Modernising Maritime Infrastructure

Design, Challenges, and Solutions for a New Millennium

Manoj Kumar

Consulting Engineer,

Port Infrastructure Planning & Design

TEXT DISCLAIMER

This book has been published with all reasonable efforts taken to make the material error-free after the consent of the author. No part of this book shall be used or reproduced in any manner whatsoever without written permission from the author, except in the case of brief quotations embodied in critical articles and reviews.

The author of this book has put forth his best effort in preparing the error-free contents of this book. However, neither the author nor the publisher makes any representations, warranties, or guarantees, express or implied, as to the fitness or relevance of the ideas or recommendations applied to any engineering problem in practice. It is the sole responsibility of the engineer-in-charge to apply the stipulated principles and their future consequences. The author or publisher disclaims any such liabilities for any loss of profit or commercial or personal damages resulting from the use of the information contained in this book.

Preface

This book is intended as a reference manual for the modernisation of existing port facilities, with particular emphasis on the deepening, strengthening, and reconfiguration of quay structures. It is primarily aimed at practising engineers and professionals engaged in the planning, assessment, and upgrading of port infrastructure.

Maritime transport is the backbone of global trade. With the growth of international commerce, ships are becoming increasingly larger. Larger vessels require deeper drafts and stronger quays equipped with cranes of greater outreach. Many existing ports, however, lack sufficient depth and facilities, resulting in reduced competitiveness. Consequently, ports worldwide are deepening basins and upgrading ageing quay structures to accommodate modern fifth- and sixth-generation vessels, creating a demand for innovative, multidisciplinary engineering solutions.

The strengthening of existing quay walls requires a thorough understanding of deterioration arising from ageing, operational demands, and environmental exposure. Rational selection of upgrade measures depends on a balanced assessment of structural strength and overall stability. This book consolidates current knowledge on stability analysis, condition assessment, investigation methods, testing techniques, and construction practices, with particular reference to conditions relevant to Indian ports.

This book presents state-of-the-art concepts, fundamental principles, and practical engineering approaches for the upgradation and capacity enhancement of quay walls. Its scope is intentionally broad, covering a wide range of relevant technical topics that are organised into three main sections, with chapters arranged in a clear and logical sequence. The content has evolved from the integration of the author's design notes, technical clarifications, specifications, and professional discussions gained through long-term involvement in multidisciplinary activities related to port planning, design, retrofitting, and augmentation. The author gratefully acknowledges the senior colleagues whose creative insights, critical thinking, and selfless contributions have been a continuous source of inspiration throughout his professional career and sincere thanks to Supratim Chanda for reviewing the manuscript and providing valuable inputs that improved the overall quality of this book.

This work draws upon a wide range of published references, acknowledged in the epilogue, and sincere appreciation is extended to all contributors. Constructive feedback from engineers, researchers, and academics is welcome to enhance future editions further.

Readers seeking detailed design methods may refer to the companion book, *A Practical Approach to Design of Steel Sheet Pile Structures*. This book aims to offer practical technical guidance to help identify shortcomings in existing structures and to develop cost-effective and sustainable solutions. These solutions are intended to ensure long-term performance while meeting environmental requirements, which are increasingly important in modern port engineering.

Manoj Kumar

mk.sheetpile@gmail.com

New Delhi

May, 2026

About the Author

Manoj Kumar is a professionally active civil engineer with over 27 years of experience in multidisciplinary civil engineering, with a strong specialisation in marine and port infrastructure systems in India and abroad. He obtained his undergraduate degree in Civil Engineering in 1997 from M.I.T., Muzaffarpur, Bihar, and completed his postgraduate studies in Computer Aided Design in 1999 at the University of Roorkee, India (now IIT Roorkee).

His technical expertise includes structural engineering, geotechnics, and marine hydrodynamics, with extensive involvement in the planning, analysis, design, and execution of complex maritime infrastructure projects. His professional experience covers wharves, quay walls, solid and liquid bulk terminals, container terminals, and desalination facilities, particularly in challenging soil and foundation conditions. He possesses specialised expertise in quay wall upgradation, RCC diaphragm wall systems, steel sheet pile structures, and heavily loaded foundation design, and has successfully led more than 15 large-scale integrated port projects from inception to commissioning, involving quay walls/wharves, breakwaters, dredging, reclamation, backup yard development, and associated support facilities.

He is the author of a technical publication on steel sheet pile structures, contributing to knowledge development and dissemination in maritime engineering.

Dedicated to my parents—

whose sacrifices have shaped my life.

अपने माता-पिता को समर्पित—
जिनके त्याग ने मेरे जीवन को आकार दिया।

Table of Contents

Definition of Key Terms

Term	Definition
ACI	American Concrete Institute
API RP 2A	API Offshore pile design guideline
ASTM	American Society for Testing & Materials
BC / BCE	Before Christ / Before Common Era
CFC	Carbon Fibre Composite
CFRP	Carbon Fibre Reinforced Polymer
DWT	Deadweight Tonnage
EC3	Eurocode 3 (Steel structures)
EPDM	Ethylene Propylene Diene Monomer Rubber
FOS	Factor of Safety
FRP	Fibre Reinforced Polymer
FS	Factor of Safety
GFRP	Glass Fibre Reinforced Polymer
GT	Gross Tonnage
LOA	Length Overall
NDT	Non-Destructive Testing
O	Pile / O-Type – Tubular (pipe) piles
O+U / HZ	Other combi-wall variants
PIANC	Permanent International Association of Navigation Congresses
PVC	Polyvinyl Chloride
RCC	Reinforced Cement Concrete
RMG	Rail Mounted Gantry
RMQC	Rail Mounted Quay Crane
SBC	Safe Bearing Capacity
SMTS	Specified Minimum Tensile Strength
STS	Ship-to-Shore (cranes)
TSD	Design Tensile Force / Design Anchor Force
ULCC	Ultra-Large Crude Carrier
UPV	Ultrasonic Pulse Velocity
VLCC	Very Large Crude Carrier

Part-1
Maritime Industry Dynamics

Understanding The Changing Landscape of Maritime Infrastructure

Chapter Summary

This chapter outlines the evolution of maritime trade and its central role in global economic growth. It explains how globalisation, containerization, and industrial expansion have driven the use of larger vessels due to improved fuel efficiency, lower transport costs, and reduced environmental impact. The chapter emphasises the strategic importance of global shipping routes and the need for continuous port modernisation to handle increased vessel sizes. It compares port development across major maritime hubs, positioning India as a rapidly advancing regional player. The chapter further discusses challenges in upgrading existing port infrastructure, highlighting modern materials, construction technologies, and the importance of fostering a culture of learning and innovation to support future-ready maritime infrastructure.

Learning Objectives

- Expansion of maritime trade became vital to global business
- Bigger ships make sea trade more economical
- Ports need upgrades to handle larger vessels
- New materials and methods support the upgradation of ports
- Learning and applying new techniques should be encouraged

1.1 Evolution of Early Maritime Navigation

The history of navigation is as old as human civilisation itself. Earliest settlements were often located along rivers, which served as the first highways of the ancient world. The great civilisations of the Indus Valley, Mesopotamia, Egypt, and China not only navigated these inland waterways but were also pioneers of open sea navigation. This legacy of maritime trade set the foundation for our globalised world, demonstrating the continuing importance of sea travel in connecting distant lands and peoples, long before the advent of modern transportation methods like road, rail, and air travel.

Maritime trade has distinct advantages in terms of cargo carrying capacity and fuel economy, making it a preferred mode of transport for bulk goods. The earliest waterborne commerce emerged from river-based civilisations such as those along the Indus, Nile, and

Tigris rivers, significantly contributing to the socio-economic development of their respective regions.

India's maritime history dates back to the 3rd millennium BC, when the Indus Valley civilisation began trading with Mesopotamia. Lothal, one of the world's earliest known ports, was a centre for trade in beads, gems, and ornaments, with connections extending to West Asia and Africa. Indian traders actively engaged with Arabs, Chinese, and Indonesians, primarily exchanging spices and dyes. These trade routes also played a key role in spreading Buddhism across Asia. India's peninsular geography made it well-suited for marine trade, leading to the rise of numerous ports, some of which were among the busiest in ancient times.

1.2 Symbiosis of Maritime Trade and Industry

Today, maritime transport has been playing a crucial role in global trade and economic development, serving as the backbone for the exchange of goods worldwide. In the modern era, no nation is completely self-reliant; each depends on maritime trade to export and import necessary commodities. Since World War II, there has been a remarkable surge in global trade, which corresponds to the expansion of the maritime industry along with industrial growth. Over the past fifty years, commercial marine transport has seen consistent growth. Countries that have adopted an open approach to maritime trade have often experienced significant economic growth, benefiting from foreign investment and trade. This phenomenon, known as globalisation, is characterised by an integrated global economy marked by free trade and access to more affordable foreign labour markets.

Each year, the shipping industry handles massive volumes of critical raw materials that are impractical to move by road, rail, or air. Transporting such materials by sea enables countries to build industries, develop cities, provide housing, support population movement, and convert natural resources into finished goods for export. This value-added process brings up prosperity and supports the advancement of developing nations.

Maritime transport also facilitates the global movement of chemicals, fuels, and manufactured products. Historical trends show that as international trade has grown, so has the global economy—highlighting trade as a key driver of worldwide economic development and prosperity. This underscores the mutually beneficial relationship between maritime shipping and globalisation in shaping trade and transport.

The evolution of maritime trade has significantly influenced industrial location strategies. The focus has shifted from proximity to raw materials to closeness to maritime routes, thanks to reduced shipping costs and enhanced global connectivity. For example, steel plants that were once located near raw material sources are now being established along coastlines, allowing easy import of inputs and export of finished products. This shift reflects the cost-efficiency of sea transport, enabling industries to source materials globally and distribute goods to international markets seamlessly. It's part of a broader trend where efficient connectivity and logistics are central to success in the global economy.

Initially, crude oil was refined near extraction sites and shipped in small tankers to consumer nations. Over time, refineries were relocated closer to consumption centres, allowing crude oil to be transported in larger volumes. This led to the development of large crude carriers, significantly lowering the cost of intercontinental energy transport.

Agricultural products also highlight global interdependence. Many items travel across continents during their production cycle. For instance, cotton grown in North America might be processed in Africa, turned into clothing in Asia, and sold back in North America. This global journey isn't unique to cotton—products like orange juice and wine are sourced from different regions due to seasonal availability or labour cost differences.

Another major shift is the widespread adoption of containerization. Container shipping has transformed global trade, making it more economical and efficient. Containers have drastically cut ocean freight and port handling costs. Today's mega container ships have further reduced transport expenses, making shipping a minor part of production costs. This efficiency supports global sourcing without significant financial strain. Container ships are now central to a sophisticated logistics network. While the ships are highly automated, land-based systems pre-plan container loading and unloading, stacking positions, and land transport arrangements. This streamlined system has brought new reliability to global supply chains, enabling manufacturers and retailers to confidently order goods from across the globe, knowing they'll arrive safely and on time.

1.3 Major Global Shipping Routes and Their Importance

Global maritime routes serve as the essential conduits of international commerce, supporting the smooth, cost-efficient transport of high-volume cargo across continents. Gaining a thorough understanding of the risks and vulnerabilities inherent in these routes is vital for robust logistics planning, comprehensive risk evaluation, and strategic geopolitical decision-making.

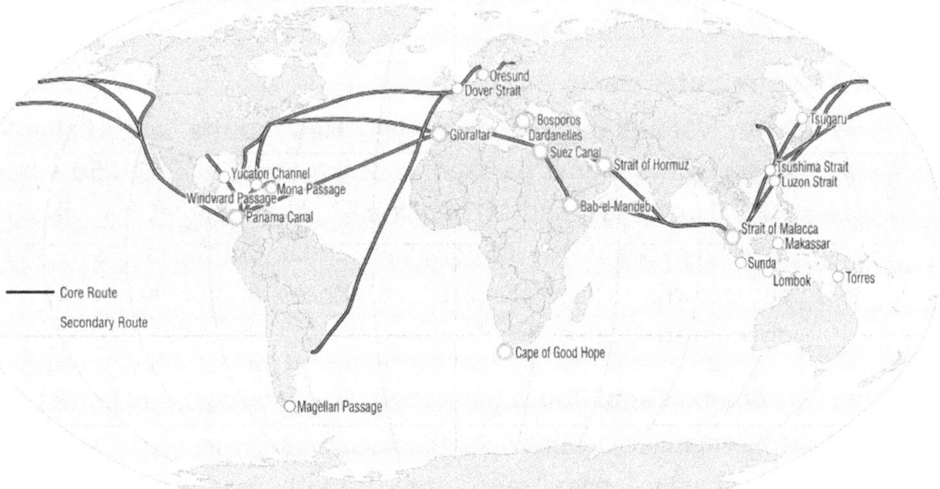

Figure 1.1: Major Global Shipping Routes

Planning the facility accounting the key strategic chokepoints—such as the Suez Canal, the Panama Canal, the Strait of Malacca, and the Strait of Hormuz- is very important as they are pivotal transit hubs whose operational security and stability exert direct influence on global inflation trends, energy supply resilience, and the overall reliability of supply chains. A list of Major Global Shipping Routes and Their Strategic Importance are given below.

i) **Asia–Europe Route (via the Suez Canal)**

This is the world's most significant container shipping corridor, connecting major Asian manufacturing centres with Europe's industrial and consumer markets. The route operates through the Suez Canal, one of the most critical global chokepoints. It carries a substantial share of global electronics, machinery, textiles, and other manufactured goods. The Suez Canal alone accounts for approximately 12% of global trade, meaning any disruption has severe economic implications. By eliminating the need to divert around Africa's Cape of Good Hope, the route saves roughly 7,000 km and reduces transit times by 10–14 days, significantly improving supply-chain efficiency.

ii) **Trans-Pacific Route (Asia ↔ North America)**

This corridor links East Asia—primarily China, Japan, and South Korea—with major U.S. West Coast ports such as Los Angeles and Long Beach. It is one of the largest and busiest shipping routes globally, handling 300–500 vessels daily. Facilitating more than $2.5 trillion in annual trade, it primarily moves consumer goods, electronics, and components vital to the U.S. retail and technology sectors, making it essential for North American supply-chain stability.

iii) **Trans-Atlantic Route (North America ↔ Europe)**

One of the oldest established maritime routes, this corridor connects North American ports with major European hubs. It carries industrial goods, vehicles, chemicals, pharmaceuticals, and various energy products. The Trans-Atlantic trade volume exceeds $1 trillion annually, supporting critical economic and industrial interdependence between the two regions.

iv) **Strait of Malacca Route**

This pivotal route links the Indian Ocean to the South China Sea, forming Asia's most important maritime passage. It accommodates 220–250 vessels daily (approximately 80,000 per year) and serves as a vital chokepoint for oil, LNG, and container traffic moving between the Middle East, India, Southeast Asia, and East Asia. The Strait is particularly crucial for Japan, South Korea, and China, whose economies depend heavily on uninterrupted energy imports.

v) **Panama Canal Route (Asia ↔ U.S. East Coast / Atlantic)**

This engineered canal links the Atlantic and Pacific Oceans, enabling ships to avoid the significantly longer route around South America. It handles 14,000–15,000 transits annually (35–40 per day). The canal's 2016 expansion to accommodate New Panamax vessels has increased capacity and strengthened its role in facilitating Asia–U.S. East Coast trade.

vi) **Strait of Hormuz (Middle East → Asia/Europe)**

A narrow but strategically critical passage between the Persian Gulf and the Gulf of Oman, the Strait of Hormuz is one of the world's most important energy chokepoints. Nearly 20% of the global oil trade passes through this strait. It is indispensable for exporting petroleum from key producers such as Saudi Arabia, the UAE, Kuwait, Iraq, and Qatar.

vii) **Cape of Good Hope Route (Southern Africa)**

This is the primary alternative to the Suez Canal, routing ships around Africa's southern tip. It typically handles 100–120 ships daily and becomes essential when the Suez Canal is disrupted or when security risks affect the Red Sea. However, this route is significantly longer—about 12,000 km—leading to higher fuel costs, longer transit times, and exposure to rough seas and challenging weather conditions in the Southern Ocean.

viii) **Bosporus Strait (Black Sea ↔ Mediterranean)**

A narrow maritime corridor passing through Istanbul, the Bosporus acts as the gateway between the Black Sea and the Mediterranean. It is crucial for transporting oil, gas, and other commodities from the Caspian region and Black Sea nations to global markets.

ix) **Northern Sea Route (Arctic Route)**

An emerging corridor along Russia's Arctic coastline, the Northern Sea Route is gaining importance as melting ice opens previously inaccessible waters. It can reduce the Asia–Europe distance by up to 40%, offering a shorter alternative to the Suez route. Its relevance continues to grow due to climate-driven Arctic accessibility.

x) **Indian Ocean Trade Routes**

These routes interlink the Middle East, East Africa, India, and Southeast Asia. Their importance stems from:

- Serving as a primary corridor for Middle Eastern oil shipments to Asian markets.
- Carrying substantial volumes of bulk commodities, LNG, and manufactured goods.
- Connecting major global chokepoints in a continuous chain, notably Hormuz → Malacca → Suez, forming the backbone of East–West maritime trade.

1.3.1 Significance of Global Shipping Routes for Port Modernisation

Major global shipping routes serve as the essential arteries of international trade, directing the flow of high-volume cargo across regions. Ports situated near busy maritime chokepoints naturally attract substantial vessel traffic, making them prime locations for transhipment operations, bunkering services, vessel repairs, and integrated logistics activities. These strategic chokepoints create a geographic concentration of maritime movement, providing ports with inherent competitive advantages.

Notable examples include the Port of Singapore near the Strait of Malacca, Port Said and Jeddah adjacent to the Suez Canal, ports within the Panama Canal zone, and the Port of Colombo positioned to efficiently serve India-bound cargo. The high density of traffic in such corridors significantly influences port modernisation needs, operational throughput, investment decisions, and the adoption of advanced technologies.

Accordingly, port modernisation and the ability to attract shipping traffic function as mutually reinforcing processes, each dependent on and strengthened by the other as they evolve in tandem.

1.3.2 Progression of Port Modernisation in India

India's port modernisation has accelerated significantly since 2014, but historically, India lagged behind leading global hubs. The timeline of port modernisation across Asia reveals that India's transformation began significantly later than several major and adjacent regional competitors.

- Singapore emerged as a global benchmark as early as the 1980s, pioneering port automation, digital operations management, and high-efficiency container handling systems, establishing itself as the world's leading transhipment hub decades before others followed.
- China, beginning in the 1990s, pursued an aggressive state-driven expansion strategy—constructing mega-ports such as Shanghai, Ningbo, and Shenzhen—supported by heavy automation, integrated logistics parks, and large-scale hinterland connectivity, quickly propelling it to the top of global port rankings.
- In the Indian subcontinent, Colombo strengthened its status as South Asia's dominant transhipment hub well before Indian ports modernised, due to its deep-draft terminals and early partnerships with global operators. Hambantota, another Sri Lankan port, albeit developing more slowly, was planned and built earlier than India's current deep-draft mega-port initiatives.

India now has multiple globally competitive ports—Visakhapatnam, Mundra, and JNPA—yet still trails Singapore and China in transformation scale and hub dominance. In the South Asian region, JNPA clearly outperforms Chittagong and Karachi, and has narrowed the efficiency gap with Colombo, though not yet its transhipment importance. JNPA's rise to 23 globally strengthens India's overall port competitiveness and supports its ambitions to emerge as a major logistics and maritime hub in the Indo-Pacific.

India is therefore positioned as a rapidly improving regional power, but not yet a global port leader.

1.4 Recent Trends in Shipping Industry

The maritime sector has increasingly shifted towards the use of larger vessels—and with good reason. These ships offer substantial economic advantages. Larger vessels are more fuel-efficient due to a better fuel-to-cargo ratio and require fewer crew members, which helps reduce overall operational costs. They are also better equipped to manage bulk and containerised cargo, a critical factor in today's global economy where the volume of seaborne trade continues to rise. This trend not only supports economic efficiency but also contributes to environmental sustainability by lowering the carbon footprint per tonne of cargo transported.

The economic logic behind this shift underscores the importance of vessel size in maritime economics. The move towards larger ships has been enabled by advancements in cargo handling technologies and the widespread adoption of containerization, which standardised cargo dimensions and streamlined port operations. These developments have made it feasible to operate mega vessels efficiently.

Modern mega container ships are designed to carry significantly more containers while distributing fuel consumption across a greater number of container miles, making them more environmentally friendly. In essence, the industry's transition to larger ships is not only a cost-effective strategy but also a step toward more sustainable global logistics. This evolution has also influenced industrial logistics and supply chain strategies.

1.5 Need for Upgradation and Modernisation of Wharves

As discussed earlier, the rapid expansion of maritime trade in recent decades has driven a steady increase in ship sizes. This growth necessitates the development of new marine terminals or the upgrading of existing ones to accommodate larger vessels. Over time, port infrastructure has undergone several transformations, with ports around the world evolving from their original configurations to meet changing demands. To adapt to these developments, ports must increase their draft capacity.

Traditionally, increasing draft involves shifting mooring arrangements further out to sea, where deeper waters are naturally available. However, the pace at which ship sizes evolve often renders quay structures obsolete much sooner than anticipated, creating a challenge in managing these changes sustainably and cost-effectively. Demolishing and reconstructing quay walls to support deeper drafts is frequently impractical due to financial and environmental constraints.

As a result, attention has shifted toward upgrading and deepening existing quay walls as a more viable alternative. This approach involves modifying and reinforcing current wharf structures to bear loads beyond their original design capacity. These loads may include increased mooring and berthing forces, heavier crane operations, and sometimes additional storage demands to accommodate larger ships.

To ensure sufficient strength and stability, constructive modifications must be made to the quay walls. This may include structural reinforcement, retrofitting with additional materials, redesigning components to better distribute loads, or applying innovative construction techniques to extend the infrastructure's service life. Such upgrades are essential to align port capabilities with the evolving demands of modern maritime logistics.

1.5.1 Structural and Functional Upgradation to Reduce Obsolescence

As structures get older, they are continuously affected by environmental exposure, normal use, changing regulations, and new technology. Because of this, deterioration and obsolescence are unavoidable over time. Deterioration affects the physical condition of a structure, while obsolescence occurs when a structure can no longer meet today's functional needs. A structure may therefore remain standing and stable but still become obsolete if it cannot support current operational requirements. This makes timely and well-planned upgradation essential to maintain safety, functionality, and long-term usefulness.

Technological progress has a major influence on whether infrastructure remains relevant. In ports and marine facilities, changes such as larger ship sizes, higher cargo capacities, and modern mooring systems often exceed the design limits of older structures. As a result, jetties and port facilities built decades ago may require upgrading of structural components, fenders, bollards, and berthing arrangements to continue operating efficiently and competitively.

However, technological obsolescence is not always absolute. A structure may still be suitable for smaller vessels, but not for modern ships. Therefore, decisions regarding obsolescence should not be based only on physical condition. They must also consider Competitiveness, economic viability, user requirements, operational goals, and future development plans and landmark global scenarios are discussed in Chapter-4. In this context, the ability of a structure to be adapted, upgraded, or expanded becomes a key factor in planning.

Before planning any upgradation of existing port facilities, it is important to clearly understand the difference between the following conditions:

- **Deterioration** refers to visible physical damage caused by age, environment, and use, such as corrosion, cracks, and concrete spalling.
- **Degradation** refers to the gradual loss of material or structural performance, such as reduced strength, stiffness, or ductility, which may not always be visible.
- **Functional obsolescence** occurs when a structure cannot meet current operational or functional needs, even though it may still be structurally sound.

Structural upgradation is mainly required when deterioration or degradation has reduced the strength or safety of the structure. Such aspects are discussed in detail in **Chapter-12**. Functional upgradation, on the other hand, is needed when functional obsolescence arises due to new requirements or advancements, and these aspects are covered in **Chapter-13**.

In practice, structural and functional upgradation are closely linked. Changes in function may increase loads on the structure and require strengthening, while structural repairs may be necessary to support new services, equipment, or layouts. Therefore, a combined and holistic assessment of prevailing site conditions, structural conditions, material performance, and functional requirements is essential before deciding on the upgradation strategy.

In short, obsolescence is mainly driven by the global evolution of the maritime world, changing needs and technological advances, while deterioration is driven by material ageing and environmental effects. A systematic approach that includes condition assessment, repair, renovation, and rehabilitation helps ensure safety, usability, durability, and cost effectiveness. By adopting a forward-looking and planned approach, existing structures can be successfully upgraded to meet both present and future demands.

1.6 Challenges in Design for Upgradation

Building on the need for upgrading quay structures, the challenge lies in bridging the gap between past designs and present requirements to meet future demands. This involves re-designing existing structures to modify, strengthen, extend their service life, and retrofit them to accommodate larger vessels. As ship sizes increase, deeper basins and longer berths are required to match their draft and length. Additionally, port equipment—especially ship-to-shore cranes—must be upgraded to reach higher decks and span the wider beams of modern vessels.

Upgrading port infrastructure is a complex and multi-dimensional process. It begins with a thorough assessment of the current facilities to identify areas needing reconfiguration or modification. This evaluation must align with contemporary design standards to ensure that the upgrades are both resilient and future-ready. The objective is not only to keep pace with evolving ship designs but also to prepare ports for the next generation of maritime vessels. By doing so, ports ensure that they remain competitive, efficient, and safe, ready to facilitate the ever-growing demands.

1.7 Assessment of the Condition of Structure

Reconfiguring quay structures is a complex engineering task that involves multiple disciplines. It requires a thorough understanding of the original design conditions, the actual operational loads experienced over time, and the new demands that the upgraded structure must withstand—both locally and globally. It's important to recognise that any man-made structure begins to deteriorate from the moment it is commissioned. Therefore, upgrading and extending the life of a wharf demands comprehensive investigations to assess its current condition and address any issues identified during inspections to prevent further degradation.

Condition surveys play a vital role in evaluating the structural health, identifying the impact of operational loads, and determining the safety and stability of the structure for its remaining service life. Non-Destructive Testing (NDT) methods are employed to inspect the structure without compromising its future usability. These techniques are highly effective in assessing strength and durability, detecting internal and external flaws, evaluating material properties, and measuring geometric characteristics of structural elements and deteriorated assemblies. The data collected through these methods is systematically analysed to assess the integrity of the structure.

Older quay constructions may feature lower-grade concrete, reinforcement steel, or rubble masonry walls. In contrast, modern quay walls typically use high-grade concrete, reinforced steel, and steel sheet piles of various geometries. Hence, understanding the materials used in the original construction is crucial. Additionally, the current condition of these materials—considering ageing and environmental degradation—must be factored into the design process.

This assessment process requires a seamless integration of geotechnical and structural engineering principles to ensure the long-term stability and performance of quay structures. Accurate evaluation of short- and long-term geotechnical parameters, along with advanced soil modelling, is essential to understand soil-structure interactions. This interdisciplinary approach, combining theoretical knowledge with practical engineering expertise, is key to successful renovation and reinforcement of quay walls, ensuring resilience against future uncertainties.

Moreover, the design for upgradation must be site-specific, as each port location presents unique environmental, geological, and operational challenges. This is not merely a design preference but a fundamental necessity in maritime engineering. By accounting for local geography, sediment behaviour, and tidal patterns, engineers can develop quay wall solutions that ensure safe and efficient berthing for vessels of varying sizes—ultimately supporting the reliability and productivity of port operations.

1.8 Construction Technology for Modern Ports

The construction industry is constantly evolving, always looking for smarter ways to upgrade, extend the life, and improve the safety of existing structures. The way quay walls are designed and built has changed a lot over time, thanks to advancements in materials and technology. In the early days, ports were built using rubble masonry walls and lime concrete, reinforced with low-strength steel. But as global trade grew and demands increased, engineers started using stronger materials and more advanced construction methods—like slurry walls, bored in-situ piles, driven piles, and various types of steel sheet pile walls.

As these structures age and face tougher environmental conditions and heavier loads, they need reinforcement and rehabilitation. Traditional methods for strengthening reinforced concrete structures include techniques like concrete jacketing, shotcreting/guniting, steel plate bonding, and micro-concreting. However, these methods often come with challenges—like long downtimes and labour-intensive work.

To tackle these issues, newer materials and techniques have come into play. Things like combi-sheet piles, composite fibre wrapping with polymers, and geopolymers are now being used. These modern solutions are less disruptive, meaning port operations don't have to stop for long periods. They're also faster to apply, which helps cut down on labour costs and speeds up project timelines.

One standout innovation is combi walls, which are a big step forward in marine construction. These walls combine two types of structural elements: regular sheet piles and high-strength king piles (like tubular steel piles or H-piles). The sheet piles form a continuous wall that resists water and earth pressure, while the king piles are placed at intervals to add extra strength and handle bending forces.

Combi walls are especially useful in tough environments—like deep waters or soft soils—where traditional retaining walls might not work well. Their design allows them to go deep into the ground, creating a solid foundation that can handle strong environmental forces. Because of their durability and flexibility, combi walls have become a go-to choice for many marine construction projects.

1.9 Knowledge and Experience of Upgradation

Civil engineering is undergoing a significant transformation with the introduction of advanced materials and modern technologies. Innovations and advancements in construction materials are enhancing structural strength, durability, and long-term sustainability. These innovations not only meet the growing demands of today's infrastructure but also help reduce construction and maintenance costs. Together, these advancements are making the industry more cost-effective and better prepared for future challenges.

That said, even though the construction industry has a long history of innovation, it often takes a cautious approach when it comes to adopting new materials and methods. This hesitation usually comes from a lack of hands-on experience or familiarity with newer technologies. Sometimes, this reluctance can slow down progress and hold back innovation.

The key to moving past this resistance is sharing the right information and building practical experience. When professionals get the chance to work directly with new materials, they become more confident in using them. Case studies and real-world success stories also help—showing how new technologies have been used effectively can encourage wider adoption.

Bridging this knowledge gap is important for the industry because it helps everyone use new tools, materials, and methods better. When people understand these improvements, construction work becomes safer, more sustainable, and more innovative. By learning and sharing technical knowledge, we build a strong foundation for future projects. Over time, the goal is to create a culture where trying and adopting new materials and technologies is encouraged. This ensures that engineering and construction decisions are based on solid data, research, and best practices—not just old habits or familiar ways of working.

Evolution of Maritime Trade in India

Chapter Summary

The study of historical advancements offers valuable insights into the evolution of human innovation and infrastructure. The principle that necessity drives invention is particularly evident in the development of port systems, which have continuously adapted to meet the changing demands of maritime trade. As globalisation and technological progress accelerated, the maritime industry underwent a significant transformation, shaping the way goods and cultures moved across regions. This chapter presents a concise overview of India's maritime history, tracing its origins to approximately 3000 BCE, when the Indus Valley civilisation initiated sea trade with Mesopotamia. It further explores the emergence of ancient seaports along the Indian coastline, which played a crucial role in facilitating early international commerce and cultural exchange.

Learning Objectives

- Identify key ancient maritime trade routes and their significance.
- Explore the historical development of Indian maritime commerce.
- Analyse the evolution of early seaports along the Indian coastline.
- Evolution of Modern Port System in India

2.1 Trade and Port Evolution

Trade has always been a key part of human progress, going all the way back to ancient times. Among all transport methods, travelling by water was the easiest and most reliable, which is why early civilisations popped up near rivers and coastlines.

Sailors in those days didn't have modern tools—they used the stars and moon to find their way, along with basic knowledge of astronomy. They also paid close attention to wind patterns, tides, and river currents to plan their routes. Boats used sails to catch the wind and relied on nature to guide them, which made long-distance trade possible even without advanced navigation systems. Today, ports are still very important for global trade and local industries. How goods are shipped often shows how trade works. Looking at how ports have evolved over time helps us understand why people settled where they did and gives us useful ideas for building better infrastructure now. With digital tools like AI, IoT, and predictive analytics, ports are becoming smarter and more efficient. Drones are even being used for monitoring operations. We're also seeing a big push toward sustainability—like creating zero-emission shipping routes. Logistics is getting a boost from automation, robotics, and better connections to inland areas. Data-sharing systems are improving coordination across the supply chain.

So, knowing the history of ports isn't just interesting—it's actually really helpful for designing smarter, greener, and more connected ports for the future.

2.2 Ancient Sea-Borne Trade

The history of navigation through rivers and seas dates back to the earliest human settlements along riverbanks, as rivers naturally became one of the first modes of transporting goods. Early on, they built simple boats, and later, more advanced ships popped up all over the world. Ancient civilisations like the Indus Valley, Mesopotamia, Egypt, and China all used rivers and seas for travel and trade. Even the Polynesians, who aren't as well-known, were sailing across open oceans way before others—some say they were the first to do so. Back then, most Europeans and others stuck to daytime sailing and stayed close to land, only reaching places like California much later.

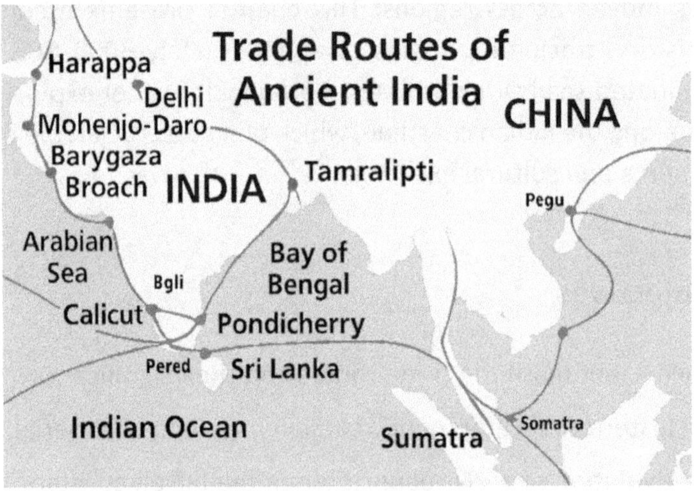

Figure 2.1: Sailing routes, 200BC-200, of Indian merchants [Nabatheans 2004]

Over time, waterborne transport proved far more efficient than road, rail, or air in terms of load capacity and fuel economy. This efficiency contributed to the early rise of river-based trade networks in major civilisations along the Indus, Nile, and Tigris, which significantly influenced regional socio-economic growth. Today, more than 90% of global trade by volume moves by sea, underscoring the central role of maritime transport in national and global economies. The link between ports, international trade, and economic development remains strong: active ports stimulate trade, and expanding trade further drives port growth and economic progress.

2.3 Indian sea trade

India's maritime history goes way back to around 3000 BC, when people from the Indus Valley started sea trade with Mesopotamia. One of the most impressive achievements from that time was the construction of a large brick dry dock in Lothal, Gujarat—considered one of the oldest docks in the world. Built around 2400 BC, it was cleverly connected to Harappan cities through the Sabarmati River, allowing the smooth movement of goods. Back then, the area we now call the Kutch desert was actually part of the Arabian Sea, making

Lothal a perfect spot for water-based trade. Lothal became a major trading hub, especially known for its exports of beads, gems, and ornaments. Its dockyard was designed to handle ships efficiently, with smart features to manage water flow and prevent silt buildup.

Goods from Lothal reached far-off regions like West Asia and Africa. The town also played a key role in trade with Arabs, Chinese, and Indonesians—mainly dealing in spices and dyes. These trade routes even helped spread Buddhism across Asia.

Figure 2.2: Photo of Lothal Dockyard (Courtesy ASI)

2.4 Old Indian Ports and Ocean Trade

India's maritime trade has a long history, beginning with the Indus Valley civilisation, which engaged in long-distance sea voyages. Since the early days of maritime activity, India has been one of the most significant centres of sea trade in South Asia. Long before the development of the Silk Road, Indian traders sailed thousands of miles across the Indian Ocean and the Arabian Sea to reach markets in West Asia, East Asia, Southeast Asia, and East Africa.

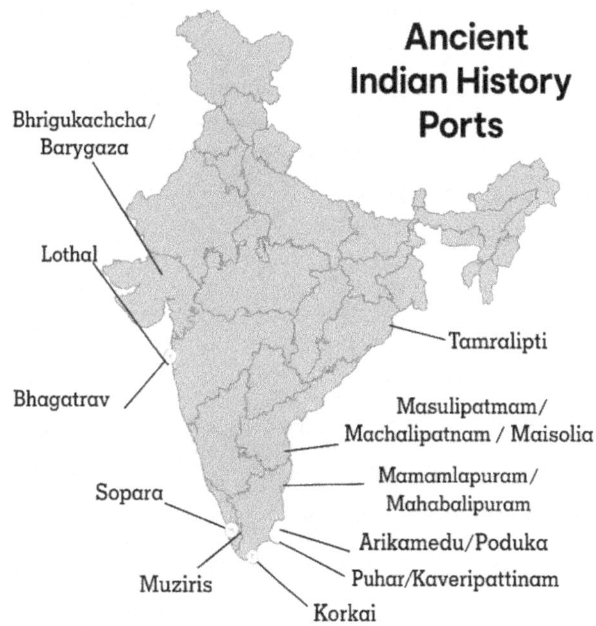

Figure 2.3: Locations of Ancient Seaports of India

Likewise, merchants from these regions—especially Arabs and Chinese—frequently visited the Indian subcontinent, trading goods such as silk, spices, porcelain, and ivory.

As ocean trade flourished, India established maritime links with the Roman Empire and other parts of Europe, well before the colonial period. Its strategic peninsular location and extensive coastline led to the early development of numerous ports, positioning India as a key player in the global maritime trade network.

Below is a list of some of the oldest ports in India that contributed to its prominent role in international maritime trade even before colonial rule.

xi) Muziris Port

Muziris Port, located in present-day Kerala, was one of India's most prominent ancient trading hubs nearly 2,000 years ago. It played a vital role in connecting India with major civilisations such as the Persians, Phoenicians, Assyrians, Greeks, Egyptians, and the Roman Empire. The port was especially known for exporting black pepper, along with other valuable items like semi-precious stones, diamonds, ivory, and pearls. Historical records suggest that goods from over 30 countries arrived at Muziris, including textiles, wine, wheat, and gold coins.

xii) Bharuch Port

Bharuch Port, located at the mouth of the Narmada River in present-day Gujarat, was a major trading hub in western India nearly 2,000 years ago. Known to international traders as Bharukaccha or Barygaza, it maintained active trade relations with merchants from Arabia, Greece, Rome, Africa, China, and Egypt.

xiii) Calicut Port

Calicut Port, also known as Kozhikode, was among the busiest trading centres in southwestern India, alongside Muziris. It was especially known for its spice trade, exporting pepper, cloves, and cinnamon. The port gained greater prominence after Portuguese explorer Vasco da Gama landed there in 1498, marking the beginning of lasting trade relations between the East and West.

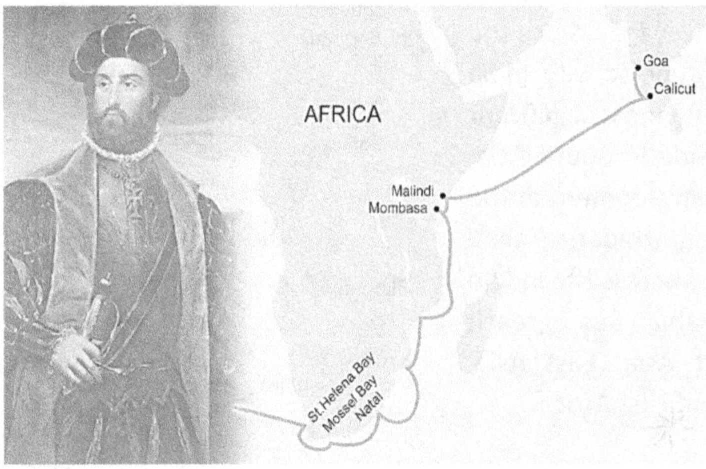

Over time, under colonial rule, the port's importance declined. However, remnants like a sea bridge, storage yards, and old cargo-handling equipment still stand as symbols of its once-thriving maritime past.

***Figure 2.4**: The navigational route of Vasco da Gama's expedition.*

xiv) Poompuhar Port

Poompuhar Port, also known as Puhar, is believed to be the Port town of the Chozha Empire, currently located in the Nagapattinam district in Tamil Nadu. It was reportedly situated at the mouth of the Kaveri River. Details about the port city have been found in several historic documents, including the Periplus of the **Erythraean** Sea. It is said that the goods that arrived in the port included horses from Gulf countries and finished goods from neighbouring regions such as Sri Lanka and Indonesia.

xv) Arikamedu Port

Arikamedu Port, located in present-day Puducherry, was historically known as Podouke. It is mentioned in ancient texts like the *Periplus Maris Erythraei* and Tamil Sangam literature, indicating its role as an active trading centre as early as the 2nd century BCE. Arikamedu is believed to have been a Chola port, primarily known for bead-making, and is considered the only port in the region with direct trade links to the Roman Empire.

xvi) Tuticorin Port

Tuticorin Port, also known as Thoothukudi, is among India's oldest seaports, established in the 6th century during the rule of the Pandya and Chola dynasties. It was renowned for its trade in fish and pearls. The port gained further importance under Dutch and British colonial rule, becoming a key centre for exporting palmyra fibres, senna leaves, salt, and dried fish, while importing cotton, pulses, grains, coal, and other goods. Today, it is known as **V.O. Chidambaranar Port** and remains one of India's major ports.

xvii) Tamralipti Port

Tamralipti Port (known in Pali as Tāmaliti) was an ancient port city and the capital of the Suhma Kingdom, located along the Bay of Bengal. The modern town of Tamluk in Purba Medinipur, West Bengal, is widely identified as its site. Positioned near the Rupnarayan River, Tamralipti is believed to have been settled as early as the 7th century BCE and maintained active trade connections with regions like Ceylon, Java, and Sumatra.

xviii) Mahabalipuram Port

Mahabalipuram Port, or Mamallapuram, was one of two major port cities in the Pallava kingdom. The town was named after Pallava king Narasimha Varman I, who was also known as Mamalla. In addition to its role in trade, the port became known for its remarkable group of royal monuments, many of which were intricately carved from solid rock.

xix) Machilipatnam Port

Machilipatnam Port, located in the present-day Krishna district of Andhra Pradesh, traces its origins back to the Satavahana period in the 3rd century BCE. Historically known as Maisolos or Masalia, the port gained prominence around the 1st century CE, particularly for its trade in muslin—a fine cotton fabric. The name Masalia or Masulipatnam is believed to be linked to this textile trade. Today, the city is also referred to as Bandar or Masulipatnam.

2.5 Evolution of Global Shipping Systems

2.5.1 Port Development with Advancements In Shipping

The development of global shipping systems has progressed in parallel with advances in shipbuilding technology. Maritime transport has transitioned over millennia from basic watercraft to highly engineered, digitally controlled vessels. This section provides a concise technical overview of the major evolutionary stages and key innovations shaping maritime engineering, navigational systems, and global logistics.

2.5.2 Early Maritime Development

Across centuries, maritime transport advanced from rudimentary boat construction to sophisticated sailing practices. Improvements in sail configuration enabled vessels to manoeuvre effectively even against prevailing winds, significantly enhancing control and ocean-voyage efficiency.

Figure 2.5: Early Maritime Practices

Prior to the Industrial Revolution, ship design and cargo handling methods remained largely static. Ports consisted of simple quays, and cargo was manually handled using basic tools. The emergence of standardised cargo vessels began in the late 18th century, illustrated by early designs such as James Brindley's Starvationer, built for coal transport.

2.5.3 Industrial Revolution and Mechanised Shipping

The shift to steam propulsion marked a fundamental transformation in maritime engineering. Iron and later steel hulls replaced wooden structures, allowing the construction of larger, stronger, and more durable ships. The screw propeller superseded paddle wheels as the primary propulsion method.

Figure 2.6: Picture of a Steam-Powered Ship

Global maritime connectivity expanded significantly with the opening of the Suez Canal in 1869, reducing transit times between Europe and Asia.

2.5.4 Early 20th Century: Advances in Shipbuilding and Navigation

Transitioning from coal-fired steam engines to diesel propulsion improved fuel efficiency, power output, and operational range. Specialised vessel types—including tankers and refrigerated cargo ships—emerged to support an increasingly diverse global trade. The introduction of radar dramatically improved navigational safety under low-visibility conditions. The commissioning of the Panama Canal in 1914 further reshaped global shipping routes.

2.5.5 Containerization and Integrated Global Logistics

A major structural shift occurred with the introduction of the standardised intermodal container in 1956, revolutionising cargo handling, minimising port dwell times, and enabling seamless multimodal logistics across maritime, road, and rail systems. Container terminals adopted ship-to-shore gantry cranes, while vessel architecture diversified into container ships, Ro-Ro carriers, and other specialised designs. Electronic systems such as GPS and AIS enhance navigation accuracy, fleet monitoring, and maritime traffic management.

The late 19th and 20th centuries also witnessed the steady replacement of steamships with diesel-powered vessels, accompanied by substantial increases in vessel size and carrying capacity. Sea transport expanded to support a wide spectrum of commodities, and coastal industrialisation accelerated as industries relocated near ports to capitalise on efficient bulk transport.

2.5.6 Modern Shipping and Digital Transformation (2000–Present)

Contemporary maritime operations are characterised by the deployment of ultra-large container vessels exceeding 24,000 TEUs, extensive automation across shipboard systems, and comprehensive digital integration throughout the global supply chain. Key technological advancements shaping this modern operating environment include:

- **LNG and dual-fuel propulsion technologies**, which support significant reductions in greenhouse-gas emissions and enable compliance with emerging decarbonization targets.
- **Ballast water treatment systems** engineered to meet IMO-mandated environmental performance standards, thereby minimising ecological impacts from ballast discharge.
- **Integrated Bridge Systems (IBS)** that consolidate and synthesise navigational, sensor, and operational data to enhance situational awareness and decision-making accuracy.
- **Blockchain-enabled digital documentation frameworks**, ensuring secure, tamper-resistant processing of trade records and logistic transactions.
- **Advancements toward autonomous and remotely operated vessels**, supported by AI-driven route-optimisation algorithms that improve voyage efficiency and reduce operational risk.

Environmental regulation and the accelerating adoption of green technologies now serve as primary drivers of innovation, influencing vessel design, port infrastructure development, and operational strategies across the maritime sector.

Containerization continued to redefine shipping economics throughout the 20th century, driving ports to upgrade infrastructure for larger vessels and rapid cargo handling. Steam power and mechanisation introduced during the Industrial Revolution laid the foundation for these advancements and reshaped global trade.

In recent decades, vessel dimensions have expanded dramatically. Ultra-large ore carriers such as the Valemax class, capable of transporting up to 400,000 metric tons, have reduced per-ton transport costs and emissions. Modern container ships now exceed 24,000 TEUs, with even larger designs under development. This trend has significantly influenced port engineering, hinterland connectivity, and global trade efficiency.

2.6 Evolution of the Modern Port System in India

2.6.1 Before 1947 (Post-Independence)

Before independence, Indian ports were not developed to support Indian trade or industry. Ports like Bombay, Calcutta, and Madras mainly worked as exit points to send raw materials to Britain and bring back finished goods. They were used to support the British economy, not as modern logistics or industrial hubs for India.

After independence, the Government of India developed these ports through nationalisation, dock expansion, mechanisation, and improved rail connectivity. Mumbai was strengthened as a major commercial port, Kolkata was expanded with new docks, and later the Haldia complex was developed to overcome river constraints, and Madras developed better harbour facilities and cargo-handling infrastructure.

2.6.2 1947 to 1965s (Early years after independence)

After independence, the government took full control of ports. The main focus was on national control, security, and replacing ports lost after Partition, such as Karachi. New ports like Kandla were developed, but efficiency and global competitiveness were not the priority at this stage. The major ports developed in this period are;

i) Kandla (Deendayal Port) – 1955

Kandla was set up in 1955 as India's first major port after independence to take over the role of Karachi after Partition. Located in the Gulf of Kachchh, it was developed to secure maritime trade for north-western India. The port mainly handled petroleum, fertilisers, food grains, and other bulk cargo, reflecting India's early focus on trade continuity, strategic control, and self-reliance.

ii) Visakhapatnam Port

Visakhapatnam was expanded after 1947 to support India's heavy-industry-based development, especially steel and mineral exports. It was closely connected to major steel plants such as Bhilai, Bokaro, and, later, the Visakhapatnam Steel Plant. With access to mineral-rich regions, the port mainly handled iron ore, coal, and bulk cargo, becoming a key industrial port on the east coast.

iii) Paradip Port

Paradip Port was developed to serve the mineral-rich eastern industrial region, particularly Odisha. Designed mainly for bulk cargo, it handled iron ore exports and imports of coal, crude oil, and fertilisers. The port supported nearby steel plants and power stations and helped reduce pressure on Kolkata port, reflecting a state-led, resource-based industrial strategy.

iv) Cochin (Kochi) Port

Cochin was expanded after independence by using its natural deep-water harbour, making it an important west-coast port. It developed as a multipurpose port for commercial shipping, ship repair, and naval activities. The Cochin Shipyard strengthened its role in shipbuilding, while its location on key sea routes added to its trade and security importance.

2.6.3 1965s to 1980s (Planned economy phase)

During this period, ports were expanded mainly to support government industries and energy needs. Ports such as Visakhapatnam, Cochin, and New Mangalore were developed to handle steel, minerals, coal, oil, and fertilisers. Most ports were run by government authorities, with limited technology and slow modernisation.

- New Mangalore Port was developed to meet India's growing petroleum and energy import needs and to support refineries and industries in Karnataka and nearby regions.
- Visakhapatnam was expanded to support India's heavy-industry-based development, especially steel and mineral exports. It was closely connected to major steel plants such as Bhilai, Bokaro, and, later, the Visakhapatnam Steel Plant. With access to mineral-rich regions, the port mainly handled iron ore, coal, and bulk cargo, becoming a key industrial port on the east coast.
- Cochin was expanded using its natural deep-water harbour as a multipurpose port for commercial shipping, ship repair, and naval activities. The Cochin Shipyard strengthened its role in shipbuilding, while its location on key sea routes added to its trade and security importance.

2.6.4 After 1991 (Economic Reforms)

Real modernisation of Indian ports began after the economic reforms of 1991, when private sector participation was introduced, and greater emphasis was placed on efficiency, capacity expansion, and improved services. A major milestone during this period was the development of **Jawaharlal Nehru Port (JNPA)** near Mumbai as India's first dedicated container port, marking a shift toward containerised and globally integrated trade. However, since modernisation started relatively late, Indian ports continued to face a development gap when compared with leading global ports.

2000s onwards (Modern commercial phase)

From the 2000s onward, Indian port development entered a modern commercial phase with the rise of major private ports such as Mundra, Pipavav, Krishnapatnam, and Dhamra. These ports introduced faster operations, advanced mechanisation, deeper drafts, and strong rail-road connectivity, significantly improving efficiency and capacity. Mundra emerged as the largest multipurpose commercial port, while Pipavav and Krishnapatnam strengthened container and bulk cargo handling on the west and east coasts. This phase marked a clear shift from government-controlled ports to commercially driven, globally integrated maritime trade hubs.

2.6.5 Global Port Powerhouses

(India in Comparison with China and Europe)

i. Port System Structures

India, China, and Europe follow distinct port development models shaped by history, governance, and trade strategy.

- **India** operates a mixed ownership system, with government-controlled major ports and increasingly important private ports. The policy focus is on port-led development to reduce logistics costs and support economic growth.
- **China** follows a state-owned, centrally coordinated model, aligning ports directly with national export, industrial, and geopolitical strategy.
- **Europe** primarily uses a landlord port model, where public authorities own land and regulate infrastructure while private operators run terminals, ensuring market efficiency and multimodal integration.

In global terms, China dominates port throughput, Europe acts as a gateway for global trade, and India mainly serves regional and domestic trade, though it is rapidly modernising.

ii. Port Capacity and Scale

- **India's** ports are largely medium-scale, with limited deep-draft capacity. Mundra is the largest Indian port at about 150 million tonnes annually, but only a few ports can handle ultra-large vessels. Capacity expansion is ongoing under Sagarmala, yet scale remains a constraint.
- **China** has the largest and busiest port system in the world. Shanghai is the world's top container port, and several Chinese ports handle over 500 million tonnes annually, designed for ultra-large container vessels and mega bulk carriers.
- **Europe's** ports—such as Rotterdam, Antwerp, and Hamburg—handle large volumes but focus more on high-value cargo, speed, and reliability rather than sheer tonnage, supported by deep drafts and strong hinterland links.

Overall capacity ranking: China > Europe > India

iii. Port Efficiency and Technology

- **Indian ports** generally have higher turnaround times, due to procedural delays, limited automation, and hinterland bottlenecks. Digitalisation and customs reforms have improved performance, but gaps remain.
- **Chinese ports** operate with very low turnaround times, extensive automation, centralised customs clearance, and highly synchronised operations. Many terminals function almost like industrial assembly lines.

- **European ports** combine advanced digital systems, automation, and standardised procedures, delivering predictable and reliable logistics performance, even if operations are more environmentally regulated.

In short, India lags in consistency, China prioritises speed and scale, and Europe focuses on logistics efficiency.

iv. **Governance and Port Models**

- **India** runs a dual system: major ports under port authorities and private ports under market-oriented management. Reforms are underway to move toward a landlord-style framework.
- **China** maintains strong state ownership and planning, integrating ports with industrial zones, export clusters, and the Belt and Road Initiative.
- **Europe** has the most institutionally mature governance, with clear regulation, operational autonomy, and environmental oversight.

Governance maturity: Europe > China > India

v. **Hinterland Connectivity and Logistics Integration**

- **India's port**-hinterland connectivity is weak to moderate, constrained by rail congestion, limited inland waterways, and developing logistics parks, despite ongoing infrastructure upgrades.
- **China** treats ports as part of a national logistics system, supported by strong rail networks, extensive inland waterways, and planned logistics hubs.
- **Europe** has highly integrated multimodal connectivity, using inland waterways, dense rail networks, and cross-border logistics corridors.

China and Europe view ports as logistics nodes, while many Indian ports still function as standalone infrastructure assets.

vi. **Role in Global Trade Routes**

- **India** plays an important role in the Indian Ocean, especially for energy imports, but largely operates on feeder routes with limited control over global shipping lanes.
- **China** is central to East–West trade routes and the Asia–Europe container corridor, with global terminal ownership strengthening its shipping influence.
- **Europe** acts as a gateway region, linking Asia, Africa, and the Americas through strong maritime-rail integration into continental hinterlands.

vii. **Strategic and Geopolitical Role**

- **India**: Emerging maritime strategy, limited global port ownership, growing maritime diplomacy.
- **China**: Ports used as strategic and geopolitical instruments, backed by centralised planning and overseas port investments.
- **Europe**: Ports serve economic integration goals, with regionally coordinated strategy.

Functional distinction: China uses ports strategically, Europe economically, and India developmentally.

Indian ports are strategically positioned and rapidly modernising, but remain behind China and Europe in scale, efficiency, and integration. China leads through centralised planning and massive investment, while Europe excels in governance and multimodal logistics. India's progress toward global competitiveness depends on hinterland connectivity, governance reform, and operational efficiency, supported by private participation and policy initiatives such as Sagarmala.

Shipping, Logistics, and Essentials of Port Planning

Chapter Summary

This chapter presents different ship types and sizes, each designed for specific roles such as transporting cargo, carrying passengers, or conducting research. It explains the ships of various classes and sub-classes, and introduces key logistics terms used in maritime operations. The chapter also explains how ship measurement parameters—covering dimensions, volume, structural components, and directional terminology are essential for planning and preparing designs for port facilities. Additionally, it outlines the basic infrastructure requirements for different types of terminals, defines performance indicators, and describes the systems that support terminal operations.

Learning Objectives

- Describes ship types and their specific roles.
- Maritime logistics terms.
- Ship dimensional parameters and design terms.
- Cargo terminals requirements and performance indicators.
- Other systems to support port operations.

3.1 Introduction

Designing marine port structures requires a multidisciplinary approach that brings together engineering, environmental science, logistics, and safety. The goal is to ensure functionality, durability, and sustainability while accommodating current and future maritime needs. To provide a brief overview, some of the key elements are discussed here.

3.2 Shipping Class

Ships are big water vehicles that travel across oceans and deep waterways, carrying goods, people, or doing special jobs. After World War II, the need for ships grew to help rebuild and move materials globally. Today's ships come in many shapes and sizes, each built for specific tasks like cargo transport, tourism, energy collection, or research.

Depending on what they carry, ships are grouped into types:

- **Dry bulk carriers** move things like coal, grain, and ore.
- **General cargo ships** carry packed goods and may follow fixed routes or go wherever needed.

- **Container ships** are designed to carry standard-sized containers, making it easy to switch between trucks, trains, and ships. These containers can hold anything from food to electronics.
- **Passenger ships** include cruise liners and ferries.
- **Liquid bulk carriers** transport oil, chemicals, and gases like LPG and LNG.
- **Research vessels** are built like passenger ships but have labs and space for scientists to study the ocean.

Cargo is usually grouped into dry, liquid, and specialised types. Dry includes bulk and container goods, liquid is mostly oil and chemicals, and specialised cargo includes things like livestock or heavy equipment. Ships are also classified by size and the routes they can handle, which helps in planning port facilities and trade logistics.

3.2.1 Bulk Carriers

Dry bulk refers to large amounts of unpacked, non-liquid goods like coal, iron ore, grain, and fertiliser. These items are loaded loose into a ship's cargo hold, not in containers or packaging. Bulk carriers are easy to spot—they have big hatches on deck that cover separate cargo holds, as shown in the figure below. At ports, cargo is loaded using conveyors, cranes, or spouts. Some ships have their own cranes to load/unload without needing port equipment. Unloading usually involves cranes with grabs, and sometimes people and vehicles help inside the holds. Cargo might go into barges, then to silos or storage areas, or directly into trucks for smaller ships.

Figure 3.1: Photo of Bulk Carrier Vessel

Bulk carriers come in many sizes, from small ships to massive ones over 360,000 tons. Their size is measured by **deadweight tonnage (DWT)**, which includes the cargo plus everything else the ship carries, like fuel, crew, and supplies.

Bulk carriers vary widely in size, from a few hundred to over 360,000 deadweight tons (DWT), which includes cargo plus fuel, crew, and supplies. Their classification is mainly based on size and cargo capacity.

3.2.1.1 Class of Bulk Carrier

i) Mini Bulk Carrier

Mini Bulk Carrier, less than 10,000 DWT, used mainly for short coastal routes and feeder services to larger ships. Common in small ports and often general cargo or specialised short-sea vessels.

ii) Handysize

Handysize bulkers range from 10,000 to 35,000 DWT. Typically has 4–5 cargo holds and can carry various bulk goods. Some are built to transport logs and are equipped with special stanchions and lashing gear. Highly versatile and suitable for many ports.

iii) Handymax

Handymax bulkers range between 35,000 and 50,000 DWT (up to 60,000 DWT in some cases). Ideal for intercontinental trade and smaller ports with draft or length limits. Commonly used for dry cargo like coal, grain, and cement. Equipped with onboard cranes for loading/unloading.

iv) Supramax

One very important size is the "Supramax" - a type which became more and more popular since 2001. Ship sizes range from 50,000 to 60,000 DWT, with some designs reaching 63,000 DWT. Features five cargo holds and deck cranes (25–40 MT capacity), often with grabs. Many are double-hulled and have wide hatch openings for easier cargo handling. These are popular for their flexibility and port accessibility.

v) Panamax Class

The term 'Panamax' (length 200-230 m, draft 13 ~ 15 m) refers to design size limitations imposed by the Panama Canal locks and adopted by the international shipping community. The beam of ships must not exceed 106 feet (32.2 m), and fully loaded vessels must not exceed 80,000 tons deadweight. The various ship sizes in Panamax class are as follows;

vi) Ultramax

Ultramax has a DWT of up to 65,000 tonnes. This class is new on the market, and these vessels are usually equipped with an 'eco main engine' that reduces the amount of fossil fuels used, has an optimised hull shape and uses solar panels and batteries. There are a few variations in terms of specification, but generally, they have been designed with sustainability in mind.

vii) Panamax

Panamax-type ships have DWT of up to 80,000 tonnes. Their name comes from the fact that they are the largest vessel able to transit the Panama Canal. The size of locks in this area restricts the size of vessels able to use the route.

viii) Post-Panamax Bulk Carrier

The dead weight tonnage of these vessels varies between 80,000 to 120,000, and they usually have 9 cargo holds. These vessels have more breadth than the Panamax and were designed for the bulkers to transit through the new locks of the Panama Canal.

ix) Kamsarmax

These ships have DWT of up to 82,000 tonnes. Similar to the Panama, this slightly larger vessel is the largest ship that can enter Port Kamsar in Guinea, West Africa. These vessels do not have their own 'gear' and so must use what is available at ports.

x) Capesize

Capesize vessels have a deadweight capacity of up to 200,000 tonnes. These ships are too large to pass through the original Panama or Suez Canals and must instead navigate around Cape Horn or the Cape of Good Hope. Due to their massive size and carrying capacity, they are primarily used for transporting bulk commodities like coal and iron ore. Only a limited number of ports worldwide can accommodate them when fully loaded. These vessels typically lack onboard cargo-handling gear and rely on port facilities for loading and unloading. Capesize ships generally feature nine cargo holds.

xi) Babycape

The Babycape sector is the newest segment of the bulk market. There were also smaller Capesizes known as mini-capes or babe-capes, ranging between about 110,000 to about 120,000 tons. They provide an opportunity to the customers to load an optimum amount of cargo to fulfill their supply chains.

xii) Suezmax

Suezmax means the largest vessel that can pass through the Suez Canal. After the modification of the Suez Canal in 2017, the maximum allowed draught is currently up to the 22 m and can fit the ships of DWT upto 240, 000 tonnes.

xiii) Ultra Large Ore Carriers

These are highly specialised vessels, primarily built for specific trades, and are larger than Capesize ships. They are mainly used for transporting iron ore. The largest among them are the Valemax (also known as Chinamax) vessels, with deadweight capacities reaching up to 400,000 tonnes.

Currently, bulk carriers represent nearly 40% of the global merchant fleet, and this share is expected to grow in the coming decades. As per the 1999 International Convention for the Safety of Life at Sea, a bulk carrier is defined as: *A ship* constructed *with a single deck, top side tanks, and hopper side tanks in cargo spaces, intended primarily to carry dry cargo in bulk, including ore carriers and combination carriers.*

xiv) Open Hatch Bulk Carrier

As the name suggests, these vessels feature open hatches that allow direct loading and unloading of cargo. Unlike conventional bulk carriers, they lack upper and lower wing tanks and instead have straight vertical sides, making them suitable for transporting forestry products.

These ships are equipped with either movable or fixed gantry cranes, which assist in cargo operations. Designed to handle a wide range of cargo types, their cranes often come with specialised attachments such as clamps, multi-lifts, and frames to accommodate different loading needs.

3.2.2 General Cargo Vessels

General cargo, also known as breakbulk, typically consists of goods packed in pallets or bags. While some vessels have specialised handling systems for specific cargoes, loading and unloading are generally done using cranes with straps for boxes and slings for bags. These ships can also carry loose or irregular cargo, which is carefully packed by the crew and port workers to prevent damage and optimise space. Though bulk and container ships have largely taken over, general cargo vessels are still in operation worldwide.

These ships are easily recognised by the cranes mounted on their decks and are capable of transporting a wide range of items—from timber to heavy machinery. Representing about 20% of the global fleet, general cargo ships remain one of the most common vessel types.

Originally mass-produced during World War II, the famous 10,000 DWT "Liberty type" ships were designed to carry general, bagged, baled, and even bulk cargo. Today's general cargo vessels typically range from 5,000 to 25,000 DWT and are often built with two decks, known as tween Deckers. Each cargo hold in a tween-decker is divided into two compartments: the upper cargo hold (between the main deck and tween deck) and the lower cargo hold (beneath the tween deck).

Tween-Deckers offer two key advantages over single-deck vessels:

1. They provide more cargo compartments, allowing different cargoes to be separated to avoid contamination.
2. They support higher stacking of bagged or baled cargo by distributing weight across two decks.

Additionally, tween decks can be folded against the hold sides to allow bulk cargo loading into a single compartment. Modern general cargo ships are also fitted with container handling systems, making them multi-purpose vessels.

Figure-3.2: General Cargo Ship

Their onboard gear, typically with a lifting capacity of 30–40 tons SWL, enables them to handle both containers and general cargo efficiently.

3.2.3 Container Ships

Container ships have become the primary mode of transporting manufactured goods globally. These vessels are specifically designed to carry large volumes of cargo in standardised shipping containers, which can be easily transferred between trucks, trains, and ships. Containers are versatile and can hold a wide range of goods—from food and electronics to vehicles, liquids, and refrigerated items. Container ships mainly operate on liner routes and are among the largest vessels in operation.

These ships are built exclusively for container transport. Their cargo holds are cellular, featuring vertical guides to slot containers securely. Container ships vary in size, from small vessels with a capacity of around 500 TEU to massive ones holding up to 22,000 TEU. They can carry both 20-foot and 40-foot containers, and their total capacity is usually specified for each type. Smaller container ships, often geared, serve as feeder vessels connecting smaller ports to major terminals. Larger ships are typically gearless and capable of speeds exceeding 25 knots, making them suitable for long-distance voyages.

Unlike other dry cargo vessels, container ships are usually chartered on time-based contracts by major liner companies, which then offer fixed rates per container to customers.

a. **Reefer vessels**, designed to carry perishable and refrigerated goods like fruits and meat, are similar to container ships. These ships have temperature-controlled holds. However, traditional reefer vessels have largely been replaced by reefer containers, which are now commonly transported on container ships. These containers require a power source and regular monitoring during transit.

In terms of service range, container ships are categorised into two main types:

a. **Feeder Ships**: Medium-sized vessels carrying 300 to 1,000 containers, used for short routes (typically under 500 nautical miles) between smaller ports and major transhipment hubs.

b. **Mother Ships**: Large container vessels used for international trade, significantly bigger than feeder ships.

Mother ships are further classified into the following categories:

i) **Early Container Ships**: Introduced in the 1960s–70s, these vessels were 140–200 meters long, with capacities up to 2,000 TEU. They were slow, less efficient, and often equipped with cranes.

ii) **Panamax**: Introduced in the 1980s, these ships carry 3,000–4,500 TEU and are designed to fit the original Panama Canal. Their dimensions are limited to 290 meters in length, 32 meters in breadth, and 12.5 meters in draught.

iii) **Post-Panamax**: Larger than Panamax vessels, these ships cannot pass through the original Panama Canal and typically carry 6,000–8,500 TEU.

iv) **New Panamax (Neo-Panamax)**: Built to fit the expanded Panama Canal (opened in 2016), these vessels can carry up to 12,500 TEU and measure up to 300 meters in length and 50 meters in width.

Figure-3.3: Container Ship

v) **Suezmax**: Designed to pass through the Suez Canal, these ships have a capacity of around 12,500 TEU, with a breadth of 50–57 meters and a draught of up to 16.5 meters.

vi) **Very Large Container Ships (VLCS)**: Introduced in 2006 with the Maersk E-series, these vessels can carry between 11,000 and 15,000 TEU.

vii) **Ultra Large Container Ships (ULCS)**: Also known as Post-Suezmax, these vessels are too large for the Suez Canal and typically carry 18,000–21,000 TEU, with a breadth of 60 meters and a draught of up to 21 meters.

viii) **Megamax-24 (MGX-24)**: The largest container ships currently in service, with capacities ranging from 21,000 to 25,000 TEU. These vessels feature 24 container bays, 24 deck rows, and can stack up to 12 containers in the hold and 12 on deck.

3.2.3.1 Container numbering

Each container ship is divided into sections called **bays**, which run lengthwise from the **bow (front)** to the **stern (rear)**. The numbering of bays typically starts at the bow (e.g., Bay 01) and increases toward the stern (e.g., Bay 40), depending on the ship's size.

3.2.3.2 Three-Dimensional Container Positioning

Container placement on a ship is defined using a three-coordinate system:

i) **Row**: Runs from the bow to the stern. (Bays)

ii) **Tier**: Indicates vertical stacking levels, starting from the bottom of the cargo hold and increasing upward.

iii) **Slot**: Refers to the horizontal position across the ship's width:

- **Odd-numbered slots** are on the **starboard side.**
- **Even-numbered slots** are on the **port side.**
- Lower slot numbers are closer to the **centreline**, increasing outward.

3.2.3.3 Bay Numbers

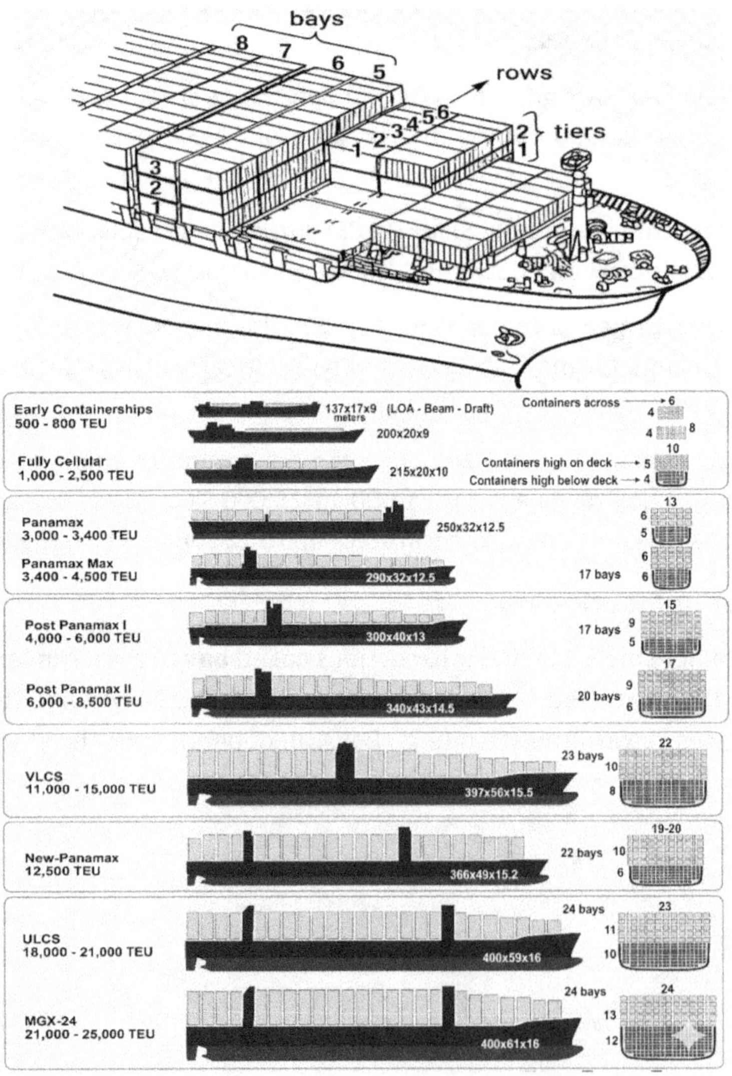

Figure-3.4: Shipboard container numbering system

- Odd-numbered bays are designated for 20-foot containers.
- Even-numbered bays are designated for 40-foot containers.
- The even number between two 20-foot bays indicates a 40-foot bay.

For example, bay numbers for 20-foot containers might be 01, 03, 05, up to 75, while 40-foot container bays would be numbered 02, 04, 06, up to 74.

3.2.4 Tanker vessels

Tanker vessels are designed to transport uniform cargoes of liquids or gases, such as bitumen, fuel oil, diesel, kerosene, gasoline, edible oils, paraffin, molasses, and even wine. These ships are not limited by space constraints and typically operate at low speeds, with a maximum cruising speed of around 15.5 knots, powered by large marine diesel engines.

Tanker vessels are categorised as follows:

i) **Handymax, Handysize, Coastal, and similar classes**

These tankers have capacities below 50,000 DWT and lengths up to approximately 200 meters.

ii) **Panamax**

Ranging from 200 to 250 meters in length, these tankers have capacities between 50,000 and 80,000 DWT.

iii) **Aframax**

With lengths around 245 meters and capacities between 80,000 and 120,000 DWT, Aframax tankers are ideal for short to medium-haul crude oil transport. They are commonly used in smaller ports that cannot accommodate VLCCs and are favoured by non-OPEC countries with lower crude production. Typical routes include South America to the U.S. Gulf, North Africa to Southern Europe, the former Soviet Union to Northern Europe, and Southeast Asia.

iv) **Suezmax**

These vessels, typically 285 meters long, are the largest tankers that can transit the Suez Canal. They have capacities between 120,000 and 200,000 DWT and beams ranging from 50 to 77.5 meters. While the pre-2017 canal allowed a maximum draught of 18.90 meters, newer vessels built after upgrades can have draughts up to 22 meters.

v) **VLCC (Very Large Crude Carriers)**

VLCCs can reach lengths up to 470 meters with beams of 60 meters, though standard dimensions are 300–330 meters in length, 58 meters in width, and 31 meters in depth. With capacities ranging from 200,000 to 320,000 DWT, these vessels—often called **Super-tankers**—are used for long-haul crude oil transport from the Persian Gulf to Europe, Asia, and North America. They are also commonly seen in regions like the North Sea, Mediterranean, and West Africa and can transit the Suez Canal.

Figure-3.5: Oil Tanker

3.2.5 Specialised vessels

3.2.5.1 Ro-Ro Vessels

Ro-Ro (Roll-on/Roll-off) vessels are designed to transport wheeled cargo such as cars, trucks, trailers, and even train cars. These ships include vehicle ferries and cargo carriers for truck trailers. The most common type is the car carrier, which features multiple vehicle decks connected by internal ramps and shore access via loading ramps. Their cargo capacity is measured in Car Equivalent Units (CEU), with the largest carriers exceeding 6,000 CEU.

Figure-3.6: Ro-Ro Ship

3.2.5.2 Passenger Vessels

This category ranges from small ferries to large cruise ships capable of carrying over 6,000 passengers. These vessels are highly specialised, offering amenities such as lounges, restaurants, shops, and entertainment facilities. Modern cruise ships include casinos, gyms, theatres, cinemas, pools, and bars. The largest cruise ships measure up to 360 meters in length and 60 meters in width, operating mainly in regions like the Mediterranean, Caribbean, and Scandinavia.

Figure-3.7:Cruz Ship

3.2.5.3 Heavy-lift/Project Cargo Vessels

These vessels are built to transport extremely heavy or oversized cargo, such as ships or industrial components. They are used for moving offshore platforms and other large structures. Some are equipped with high-capacity cranes for ports lacking heavy-lift infrastructure, while others are semi-submersible, allowing cargo to be floated into position before being lifted. Common cargo includes wind turbine components, quay cranes, and machinery. Some are adapted with jack-up legs for stability during offshore installations. Heavy-lift ships can handle project cargoes up to 2,000 tons, which conventional vessels cannot accommodate.

Figure-3.8: Heavy Lift Cargo Ship

3.2.5.4 Wood Chip Carriers

Used for transporting wood chips to pulp mills, these vessels typically range from 40,000 to 60,000 DWT, though capacity is often measured by hold volume. They are equipped with cranes and conveyor belts for efficient discharge.

3.2.5.5 Livestock Carriers

These vessels transport animals, mainly sheep and cattle. They require specialised features such as fodder storage, ventilation, water supply, waste disposal systems, ramps, and crew accommodations. Most are converted from other vessel types, though some are purpose-built, especially for cattle. Some can carry up to 120,000 sheep, with common routes from Australia and New Zealand to the Middle East.

3.2.5.6 Product Carriers

Smaller than crude oil tankers, product carriers transport refined petroleum products like jet fuel, diesel, asphalt, lubricants, and tar. They also carry non-petroleum liquids such as molasses and palm oil, typically from large terminals to smaller ports.

3.2.5.7 Chemical Carriers

These vessels, with deadweight ranging from 5,000 to 40,000 tons, are equipped with specialised systems tailored to the cargo type. These may include heating/cooling systems and advanced cleaning mechanisms to maintain cargo purity in multi-purpose tanks.

3.2.5.8 Liquefied Gas Carriers

Originally converted from oil tankers, these vessels are now purpose-built to transport Liquefied Petroleum Gas (LPG) and Liquefied Natural Gas (LNG) under pressure. Their spherical tanks provide structural strength. LNG carriers are generally larger than LPG carriers, with the largest being Q-Flex vessels, capable of carrying up to 266,000 cubic meters of gas.

Figure-3.9: LNG Carrier Ship

These are highly specialised tankers designed for transporting liquefied gases. They are primarily classified into two types based on the cargo they carry.

3.2.5.9 LNG Carriers (Liquefied Natural Gas)

These vessels are built to transport methane, the main component of natural gas, in its liquefied form. To maintain methane as a liquid, it must be stored at extremely low temperatures of −162°C under atmospheric pressure. LNG carriers are equipped with insulated tanks, often spherical in shape, to withstand the pressure and maintain the required temperature.

- Size Measurement: LNG carrier capacity is measured in cubic meters (CBM).
- Typical Capacity: Ranges from medium-sized vessels to the largest, such as:

- **Q-Flex**: 210,000–217,000 CBM
- **Q-Max**: 261,700–266,000 CBM

These two types form the Qatari Q-Class, which accounted for 18% of global LNG carrying capacity as of 2016.

3.2.5.10 LPG Carriers (Liquefied Petroleum Gas)

LPG carriers transport gases like propane, butane, ammonia, and other chemical gases. These vessels use two main methods to keep the cargo in liquid form:

- **Fully Pressurised Vessels**: Maintain cargo at high pressure (10–12 bars) without refrigeration.
- **Refrigerated Vessels**: Cool the cargo to temperatures as low as –50°C, allowing it to remain liquid at lower pressures.
- **Size Measurement**: Also measured in CBM, with capacities ranging from:
  - Small LPG Carriers: Around 500 CBM
  - VLGC (Very Large Gas Carriers): Up to 85,000 CBM

LPG carriers are versatile and used for transporting both petroleum-based gases and chemical gases. Their design varies depending on the cargo type and required containment method

3.3 Vessels Dimension For Design

Defining the dimensions of vessels is essential for designing navigational channels, berthing basins, and calculating the forces exerted on quay walls during berthing and mooring. The key parameters used to describe a vessel’s size and load capacity include:

- **Draught**: The vertical distance between the waterline and the bottom of the hull (keel), indicating how deep the vessel sits in the water.
- **Length**: The overall length of the vessel from bow to stern.
- **Beam**: The maximum width of the vessel, typically measured at the midsection.

- **Wind-exposed surfaces**: Areas of the vessel that are subject to wind forces, affecting manoeuvrability and mooring loads.

3.3.1 Load and Volume Metrics

a. **Dead Weight Tonnes (DWT)**: The total weight a vessel can safely carry, including cargo, fuel, water, provisions, crew, and other supplies.

b. **Vessel's Gross Tonnage (GT)**: A measure of the vessel's total internal volume, calculated according to the IMO's 1969 International Convention on Tonnage Measurement.

c. **Gross Registered Tons (GRT)**: An older metric representing the internal volume in Moorsom tons (1 Moorsom ton = 100 cubic feet or 2.83 m^3). This has largely been replaced by GT.

Some vessel types use alternative capacity indicators:

- **Gas carriers** (e.g., LNG and LPG vessels) are measured by their **cubic meter (m^3)** capacity.
- **Container ships** are classified by **TEUs (Twenty-Foot Equivalent Units)**, representing the number of standard containers they can carry.

It's important to understand that although terms like DWT, GT, and GRT are widely used, they do not truly indicate how well a vessel can manoeuvre, so they are not used alone for design decisions.

3.3.2 Key Structural Components of a Vessel

a. **Bow**: The forward-most part of the ship that cuts through water. It should minimise drag and be tall enough to prevent excessive water splashing.

b. **Thrusters**: Propeller-like devices installed on both sides of the bow to aid manoeuvring in tight spaces, especially at low speeds near ports or in canals.

c. **Hull**: The main body of the ship below the waterline, acting as a protective shell against external elements.

d. **Keel**: The structural backbone of the ship, running along the bottom centreline, distributing stress and providing strength.

e. **Freeboard**: The vertical distance between the waterline and the upper deck. It varies with cargo load and must meet minimum safety standards as per the International Convention on Load Lines (ICLL 1966) to prevent sinking. Thus, to ensure the safety of ships at sea, load line marking is used to ensure the maximum possible draft for a safe voyage.

3.3.3 Directional and Positional Terms

a. **Stern (After End)**: The rear part of the ship. When moving backwards, the vessel is said to be moving astern.

b. **Fore End (Bow)**: The front part of the ship. Movement in this direction is referred to as ahead.

c. **Fore and Aft**: Terms used to describe directions along the length of the vessel.

d. Amidships: The central area between the bow and stern.

The rear portion of the ship is called the after end or **stern**. When moving stern first, the vessel is said to be moving astern. The front portion of the ship is called the fore end, whilst the extreme forward end is called the **bow**. When moving bow first, the vessel is said to be moving ahead. Fore and aft are generally used for directional purposes. The area between the forward and aft portions of the vessel is called **amidships**.

3.3.4 Special Design Features

i. Bulbous Bow

A protruding bulb at the front below the waterline, designed to reduce pitching and improve fuel efficiency. Commonly found on tankers, bulk carriers, and modern cargo ships. The bulbous bow can improve passenger and crew comfort, as it can reduce pitching in heavy seas and has been provided in tankers, bulk carriers, and modern cargo liners to increase speed when in ballast.

ii. Beam :

The widest part of the vessel, typically located amidships, crucial for stability and cargo capacity.

3.3.5 Shipboard Measurements

A ship is a three-dimensional structure defined by its length, breadth (beam), and depth. Understanding these dimensions is essential for safe navigation, docking, and operational planning.

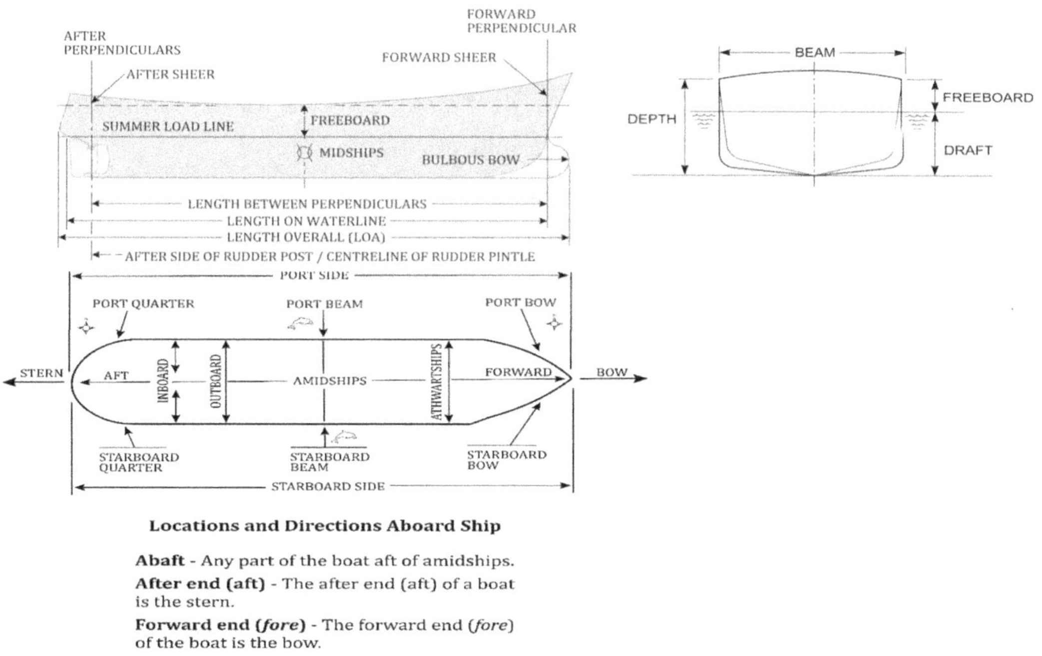

Figure-3.10: Directional and Positional Terms of Ship

3.3.5.1 Length Measurements

i) **Length Overall (LOA)**: The total length of the ship from the foremost point of the bow to the rearmost point of the stern. This is a key measurement used in ship specifications and manoeuvring.

ii) **Length Between Perpendiculars (LBP):** The distance between the forward perpendicular (where the stem meets the waterline) and the aft perpendicular (passing through the rudder stock).

iii) **Length at Waterline (LWL)**: The length of the ship measured along the waterline. This dimension is crucial for calculating speed, resistance, and hydrodynamic performance.

3.3.5.2 Breadth

Beam: The widest part of the ship, measured from the outermost point on one side to the outermost point on the opposite side, typically at the midsection (amidships).

Figure-3.10: Directional and Positional Terms of Ship

3.3.5.3 Depth and Vertical Measurements

i) **Depth**: The vertical distance from the bottom of the keel to a selected reference deck. It includes several key vertical parameters:

- **Waterline**: The line where the hull meets the surface of the water when the ship is afloat.
- **Freeboard**: The vertical distance from the waterline to the upper deck edge. It varies with cargo load and must meet minimum safety standards under the International Convention on Load Lines (ICLL 1966).
- **Draft (Draught)**: The vertical distance from the waterline to the bottom of the keel, indicating how deep the ship sits in the water.
- **Air Draft**: The vertical distance from the waterline to the highest point of the ship (e.g., mast or funnel). This measurement is important for passing under bridges or overhead structures.

These vertical dimensions change depending on the ship's load, including cargo, fuel, and provisions. Draft is typically measured in meters, and draft marks are painted on the hull near the bow and stern for visibility.

3.4 Port Terminal

A **port terminal** is a designated facility within a port where cargo and passengers are transferred from ships to shore, and subsequently transported via land-based systems such as trucks, trains, or pipelines. It functions as a key interface between maritime and inland logistics, facilitating the efficient handling, storage, and movement of goods, cargoes, and people.

3.4.1 Container Terminals

A container terminal is a specialised facility designed for the **transhipment, handling, and temporary** storage of shipping containers between at least two modes of transport—typically sea and land. It consists of several key components as given below:

i) **Quays**: Where ships dock for loading and unloading. The wharf and terminal facilitate the transhipment, handling, and temporary storage of containers between at least two transportation modes.

ii) **Container Yard**: An area for storing containers temporarily.

iii) **Equipment**: This includes Ship-to-Shore (STS) gantry cranes, which come in different sizes depending on the type of container ships they serve, as well as yard-handling equipment and horizontal transport systems.

On the quayside, STS cranes are used to load and unload container ships. Their size varies based on the ship class they can handle. A Panamax STS can work with ships up to 13 containers wide, while a Post-Panamax STS can handle up to 16 containers. The newest type, called the 'Ultra Post-Panamax,' can serve the latest 24,000-TEU container ships by spanning up to 24 containers.

iv) **Support Infrastructure**: Administrative buildings, maintenance facilities, and warehouses.

3.4.2 Dry Bulk Terminals

Dry bulk cargoes include iron ore, coal, limestone, alumina and grains. They are not homogeneous, as there are several grades of coal, iron ore, and limestone used for making steel and energy generation (steam coal). Each type of raw bulk material has a different market, implying a segmentation of the bulk trades.

When handling these types of cargo in a port terminal, directionality in bulk trade is an important factor. This means export bulk terminals are designed quite differently from import bulk terminals.

3.4.2.1 Export terminals

Export terminals are sophisticated systems designed for the efficient handling and storage of bulk cargo. The process typically begins with cargo arriving via various transport modes, such as rail (rake) or road. Upon arrival, the cargo is moved to a storage yard, which is equipped with specialised machinery like stacker reclaimers that stack and retrieve cargo from stockpiles.

The final stage involves loading the cargo onto ships using a ship loader, which is connected to the storage yard through a conveyor system. This integrated setup ensures a seamless transition of cargo from land-based transport to maritime vessels, ready for export. Effective terminal design requires a balanced interface between hinterland connectivity and maritime access.

3.4.2.2 Import terminals

Import terminals handling commodities like coal and iron ore utilise ship unloaders equipped with large grabs to discharge cargo. The selection of crane type is based on the terminal operator's operational requirements, including the size of vessels to be served and the minimum cargo handling rate in tons per hour. Once unloaded, the cargo is moved via conveyor systems along the quay to the storage yard.

Since import terminals typically manage multiple types and grades of bulk materials, the yard is organised with distributed stockpiles to accommodate this variety.

Terminal planning is complex due to the stochastic (unpredictable) arrival of vessels and the fact that consignees choose their preferred landside services and inland transport modes, adding variability to operations.

Dust control is a critical concern at dry bulk terminals, especially during the handling of materials like coal, limestone, and iron ore. During transfer operations, these materials can fragment, generating dust that may be dispersed by crosswinds. Even after stockpiling, dust remains a challenge. If the material arrives wet, wind over the stockpile can evaporate the moisture, causing dust particles to lift into the air.

The transloading capacity of a dry bulk terminal—particularly for coal and iron ore—is influenced by multiple factors, including:

- **Quay length**: Determines how many vessels can be accommodated simultaneously.
- **Yard dimensions**: Affects the volume of cargo that can be stored.
- **Equipment**: Includes cranes, stackers, reclaimers, and conveyor systems at both quay and yard.
- **Stockpile layout and height**: Impacts storage efficiency and dust control.
- **Storage time distribution**: Varies based on cargo turnover rates.
- **Number and types of bulk grades handled**: Requires segregation and specialised handling.
- **Operating hours**: Both waterside (ship operations) and landside (inland transport).
- **Arrival patterns and vessel types**: Influence scheduling and resource allocation.
- **Land transport modes**: Rail, road, or conveyor systems affect cargo movement speed and efficiency.

3.4.2.3 Grain Terminals and Quay Structures

Grain terminals are specifically designed to handle and store agricultural commodities such as rice, wheat, soybeans, and other grains and oilseeds. These terminals utilise a variety of unloading systems:

- Discontinuous systems, such as grabs, which operate in cycles.
- Continuous Ship Unloaders (CSUs), which include pneumatic chains, screw conveyors, and twin-belt machines, allowing for uninterrupted cargo flow.

Grains are typically stored using grain elevators, which employ bucket elevators or pneumatic conveyors to lift grain from lower levels and deposit it into silos or other storage units. Large grain terminals often feature multiple silos arranged side by side to accommodate high volumes.

Semi-mechanised terminals are equipped with mobile handling equipment like harbour cranes and travelling hoppers, offering flexibility in cargo operations.

3.4.2.4 Unitised Cargo Berth

Unitisation refers to the process of consolidating general cargo into standardised units to improve handling efficiency. This approach is widely used across the transport system to enhance the performance of primary transport vehicles and to minimise the need for handling individual cargo items.

The practice of unitising general cargo began in the mid-1960s, initially through pre-slinging methods. Today, there are three main methods used for unitisation and cargo consolidation:

i) Freight Containers

ii) Pallets

iii) Pre-slung Cargo

These methods streamline cargo operations, reduce handling time, and improve safety and logistics efficiency at unitised cargo berths.

3.4.2.5 General Cargo Berth

General cargo, also referred to as break-bulk cargo, describes non-unitised goods traditionally transported on general cargo vessels. These items are loaded and unloaded individually, often manually, and are typically of a size and weight that allows for human handling.

This category includes loose or packaged goods consolidated into forms such as crates, cartons, bags, boxes, bales, and drums. To improve handling efficiency, general cargo may be unitised using pallets or containers.

Additionally, heavy lift cargo, out-of-gauge cargo, and small bulk items may also fall under general cargo classification.

The equipment used for consolidating and moving general cargo is known as stevedoring equipment, which varies depending on the cargo's characteristics. Common tools include slings, strops, nets, and lifting boards, designed for lift-on/lift-off operations, typically handling loads between 1.5 to 5 tonnes.

These tools are made from various materials such as natural fibre rope, synthetic fibre rope, steel wire rope, and chains, selected based on the cargo type and handling requirements.

Breakbulk cargo, also known as general cargo, refers to goods that are loaded onto a vessel or transport mode as individual items or bundled units, rather than being placed in containers or shipped as large-volume liquid or dry bulk loads.

This category includes a wide range of commodities, such as:

- **Project cargo**: Heavy and oversized equipment like generators, turbines, wind energy components, oil and gas machinery, cable reels, gas tanks, modular units, petrochemical plant components, mining equipment, construction machinery, brewery tanks, and silos.
- **Iron and steel products**: Coils, plates, bars, slabs, wires, pipes, and tubes.
- **Forest products**: Lumber and paper goods.
- **Bagged and parcel cargo**: Items like malt, fertiliser, sugar, and rice.
- **Reefer cargo**: Perishable goods such as fruits and meats are transported via refrigerated vessels.
- **Smaller breakbulk shipments**: Large bags, skidded goods, and palletised cargo.

Breakbulk cargo is typically handled using quay cranes, floating cranes, or the ship's own gear, such as deck cranes and derricks. Onshore, various types of lifting equipment are used, including dockside cranes, level-luffing cranes, and mobile cranes.

All vertical cargo movements are performed using lift-on/lift-off gear, which includes a shackle connecting the crane or derrick to the cargo-handling equipment. Most lifts use a hook, and a variety of loose gear or tools can be attached to the lifting system depending on the cargo type.

3.4.2.6 Liquid Bulk Terminals

The transfer operations at liquid bulk terminals, particularly for oil and gas products, involve complex mechanical systems engineered for high safety, operational efficiency, and environmental compliance. A key component in these operations is the marine loading arm, which comprises articulated steel piping systems. These arms are designed with counterweight balancing mechanisms to manage structural loads and ensure smooth movement during product transfer. They are typically equipped with remote control capabilities and integrated safety interlocks to prevent accidental releases or mechanical failures.

Loading arms are configured to handle single or dual liquid lines, with optional vapour return lines to manage displaced vapours during loading/unloading. Their design allows for flexibility in handling various liquid hydrocarbons and chemical products.

To mitigate environmental impact and comply with regulatory standards (e.g., MARPOL Annex VI, EPA VOC guidelines), terminals implement Vapour Recovery Units (VRUs). These systems are engineered to capture and process volatile organic compounds (VOCs) and hydrocarbon vapours emitted during cargo operations. The captured vapours are routed through a recovery process that includes desulfurization and other purification steps, tailored to the chemical composition of the stored products.

The integration of VRUs is critical for:

- Reducing atmospheric emissions,
- Enhancing terminal safety,
- Meeting international and local environmental regulations.

The design, specification, and operational protocols of these systems must align with industry standards such as **API RP 1004**, **OCIMF guidelines**, and **ISO 28460**, ensuring robust and compliant terminal infrastructure.

3.4.2.7 RO-RO Terminals

The layout of a RO-RO (Roll-on/Roll-off) terminal depends on the kind of ramp the ship uses. Most ships just need a straight dock, but if they have ramps at the front (bow) or back (stern), the dock might need a small extension to help the ship line up properly for loading and unloading.

Modern car terminals are massive. Workers usually arrive in vans, board the ship, and each person drives a car off the ramp and parks it in a specific spot in the yard. These yards are like giant parking lots, designed to fit as many vehicles as possible while still allowing easy access. The parking layout is carefully planned to make the best use of space without slowing down operations.

Many RO-RO terminals also handle other types of cargo, like containers or break-bulk items. This is especially common when **ConRO vessels** (which carry both containers and vehicles) are involved. These hybrid ships make terminals more flexible and efficient.

Some Major RO-RO Terminals

- **Bremerhaven (Germany)** – A key hub for German car exports, especially for brands like BMW and Mercedes.
- **Yokohama (Japan)** – Central to Japan's automotive exports to Southeast Asia.
- **Jebel Ali (UAE)** – Handles RO-RO cargo from South Korea and Europe to the Middle East.

3.5 Performance Indicators of Port Facility

3.5.1 Port Performance and Vessel Turnaround Time

How well a port performs is super important in the shipping world. It's usually measured by how fast ships are handled, how quickly cargo is moved, and how long cargo stays in the port. These things help figure out how smoothly and efficiently the port is running—and where it might need to improve.

One of the biggest indicators of port efficiency is **Vessel Turnaround Time (VTT)**. That's the total time a ship spends in port—from the moment it arrives to when it leaves. Keeping this time short is crucial for ports and terminal operators to stay competitive. If ships are stuck too long, it costs more money and causes delays for shipping companies and everyone else involved in the supply chain. That's why ports are always trying to track and improve VTT.

3.5.2 Key Components of VTT

- **Waiting for Berthing** – This is the time a ship waits to be assigned a spot at the dock. If the port is busy, this can take a while.
- **Berthing Process** – The time it takes for the ship to move into position and tie up at the dock.

- **Cargo Operations** – How long it take to load or unload cargo? This depends on the type of cargo and how efficient the equipment is.
- **Idling Time** – Any time the ship isn't doing anything, like waiting for cargo to be ready or for crew changes.
- **Other Activities** – Things like inspections, paperwork, or crew-related tasks that also add to the total time.

All these steps add up to the total time a ship spends in port. If each part is made more efficient, the whole process speeds up. You can think of VTT in three main stages:

i) When the ship arrives at the port,

ii) When it's getting ready for cargo operations,

iii) And when it's actually loading or unloading at the dock.

Each of these steps counts toward the total turnaround time, and improving them helps ports run better and faster. Each of these components is essential for calculating the total time a vessel spends in port from arrival to departure. Optimising these elements can lead to improved turnaround times and, by extension, greater port efficiency. Thus, the turnaround time of a vessel starts after it has arrived at the port and ends when it has departed. All activities between the arrival and the departure count towards the turnaround time, and it can be identified in three stages. Firstly, the arrival of the vessel from within the port, secondly, the time required to service the vessel for voyage and to make it ready to take on cargo, and thirdly, the actual time used to load and discharge cargo at the berth, as illustrated in Figure 3.11.

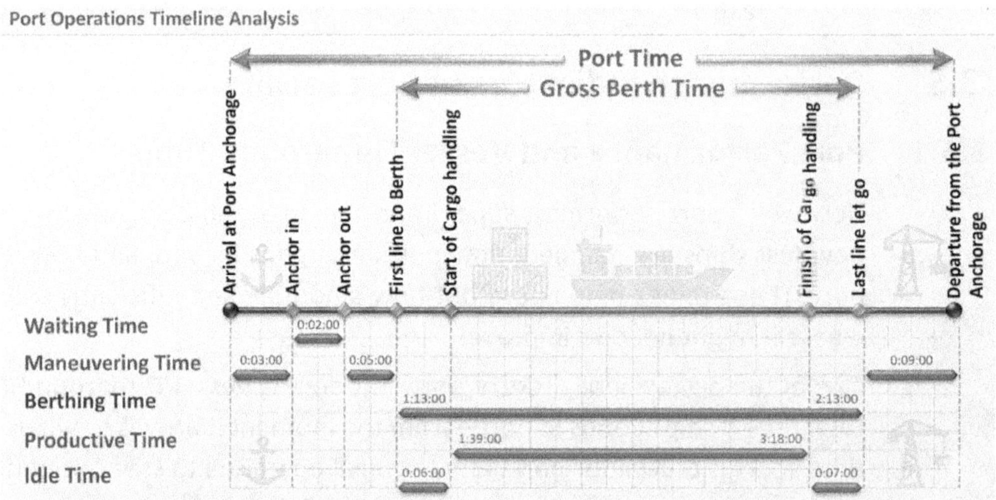

Figure-3.11: Components of Vessel Turnaround Time

3.5.3 Berth Performance Indicators

To just figure out how well a berth will perform, there are a few key things people usually look at:

- Berth Throughput
- Waiting Time

- Service Time
- Time in Port
- Grade of Waiting
- Berth Occupancy Ratio
- Berth Working Time Ratio

3.5.3.1 Berth Throughput

The berth throughput is the total tonnage or number of units "handled at one berth in a given period of time (weeks, months or years). Berth throughput is a measure of berth activity. Units of berth throughput include TEU throughput at container berths, tonnes moved at bulk and general cargo berths, and the number of vehicles handled at RO-RO berths. Some port operations, such as transhipment and re-stowage, involve double handling and double counting.

Berth Throughput (BT) = Total units handled in a period of time

3.5.3.2 Waiting Time

Waiting Time is defined as the time a ship spends waiting for an available berth, that is, the delay between a ship's arrival in' the port and its tying up at the berth. A performance indicator used by port management is the Waiting Ratio, which for an individual ship can be established by the formula:

Waiting Ratio =Time waiting for a berth / Service Time

3.5.3.3 Service Time

Service Time is defined as the time a ship stays at a berth, whether it is working or not. The service time is established from the first line ashore to the time the last line is let go. A performance indicator, the Service Time Ratio, can be established for a berth, terminal or port using the formula

Service Time Ratio = Cumulated Service Time /Total Number of Ships

3.5.3.4 Turnaround time

Turnaround time or Time in Port is the total time the ship spends in the port from arrival at the port to final departure. A performance indicator, the Time in Port ratio, can be established for a berth terminal or port using the formula

Time in Port Ratio = (Cumulated Waiting Time + Service Time) / Total number of ships

3.5.3.5 Grade of Waiting

Grade of Waiting. For commercial reasons, it is very frequent to calculate a Grade of Waiting (GW) ratio. The comparison of the waiting time with the service time provides a good indicator of what is acceptable to shipowners. Shipowners usually accept a 10% Grade of Waiting ratio, beyond which the port is considered inefficient or of low quality.

Grade of Waiting Ratio = Cumulated Waiting Time / Cumulated Service Time

3.5.3.6 Berth Occupancy Ratio

Berth Occupancy Ratio (BOR) as a measurement of the level of port infrastructure utilisation depends on the number of moorings (Layaa dan Dullaert, 2014), so the Berth Occupancy Ratio (BOR) can be calculated with different approaches according to the type of mooring in a port. The Berth Occupancy Ratio (BOR) is a critical metric in port management, reflecting the efficiency and utilisation of berth space. It's influenced by various factors such as the type of cargo, ship size, and productivity of loading and unloading operations. A standard BOR of around 65% is considered optimal, ensuring a balance between high usage and availability for incoming ships. Exceeding a 70% BOR often signals overutilization, leading to congestion and potential delays in cargo handling, while a BOR below 50% may suggest underutilization of resources. Efficient management of BOR is essential for maintaining smooth port operations and avoiding bottlenecks that can disrupt the entire supply chain. Understanding and monitoring this ratio helps in strategic planning and resource allocation to optimise port performance.

Berth Occupancy Ratio is the ratio determined by dividing the time a berth has been occupied (in hours per year) by the total number of hours in a year (8760). It shows the level of demand for port services.

Berth Occupancy Ratio = Total Service Time (per berth) /Hours in a Year (8760)

3.5.3.7 Berth Working Time

Berth Working Time Ratio is a ratio that relates the number of hours during which a ship is worked in port during the Total Service time on the berth. It informs whether or not there is a long idle time during berth operations. For example, a berth working time ratio of 50% means that the port works only 12 hours a day and the ship is idle for 12 hours.

Berth Working Time Ratio = Total Time Worked / Total Service Time

3.5.3.8 Handling Operation Indicators

Ports are basically service providers, and the efficiency of handling cargo is also a very important indicator for port performance. In cargo handling operations at a port, three sets of resources are considered, namely Ship/Shore Handling Equipment. Yard Transfer-Equipment and the Labour Force. The indicators below are frequently used for the measurement of the performance of the handling operations.

Ship-out Indicators measure the rate at which the cargo is handled to and from a vessel. There are three ship output indicators, namely: working ship output, berth ship output and port ship output:

Working Ship Output = Total* tonnage *handled* (tonnes or ship hours worked) / *Total hours worked

Berth Ship Output = Total tonnage handled* (tonnes or ship hours on berth) */Total hours on the berth

Port Ship Output = Total tonnage handled* (tonnes/ship hour in port) / *Total hours in port

Gang Output: The ship output depends, among others, on the number and efficiency of gangs used for handling. Except in small old-fashioned ports, the gang includes not only workers but also the handling equipment for ship-to-shore operations and quay transfer operations. It follows that gang output varies by type and capacity of the equipment used, the cargo type and the onboard position of cargo. In some ports, the average output is calculated per man-hour. The following Gang Output indicators are used:

Average Output/Gang/Hour = Tonnage handled / Total gang X hours worked

Average Number of Gangs/Ship = Total number of gangs / Number of ships

Utilisation Ratios refer to the effectiveness of resource utilisation and include both machinery and the human resource:

Rate of Utilisation of Cranes = Number of worked hours / Number of available hours

Rate of Utilisation of Workers = Number of man-hours worked / Number of man-hours available

('available' time refers to the time a machine or resource is available for use, less periods for maintenance and bad weather).

3.5.3.9 Storage Operation Indicators

Most ports of the world have to provide efficient and safe storage space or yard for containers, break-bulk cargo, tanks for liquid bulk storage and open storage space, warehouses for long-term storage. The cost involved in providing the outlay, development and maintenance is substantial; therefore, Dwell time of cargo is used as another performance indicator of the port.

Dwell Time is the time spent by a cargo in port and usually refers to containers. It can be applied to other cargo types.

Average Dwell Time is a utilisation ratio that does not concern port layout, capacity, and space requirements. The formula for establishing the average dwell time is valid for both open and covered storage.

Average Dwell Time = Tonnage x dwell time / Total tonnage stored per day

3.6 Hinterland connectivity

Hinterland connectivity refers to the efficiency and quality of transportation and logistics infrastructure that links a **port or terminal** to its **inland regions**—such as industrial zones, cities, and distribution centres. It plays a critical role in the overall performance of maritime logistics and supply chains.

Hinterland connectivity ensures that goods arriving at or departing from a port can be moved quickly, reliably, and cost-effectively to and from inland destinations. It is essential for:

- Reducing transit time
- Lowering transportation costs
- Improving supply chain reliability

- Enhancing port competitiveness

Key Components of Hinterland Connectivity are;

i. **Road Networks**

Highways and expressways connecting ports to industrial areas and cities. These are vital for short- to medium-distance cargo movement, especially for time-sensitive goods.

ii. **Rail Infrastructure**

Railways provide cost-effective and high-capacity transport for bulk and containerised cargo over long distances. Inland container depots (ICDs) and dry ports are often linked by rail to seaports.

iii. **Inland Waterways**

Rivers and canals are highly desirable for transporting goods to and from ports. They are environmentally friendly and suitable for bulk cargo, though limited by geography and infrastructure.

iv. **Intermodal Facilities**

Intermodal Facilities are locations where cargo is transferred between different modes of transport (e.g., from ship to train or truck). It includes container freight stations (CFS), logistics parks, and multimodal hubs.

v. **Digital Connectivity**

Use of logistics software, tracking systems, and port community systems (PCS) to streamline cargo movement and documentation.

3.6.1 Role of Hinteland Connectivity in Port Operations

i) **Attracting Shipping Lines and Cargo Volumes**

Ports with strong hinterland connectivity are more attractive to global shipping lines and logistics providers. Efficient inland transport options—such as rail, road, and inland waterways—ensure that cargo can move quickly and reliably between the port and inland destinations. This reliability encourages shippers to route more cargo through such ports, increasing overall throughput and competitiveness.

ii) **Reducing Port Congestion and Improving Turnaround Times**

When inland access is smooth and well-integrated, cargo can be cleared from the port area faster. This reduces **dwell time** (the time cargo stays in the port), minimises **yard congestion**, and allows ships to be loaded and unloaded more efficiently. As a result, **vessel turnaround times** improve, which is a key performance indicator for port efficiency.

iii) **Supporting Export and Import Flows**

For exports, hinterland connectivity links manufacturing zones, industrial parks, and agricultural regions to the port, enabling the timely delivery of goods for international shipment.

For imports, it ensures that goods arriving at the port can be distributed efficiently to consumption centres, warehouses, and retail hubs inland.

iv) Enhancing Supply Chain Integration

Ports act as critical nodes in global supply chains. Strong hinterland connectivity allows for better integration with logistics networks, including intermodal transport hubs, dry ports, and distribution centres, enabling seamless cargo movement from origin to destination.

v) Reducing Logistics Costs and Environmental Impact

Efficient hinterland transport reduces fuel consumption, transit time, and handling costs. It also supports the use of sustainable transport modes like rail and inland waterways, which have a lower environmental footprint compared to road transport.

3.7 Essentials in Planning of Port Facilities

i) Identification of Key Variables Influencing Demand

The first step in port planning is to determine the parameters that significantly affect cargo throughput and vessel traffic. Typical demand drivers include:

- Annual cargo volume growth linked to production and consumption trends
- Vessel size evolution and fleet composition
- Seasonal fluctuations and impacts of trade policies
- Distribution of cargo evacuation modes (rail, road, pipeline)

These variables form the basis for scenario development and sensitivity analysis.

ii) Demand and Vessel Traffic Forecast

Vessel traffic forecasting shall be based on the development of multiple demand scenarios. Three primary demand scenarios must be prepared:

3.7.1 Base Case Scenario

A realistic projection derived from historical cargo throughput trends (typically 5–15 years of data). This scenario case is advantageous and:

- Provides a performance-based baseline
- Reflects observed economic and logistics conditions
- Minimises dependence on overly optimistic estimates

3.7.2 Conservative Case Scenario

This case generally represents 5–15% lower demand than the Base Case. Influencing factors may include:

- Economic slowdown
- Regulatory restrictions (e.g., coal usage limits)
- Increased competition or logistics constraints

- Delays in industrial project execution
- Geopolitical or trade uncertainties

The purpose of this scenario is to stress-test the infrastructure against downside risks.

3.7.3 Optimistic Case Scenario

This case generally represents 10–20% higher demand than the Base Case, potentially driven by:

- Stronger economic growth
- New industrial corridors and major investments (power plants, refineries, renewable parks)
- Enhanced hinterland connectivity (rail corridors, highways, ICDs)
- Policy incentives (tariff reductions, SEZ benefits, export incentives)
- Cargo diversion from competing ports

The purpose of this scenario is to assess capacity requirements under accelerated growth conditions.

iii) Purpose for Scenario-Based Planning

Scenario-based planning helps port planners to:

- **Handle market uncertainties:** It allows planners to assess how fluctuating or unpredictable economic conditions may affect future cargo demand.
- **Prevent incorrect sizing of infrastructure:** It reduces the risk of designing facilities that are either too large (leading to unnecessary cost) or too small (causing congestion and delays).
- **Test the sensitivity of key operational elements**, including:
- **Berth capacity** and its ability to handle varying vessel traffic volumes
- **Yard and storage requirements** under different cargo flows
- **Equipment fleet sizing**, ensuring adequate cranes, unloaders, and support machinery
- **Financial performance indicators** such as IRR and NPV to evaluate project viability under each demand condition

iv) Vessel Traffic Growth Estimation (Fleet Upsizing Considerations)

Cargo growth does not directly translate into equivalent vessel call growth due to changes in vessel sizes and service frequencies. A practical planning rule is:

- Vessel Traffic Growth = 0.8 to 1.2 × Cargo Growth

Typical conditions:

- If vessel sizes increase over time: Vessel traffic growth ≈ 0.8× cargo growth
- If fleet composition remains stable: Vessel traffic growth ≈ 1.0× cargo growth

- If smaller vessels or more service frequency develop: Vessel traffic growth ≈ 1.2× cargo growth

This range is widely adopted across Australia, Europe, the Middle East, and Asia.

v) Importance of Vessel Traffic Forecasts

Vessel projections are essential for:

- Berth capacity estimation
- Quay length determination
- Tug and pilotage resource planning
- Navigational simulation and marine safety assessments

3.8 Dry Bulk Terminal Planning Guidelines (TBA/PTI Practice)

Based on methodologies from Terminal Business Analytics (TBA) and Port Technology International (PTI), dry bulk import terminals generally face higher operational variability due to the handling of multiple commodity grades, each with distinct physical properties and handling requirements. This variability affects storage planning, equipment usage, and throughput forecasting. The main observations are;

i) Multiple Commodity Grades

Dry bulk terminals typically manage a diverse set of cargoes such as coal, iron ore, gypsum, limestone, bauxite, clinker, and fertilisers. Each commodity behaves differently in terms of flow characteristics, moisture sensitivity, and storage needs.

ii) Operational Implications

The presence of multiple grades influences several operational aspects, including:

- **Storage yard layout and size**, especially for segregated stockpiles.
- **Re-handling frequency**, which increases when grades must be relocated or blended.
- **Equipment scheduling**, as different materials require different loading and reclaiming cycles.
- **Blending and quality control operations**, particularly for coal, limestone, and iron ore.
-

iii) Forecasting Requirements

Demand forecasting must be carried out **commodity-wise**, not as a combined tonnage, because each grade imposes unique operational constraints. Forecasts should evaluate:

- **Peak handling rates** for each commodity.
- Storage turnover cycles, which vary significantly between fast-moving and slow-moving grades.

- **Required stockpile area** based on material bulk density, stacking method, and dwell time.

iv) **Operational Planning Considerations**

Effective planning requires:

- **Applying peak demand multipliers** to account for seasonal fluctuations and vessel bunching.
- **Designing multiple stockpile zones or silos** to ensure grade segregation.
- Accommodating **wide variations in vessel parcel sizes**, ranging from 10,000 tons to 180,000 tons, which impact berth-day requirements and yard usage.

v) **Benefits of Applying TBA/PTI Guidelines**

- **Accurate berth-day and yard occupancy calculations**, improving overall terminal efficiency.
- **Reduced re-handling and congestion**, leading to smoother cargo flow and lower operating costs.
- **Optimized capital expenditure (CAPEX)** for conveyor systems, stacker–reclaimers, unloaders, and storage infrastructure.

3.8.1 Berth Capacity and Occupancy Guidelines

It defines the recommended berth utilisation levels and throughput capacities required to ensure efficient vessel turnaround, minimise waiting times, and maintain reliable terminal performance under varying operational conditions.

Target Berth Occupancy Levels for various cases are;

- **Dry Bulk Berths:** 55–65% (to maintain efficient vessel turnaround for grab unloader operations)
- **Liquid Bulk Berths:** 40–50% (due to safety requirements and longer pumping cycles)
- **General Benchmark:** Occupancy should remain below ~70% to avoid exponential increases in vessel waiting time (UNCTAD reference). Above 70%, vessel waiting increases sharply.

Typical capacity ranges for berths are;

- A single Panamax-class dry bulk berth typically handles 3–5 MTPA, depending on the productivity of the unloader and vessel size distribution
- Grab-type unloaders generally deliver 800–2,000 TPH, aligned with international benchmarks for Panamax and Capesize operations.

Planning Rule: Oversize berth capacity by 20–30% over projected demand to maintain operational resilience.

3.8.2 Capacity Planning for Bulk Terminals

Dry Bulk Storage Requirement: Storage requirements directly depend on vessel size, berth productivity, and downstream evacuation capacity.

i) Import Terminals

These require 10–20 days of throughput as a storage buffer.

The need increases when handling multiple grades, since individual piles must be segregated to prevent cross-contamination.

Larger vessels (e.g., Cape-size) arriving at longer intervals also imply more pronounced surges, requiring adequate storage accommodation.

ii) Export Terminals

Export-oriented facilities may require large buffer stockpiles not just for operational continuity but also to take advantage of market timing and price fluctuations. The blending of ores/coal to meet contractual quality specifications also influences yard size.

3.8.2.1 Open Yard Storage Guidelines

i) Open Storage Yards

A typical bulk stockyard can accommodate 40,000–60,000 tonnes per hectare. Actual capacity depends on:

- Material density (e.g., coal vs iron ore),
- Maximum permissible pile height (linked to stability and equipment reach),
- Yard drainage and environmental controls (dust suppression, runoff channels).

ii) Implications for Panamax/Cape-Size Cargoes

A single Cape-size ship carrying 150,000 tonnes of cargo may require 2.5–4 hectares for storage if piles are not immediately evacuated. Efficient conveyor evacuation systems significantly reduce yard footprint requirements.

A. Liquid Bulk Storage (For Multipurpose Ports)

Though not central to dry bulk operations:

- **7–15 days of throughput** is typical for petroleum and chemical terminals, accounting for safety spacing, firefighting envelopes, and pumping cycle constraints.

B. Integrated Operational Perspective for a Panamax/Cape-Size Terminal

A bulk terminal designed for Panamax and Cape-size vessels must balance:

a. **Berthing Efficiency: D**ictated by occupancy < 65% and adequate unloader capacity.

b. **Ship-to-shore transfer efficiency:** Grab-type unloaders sized to minimise vessel turnaround.

c. **Stockyard buffer capacity**: 10–20 days of throughput for imports, more for exports or multi-material terminals.

d. **Evacuation logistics:** conveyors, rail, or trucks must match unloader discharge rates to prevent bottlenecks.

By integrating these parameters, the terminal ensures smooth operations, minimised vessel queues, and optimised lifecycle cost performance.

C. **Conveying & Cargo Handling Systems**

a. **Grab unloaders:** 600–1,200 TPH for multipurpose terminals.

b. **Continuous ship unloaders:** 1,500–3,000+ TPH for high-volume ores.

c. **Belt conveyors:** Typical speeds **2.5–4.0 m/s**, capacity **1,000–4,000 TPH** depending on belt width (international mechanical spec norms).

d. Multiple grades → **more conveyors + transfer points** (PTI reference).

D. **Hinterland Evacuation (Rail / Road / Pipeline)**

a. **Rail evacuation share** should ideally be **>60%** for minerals/coal terminals above 10 MTPA.

b. **Truck turnaround target: ≤ 30 minutes** inside terminal (global benchmark).

c. **Liquid bulk pipelines:** sized for **1.5× ship discharge rate** to keep berth free faster (common OCIMF recommendations).

E. **Environmental & Safety Norms**

UNCTAD & OCIMF emphasise:

- **Dust control:** Dry fog or water mist with **90%+ suppression efficiency** for fine ores.
- **Noise control:** Maintain operational noise < **85 dB** at equipment source (global HSE standard).
- **Liquid bulk:** Follow **ISGOTT** & OCIMF guidelines for emergency shutdown, containment bunds, H_2S precautions.

F. **Operational Efficiency**

- **Equipment availability target:** 85–90%.
- **Dry bulk unloader utilisation:** 60–70% achievable for continuous operations.
- **Truck productivity:** 12–16 cycles/hour depending on terminal layout.
- **Berth productivity:**
  - Coal: 15,000–40,000 T/day
  - Ore: 30,000–60,000 T/day

G. **Financial & Phasing Guidelines**

- UNCTAD recommends **phased development** linked to demand triggers (avoids sunk cost & underutilization).
- **Contingency allowance:** 10–20% of project cost in developing-country ports.

- **Breakeven throughput:** Typically **40–60%** of design capacity, depending on tariff regime & CAPEX intensity.

3.8.2.2 Vessel Handling & Unloader Selection

i) Unloader Capacity Sizing

- **Panamax vessels (60,000–75,000 t)**
  → Recommended grab-unloader capacity: **1,000–1,500 TPH**.
- **Cape-size vessels (120,000–180,000 t)**
  → Recommended dual-unloader configuration: **1,500–2,500 TPH each**, depending on berth design.

Design Thumb Rule : *Size the unloading system so that a Panamax can be discharged in **36–48 hours**, and a Cape-size in **60–72 hours** to avoid berth congestion.*

3.8.2.3 Conveying & Cargo Handling Systems

- **Grab unloaders:** 600–1,200 TPH for multipurpose terminals.
- **Continuous ship unloaders:** 1,500–3,000+ TPH for high-volume ores.
- **Belt conveyors:** Typical speeds **2.5–4.5 m/s**, capacity **1,000–4,000 TPH** depending on belt width (international mechanical spec norms).

a. Conveyor Capacity

- Conveyor TPH ≥ **Unloader TPH × 1.1 (10% surge buffer)**
- Stacker/reclaimer capacity ≈ **70–80%** of conveyor capacity.
- Multiple grades → **more conveyors + transfer points** (PTI reference).

b. Typical Conveyor Belt Widths

Table-3.1: Typical Conveyor Belt Width vs Capacity

Belt Width	Capacity Range
1000 mm	1000–1200 TPH
1200 mm	1500–1800 TPH
1400 mm	2000–2400 TPH
1600 mm	2500–3200 TPH

c. Stacker & Reclaimer Capacity Planning

- Stackers: **1500–4000 TPH**
- Reclaimers: **1000–3000 TPH**

d. Equipment Mix

Table-3.2: Overview for Equipment Planning

Terminal Size	Recommended Machines
< 5 MTPA	1 stacker + 1 BWR
5–10 MTPA	1 SCR + 1 BWR
> 10 MTPA	Multiple dedicated stackers & reclaimers

e. Stockpile Geometry

- Pile height: **8–15 m**
- Pile width: **30–50 m**
- Stacker boom: determines pile width & yard utilisation efficiency.

3.8.2.4 Hinterland Evacuation (Rail / Road / Pipeline)

a. Rail Evacuation

- One rake: **3500–4000 t**
- Load/unload time: **45–90 min**

Rail evacuation share should ideally be **>60%** for minerals/coal terminals above 10 MTPA.

b. Truck Loading

- Truck loading rate: **600–1200 TPH**
- One lane handles **25–40 trucks/hr**

Truck turnaround target: ≤ 30 minutes inside terminal (global benchmark).

c. **Liquid bulk pipelines:** sized for **1.5× ship discharge rate** to keep berth free faster (common OCIMF recommendations).

3.8.2.5 Power Supply Planning

- Grab unloaders: **1–1.5 MW each**
- Stackers: **300–700 kW**
- Reclaimers: **500–1000 kW**
- Conveyors: **5–10 kW/m** depending on loading
- Substation capacity: Add **15–20% spare** for redundancy & future expansion.

3.8.2.6 Dust Suppression Planning

Bulk terminals generate dust from unloading, conveyors, transfer towers, and stockyards. Effective dust suppression is essential for safety, visibility, environmental compliance, and neighbourhood impact.

a. **Stockyard Dust Suppression**

- Install **3–5 mist cannons per hectare** (water-based atomized fog).
- Use **programmable sprinkler lines** around pile perimeters.
- Install **wind fences (8–12 m high)** around stockyard boundaries.
- Cover high-wind belt sections or use enclosed galleries.

b. **Conveyor Dust Control**

- Enclose conveyors, especially at:
  - Transfer towers
  - Surge points
  - Discharge chutes
- Use **dust skirts**, **hoods**, and **bag filters** at critical transfer points.
- Install **dry fog systems** at hoppers and unloader grabs.

c. **Unloader Dust Suppression**

- Fit unloaders with grab-closing dust reduction systems.
- Use **mobile water fogging units** beneath discharge hoppers.

Thumb Rule: Dust control system capacity should match at least **10–15%** of peak material-handling rate (in terms of dust load reduction).

3.8.2.7 Fire-Fighting System Planning

Even dry bulk terminals require extensive fire protection—particularly for coal, petcoke, fertilisers, or flammable cargo residues.

a. **Stockyard Fire Protection**

- Provide a **hydrant network** with hydrants every **45–60 m** along yard roads.
- Install **monitors** (water/foam) covering:
  - Conveyor take-up zones
  - Hopper & transfer tower areas
  - Pile surfaces (for coal/pet-coke, risk of spontaneous combustion)
- Maintain fire-water storage capacity equal to:
  - **4 hours of continuous firefighting**
  - Flow rate depends on terminal size (typically **600–1200 m^3/hr**)

b. **Electrical Room & Substation Safety**

Equip MCC rooms and substations with:

- **Clean agent systems (FM-200 / Novec-1230)**

- **Smoke**/heat detectors
- 2-hour fire-**rated** enclosures

c. Unloader Fire Safety

Provide:

- Fire extinguishers in cabin & machinery house
- Hydraulic oil fire detection
- Emergency stop + fail-safe parking mechanism

Thumb Rule: *Firewater pumping capacity = 30–40 l/s per hydrant stream + additional foam monitors depending on cargo hazard.*

3.9 Planning for Container Terminals

Container terminal planning depends heavily on yard footprint, quay length, nautical profile, and equipment strategy, as highlighted in global port-planning literature.

i) Berth Occupancy Targets

- Optimal berth occupancy: 60–70%, ensures incoming vessels rarely queue, preserving schedule integrity.
- **Critical threshold:** Above **70–75%**, waiting times rise sharply due to fixed crane productivity and draft/tide restrictions.
- Because modern terminals face severe space and performance constraints, occupancy must be kept well below saturation levels.

ii) Quay Length Planning

- Panamax: 250–300 m LOA → 300 m berth
- New-Panamax: 335 m LOA → 350–400 m berth
- ULCVs (18k–24k TEU): 400+ m LOA → 450 m berth

iii) Quay Crane Allocation

- **1 QC per 65–80 m of vessel length** → aligns with modern operator planning behaviour.
- QC deployment by vessel size:
  - Panamax: **3–4 QCs**
  - New-Panamax: **4–5 QCs**
  - ULCV: **6–8 QCs**

This aligns with major operator practices and the planning behaviours observed in mega-terminals.

iv) Quay Crane Productivity

Productivity Benchmarks

- Gross crane rate (GCR): **25–35 moves/hr**

- Net crane rate (NCR): **30–40 moves/hr** (well-optimised terminals)
- Vessel productivity = **80–150 moves/hr** depending on number of QCs.

QC Technical Specs

- Outreach:
  - Panamax: **13–16 rows**
  - Post-Panamax: **18–22 rows**
  - ULCV: **22–26 rows**
- Hoisting speeds (loaded): **60–90 m/min**
- Trolley speed: **150–200 m/min**
- Gantry travel: **45–70 m/min**

3.9.1 Container Yard Handling Equipment Planning

Container yard design is heavily influenced by **space availability, automation level, and yard strategy**, as emphasised in literature on yard planning and container yard management.

i) Equipment Options

Table-3.3: Equipment for Container Yard Handling

Equipment	Stack Height	Productivity	When Used
RTG (rubber-tyred gantry)	5–8 high	20–30 moves/hr	Most common, high density
RMG (rail-mounted gantry)	6–10 high	25–35 moves/hr	Automated or rail-integrated yards
Straddle Carriers	2–3 high	20–30 moves/hr	High-speed horizontal transport
Reach Stackers	3–5 high	12–18 moves/hr	Low density, overflow yards

ii) Yard Density Thumb Rules

- RTG yards: 800–1,200 TEU/ha
- RMG yards: **1,200–1,600 TEU/ha**
- Straddle carrier yards: **500–750 TEU/ha**

iii) Crane Fleet Sizing

- 1 RTG per **25,000–35,000 TEU/year** of throughput handled in a block, or
- **6–8 RTGs per 100,000 TEU** throughput.

iv) Container Yard Storage Planning

- **Dwell Time Assumptions**

- Transhipment cargo: **2–4 days**
- Imports: **3–7 days**
- Exports: **3–10 days** (Supported by typical yard dwell patterns in modern terminals).

- **Yard Layout**

Key layout considerations:

- Storage blocks shall be divided into import, export, reefer, OOG (out-of-gauge), and empty sections, and the yard must allow **smooth circulation lanes** for ITVs and yard cranes.

- **Gate & Landside Planning**

- Gate Capacity

  OCR/automated gate lane: **60–80 trucks/hour**

  Manual lanes: **30–50 trucks/hour**

- Truck Turnaround Time (TTT)

  Target TTT: **25–45 minutes**

- **Gate Mix**

Separate lanes for imports, exports, reefers, empties, and hazardous cargo.

- **Integrated Container Terminal Planning Flow**

1. **Determine TEU throughput** (base + 20-year projection).
2. **Select berth length & crane count** based on vessel mix.
3. **Dimension yard** using density, dwell time, and automation strategy.
4. **Choose yard equipment system:** RTG/RMG/SC based on footprint and productivity.
5. **Plan internal transport:** ITVs/AGVs fleet sizing.
6. **Design the gate complex** for peak truck volumes.
7. **Integrate firefighting & dust suppression** systems at the conceptual stage.
8. **Plan reefer, OOG, IMDG zones** with safety spacing.
9. **Electrical & IT backbone**: include OCR, TOS, YMS, AGV control.
10. **Evaluate berth occupancy** (<70%) and yard crane productivity via simulation models.

Simulation and modelling approaches are crucial for final dimensioning, as emphasised in terminal planning literature.

3.9.2 Power Supply & Utility Planning

I. Equipment Power Demand

Table-3.4: Power Requirement for Container Yard Handling Equipment

Equipment	Power Demand
Ship-to-shore QC	1.0–2.5 MW each
RMG	300–600 kW
E-RTG	300–500 kW
Reefer plug	5–7 kW per container

II. Electrical Redundancy

- 20–25% additional substation capacity for expansions.

III. Yard Dust Management

- **2–3 mechanical sweepers per 50 ha**
- **Paved surfaces** in all operational areas (reduces dust at source)
- **Periodic water spraying** on:
  - Empty yards
  - Access roads
  - Heavy-vehicle corridors

IV. Wind & Environmental Control

- **Wind fences (6–10 m)** at exposed yard boundaries.
- **Green belts** where space permits.

Thumb Rule: Dust suppression should cover **≥80%** of the total paved operational area, especially in dry/windy climates.

3.9.3 Fire-Fighting & Emergency Planning

Container terminals face complex risks, including reefer faults, dangerous goods (IMDG), rubber tyres, hydraulic systems, cable fires, and yard equipment ignition.

i) Fire Hydrant System

- Hydrants every **45–60 m** along quay and yard perimeters.
- Hydrant pressure: **7–10 bar**
- Flow: **1,000–1,500 L/min**
- Firewater storage for **4 hours** of continuous firefighting.
- Pump capacity: **2,000–6,000 L/min** depending on terminal scale.
-

ii) **Hazardous & Reefer Areas**

- Dedicated **IMDG zones** with separation distances.
- Continuous **thermal imaging** of reefer stacks.
- Quick-disconnect power supply for emergency isolation.

iii) **Quay Crane Fire Protection**

- Automatic fire detection in the machinery house.
- CO_2 / clean-agent suppression in electrical rooms.
- Hand-held extinguishers in the operator cabin and gantry.
- Emergency "safe-park" position in case of fire.

iv) **Yard Equipment Fire Safety**

- Straddle carriers, RTGs, reach stackers equipped with:
    - Engine-bay thermal detectors
    - Onboard extinguishers
    - Spark arrestors
- Fuelling stations: foam-suppression systems + spill containment.

v) **Substation & Control Room Fire Protection**

- 2-hour fire-rated construction
- FM-200 / Novec-1230 gaseous suppression systems
- Heat and smoke detection arrays

Thumb Rule: Fire system must be designed to combat one stack fire + one equipment fire simultaneously, matching global operator safety practices.

3.10 Building Facility Planning of Port Terminal

3.10.1 Bulk Terminal – Building Planning

Bulk terminals require buildings that support operations, control, maintenance, safety systems, and workforce functions. Planning mainly depends on throughput (MTPA), yard size, and equipment fleet.

i) **Operations Building (Control Room + Admin + Port Ops)**

Size:

- Small terminal (<5 MTPA): **500–1,000 m²**
- Medium (5–15 MTPA): **1,000–2,000 m²**
- Large (>15 MTPA): **2,000–4,000 m²**

Must be located with **line-of-sight to berths** or elevated camera coverage, and it shall include:

- Vessel operations room
- Conveyor control & monitoring centre
- SCADA-based utility control
- HSE office, meeting rooms

Additional Rule: Operations building area ≈ 0.03–0.05% of total terminal area.

ii) **Maintenance Workshop & Stores**

- **Area:** For unloaders, stackers, reclaimers, loaders → **1,500–3,000 m²** covered space
- **General planning:**
    - Provide 1 workshop bay / 4 major equipment units
    - Include machine shop, fabrication area, and lubrication centre
    - Critical spare parts store = **100–200 m²**

Ideally, these buildings shall be located adjacent to the stockyard with safe access for heavy equipment movement.

iii) **Electrical Substations / MCC Rooms**

As a thumb rule, there shall be 1 substation per 300–500 m of conveyor and the following equipment shall be facilitated with dedicated substations:

- Ship unloaders
- Stackers/reclaimers
- Major conveyor junctions

- Substation size shall be 150–250 m² (air-conditioned for drives)

Power Redundancy Rule: Provide 20–25% spare transformer & switchgear capacity.

iv) **Fire Station & Emergency Response Centre**

For terminals with combustible materials (coal/petcoke/grain) shall have;

- Fire station: **150–300 m²**
- Include foam tender bay + emergency equipment store

Firewater tanks (buildings/utilities) shall have capacities for;

- **2–4 hours** of storage
- Minimum **600–1,200 m³/hr** pump capacity

v) **Laboratory (For cargo quality testing)**

Tor coal, ore, fertiliser, grain, the required size is **80–150 m²,** depending on sample volume. It shall include sample prep, chemical analysis, and moisture measurement.

vi) **Drivers' Amenities & Gate Complex**

Generally, the average truck gate processing rate:

30–50 trucks/hr/lane (manual)

60–80 trucks/hr/lane (automated)

Size of gate building: **100–300 m²; and driver rest facility: 100–150 m²**

3.10.2 Container Terminal – Building Planning

Container terminals require more structured administrative, operational, and automated systems buildings due to complex equipment interactions.

I. **Terminal Operations Building (TOB)**

Size:

- Médium terminal (0.5–2M TEU): **2,000–3,500 m²**
- Large terminals (2–6M TEU): **3,500–6,000 m²**
- Mega hubs (>6M TEU): **6,000–10,000 m²**

Functional zones:

- Vessel planning, yard planning, gate planning
- Control tower with panoramic or CCTV-based view
- IT server room, training & meeting rooms
- Customs & documentation section (if required)

Placement Rule: Ideally **50–80 m** from the quay edge with full CCTV/RTK coverage.

II. **Gate Complex & Truck Processing Area**

- 1 lane = 60–80 trucks/hr (OCR)
- Gate building: 150–300 m²
- Provide:
  - Weighbridges (in/out)
  - OCR portals
  - Customs booths
  - Driver amenities (100–150 m²)

III. **Maintenance & Equipment Workshop**

- For RTGs, Reach Stackers, Straddle Carriers: **3,000–6,000 m²** depending on fleet size
- Workshops sized as:
  - 1 major workshop bay per **8–10 RTGs**
  - 1 straddle-carrier bay per **6–8 units**
- Spare parts store: **150–300 m²**

IV. **Reefer Control Building / Substations**

- Provide **1 reefer management room per 400–600 plugs**
- Reefer racks require:

    - Reefer panel room: **50–100 m²**
    - Backup generator enclosure (if grid is unstable)

V. **Customs, Inspection & Examination Facilities**

- Examination area:
    - **150–300 m² indoor** (cold chain if required)
    - **Hardstand of 800–1,500 m²**
- Scanning (X-ray/drive-through):
    - **25 m scanning tunnel** + control room **30–50 m²**

VI. **IT / TOS / Data Canter Facilities**

- Data canter room: **30–50 m²** per 1M TEU
- UPS room: **20–40 m²**
- Redundant server racks + backup fibre loops

VII. **Fire Station & Emergency Response (Container Terminal)**

- Fire station: **150–250 m²**
- Firewater storage:
    - **4 hours** reserve
    - Pump capacity: **2,000–6,000 L/min**
- Additional requirements:
    - IMDG storage area: **50–100 m²** (isolated)
    - Reefer monitoring rooms with thermal imaging
    - Emergency muster points every **200–300 m**

VIII. **System Support Facilities**

- Driver training building: **50–100 m²**
- Security building: **80–150 m²**
- Canteen & welfare building: **200–500 m²**
- Administration building (if separate from TOB): **1,000–3,000 m²**

3.10.3 General Provision for Buildings

Table-3.5: Area Benchmark for various Container Terminal Types

Terminal Size	Total Terminal Area	Building Area (All Uses)
Small	<50 ha	**0.2–0.4% of area**
Medium	50–150 ha	**0.3–0.6% of area**
Large	150–400 ha	**0.4–0.8% of area**
Mega	>400 ha	**0.5–1.0% of area**

- Key Ratios
    - **Operations building:** 1 m^2 per **6–10 employees** in operations
    - **Maintenance workshops:** 1 workshop per **6–10 major equipment units**
    - **Substations:** 1 per 300–500 m conveyor or 4–6 RTGs
    - **Parking:**
        - Staff: 1 space per 1.2 employees
        - Visitors: 1 space per 50 TEU/hr gate capacity

3.10.4 Electrical Sub-Station (ESS) Planning

3.10.4.1 Sub-Station Capacity Planning Guidelines

i) Transformer Rating Guidelines

Typical terminal-wide load mix:

- Bulk terminal: **5–20 MVA** total
- Container terminal: **10–50 MVA** (high crane + reefer demand)

General Guidelines:

- 1 MVA per 4 ship-to-shore cranes, or
- 1 MVA per 125–150 reefer plugs, or
- 1 MVA per 800–1,200 m of conveyor (bulk terminals)

ii) Sub-Station Spacing & Location Planning

a. Horizontal Spacing

- **1 substation every 300–500 m** along conveyor corridors (bulk terminals).
- **1 substation per 4–6 RTGs** in RTG-operated container blocks.
- For rail yards or long evacuation belts → **1 substation per 700–900 m** (max).

b. Strategic Placement Rules

- Place substations close to major load centres:
    - Ship unloaders / STS cranes
    - Yard cranes (RTG/RMG blocks)
    - Reefer areas
    - Conveyor drives & transfer towers

c. Avoid locating substations in flood-prone or dusty zones—raise floor level by 0.5–1.0 m above the surrounding grade.

d. ESS Building Area

- **Compact substation:** 120–180 m^2

- **Medium:** 200–300 m^2
- **Large terminal ESS:** 350–500 m^2

3.10.4.2 Cable Routing & Voltage Level Guidelines

a. **Voltage Levels in Ports**
   - MV Distribution: 11 kV / 22 kV / 33 kV typical.
   - STS cranes: often fed at 11 kV.
   - E-RTG: 400–480 V or 690 V.

b. **Cable Route Design**
   - MV cable max run without midway substation: ~1 km.
   - Use underground cables in the yard, cable trays on buildings/conveyors.

c. **Cable Sizing**

   Use a cable sized at 125–150% of the continuous current to handle:
   - High ambient temperatures
   - Dust loading
   - Voltage drop over long distances

3.10.4.3 Substation Safety Distances

- Safe distance from fuel/oil storage: ≥ 15 m
- From dusty stockyards (bulk terminals): ≥ 20–25 m
- Avoid placing below conveyor transfer towers to reduce dust ingress.

3.11 Dry Bulk Receiving Systems

(Rail – Wagon Tippler & Track Hopper)

The following guidelines may be adopted to choose the receiving system:

Wagon Tippler is preferred when:

- Material is not self-flowing (e.g., sticky ores, coal lumps)
- Throughput > 3–4 MTPA
- Handling BOXN/BOBR wagons
- High-volume, continuous receiving is needed

Track Hopper is preferred when:

- Material is free-flowing (limestone, gypsum, coal)
- Throughput < 3 MTPA or moderate
- Faster rake turnaround is required
- Space is available for a long hopper pit & conveyor system
-

3.11.1 Wagon Tippler System

i) Capacity Planning

- 1 Wagon Tippler (Rotary / O-type) handles 1.5–2.5 MTPA
- Cycle time: 3–5 minutes per wagon
- Discharge rate: 1,200–1,800 TPH (effective)
- Rake unloading time: 45–75 minutes per rake
    - Provide 2 tipplers if throughput exceeds 4–5 MTPA
    - Hopper below tippler: 100–200 tons surge capacity

ii) Space Requirements

- Typical tippler house footprint: 30–40 m long
- Hoppers: 8–12 m deep
- Rail approach should allow pulling & pushing operations

3.11.2 Track Hopper System

- Throughput: 2–3 MTPA per track hopper
- Hopper discharge rate: 1,000–2,000 TPH
- Rake unloading time: 30–45 minutes (faster than tippler)

i) Hopper Geometry

- Hopper length: 60–120 m (to accommodate 10–20 wagons over hopper)
- Hopper depth: 6–8 m
- Hopper width: 6–10 m
- Divided into multiple "pockets" (4–8 pockets typical)

ii) Throughput Planning for Wagon Tippler & Track Hopper

Table-3.6: Salient Points for Planning of Wagon Tippler & Track Hopper

Parameter	**Thumb Rule**
Rake size	58–59 BOXN wagons (≈ 3,900–4,000 tons)
Rakes/day	*(Daily receiving volume / 4,000 tons)*
Conveyor capacity	≥ 1.1 × receiving system discharge rate
Yard buffer	5–10 days throughput for imports
Minus grade material	Track hopper preferred
Lump/large ore	Wagon tippler preferred

iii) Rake Turnaround Time (RTT)

Wagon Tippler - **60–90 minutes/rake**

Track Hopper - **30–45 minutes/rake**

RTT < 1 hour is desirable for high-volume terminals.

iv) Key Decision Matrix

Table-3.7: Decision Matrix for Wagon Tippler & Track Hopper

Parameter	Wagon Tippler	Track Hopper
Material type	Sticky, lump ore, coal	Free-flowing coal, limestone, gypsum
CAPEX	Higher	Lower
Space requirement	Less	More
Operating cost	Medium	Low
Unloading time	Slower	Faster
Automation	High	Moderate
Dust	High → needs extraction	Medium → needs suppression

- 1 Wagon Tippler → 2 MTPA
- 1 Track Hopper → 2–3 MTPA
- Track Hopper is 30–40% faster in rake unloading
- Tippler needed if the material does not self-discharge
- Conveyors = 1.1× hopper/tippler TPH
- Provide 5–10 days storage buffer for receiving terminals

v) Layout Planning for Wagon Tippler & Track Hopper

General Flow Diagram of the bulk receiving system using Wagon Tippler and Track hopper is indicated below.

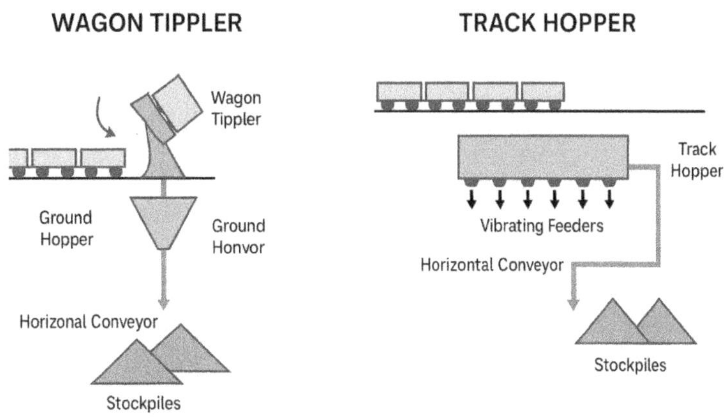

Figure-3.12: Dry Bulk Receiving System Options by Rail

i. Wagon Tippler Layout: Inbound rake → Tippler + Side Arm → Ground Hopper → Belt Weigh Feeder → Yard Conveyor → Stacker → Stockpiles

ii. Track Hopper Layout: Inbound rake over hopper → Bottom doors open → Pocket feeders → Yard Conveyor → Stacker → Stockpiles

3.12 Planning of Railway System for Port Terminals

3.12.1 Railway Sidings for Bulk Terminal

Efficient railway siding design is critical to the performance of bulk port terminals handling coal, iron ore, limestone, fertilisers, and similar commodities. While unloading systems such as tipplers or bottom-discharge hoppers receive attention, in practice, yard shunting time frequently becomes the limiting factor for port throughput. This section consolidates planning guidelines, shunting time calculation methodology, and design/operational measures to reduce shunting time, following practices aligned with Indian Railways (IR), RDSO guidelines, and global heavy-haul port terminals.

i) Design Basis (Mandatory Inputs)

Before planning the yard, the following shall be defined;

- Annual throughput (MTPA)
- Peak day factor (1.2–1.4 × average)
- Commodity type
- Wagon type (BOXN / BOBRN / BCNHL, etc.) for Bulk Cargoes
- Rake size (58 or 61 wagons)
- Unloading system (tippler / bottom discharge)

ii) Track Length Requirements

Table-3.8: Track Length Requirements

Item	Requirement
58–61 wagon rake	750–790 m
Design clear length	**≥ 900 m**
Preferred (future-proof)	**1,000 m**

Non-negotiable rule: No rake splitting shall occur inside the port yard.

iii) Yard Zoning Philosophy

A proper bulk terminal siding shall be divided into:

Main Line → Reception Yard → Marshalling / Holding → Unloading Yard → Dispatch Yard → Main Line

Golden Rule: Reception, unloading, and dispatch shall not share the same operational track.

iv) Number of Yard Track (Guidelines)

Table-3.9: Track Length Requirements

Rakes/day	Minimum Reception Lines
≤ 4	1
5–8	2
9–14	3
> 14	4+

Capacity Rule: One yard line efficiently handles 4–5 rakes/day.

i) Geometry Constraints

a. Gradients

- Tippler/unloading area: Level only
- Yard tracks: ≤ 1 in 1,000
- Lead tracks: ≤ 1 in 400
- Avoid > 1 in 200

b. Curves

- Yard tracks: ≥ 175–200 m
- Tippler approach: Straight preferred
- Main line connection: ≥ 600 m

ii) Definition and Components of Shunting Time

Shunting time is the total time consumed by internal rake movements within the port siding, excluding:

- Mainline haulage
- Mechanical unloading/loading time

It includes:

- Loco attach/detach
- Direction changes (run-around)
- Yard movements
- Point operation and signalling delays
- Brake continuity testing
- Wagon positioning
- Operational buffers

iii) General Equation for Shunting Time Calculation

$$T_{shunt} = T_{attach/detach} + T_{runaround} + T_{movement} + T_{points} + T_{brake} + T_{spotting} + T_{buffer}$$

Table-3.10: Standard Shunting Time Values

Activity	Time (min)	Planning Norm
Loco attach or detach	10	Each occurrence
Run-around	20–25	Adopt 25
Yard movement	8–10 min/km	Loaded rake
Point operation	5	Per shunt movement
Brake continuity check	7	Mandatory
Tippler positioning	15	Average
Bottom discharge positioning	8	Average
Operational buffer	10–15	Mandatory

- Simplified Shunting Time Formula

$$\boxed{T_{shunt} = (A \times 10) + (R \times 25) + (D \times 10) + 15}$$

Where:

- **A** = number of loco attach/detach
- **R** = number of run-arounds
- **D** = internal travel distance in km
- **15 min** = spotting + contingency

Table 3.11: Typical Shunting Time Benchmarks

Layout Quality	Shunting Time
Best practice	1.0–1.25 hr/rake
Average yard	1.5–2.0 hr/rake
Poor/congested	2.5–3.0 hr/rake

3.12.2 Railway System for Container Terminals

This section provides systematic planning guidelines, capacity sizing, layout principles, operating rules, and optimisation strategies for railway systems serving port-based container terminals.

i) Role of Rail System in Container Terminal Capacity

Rail system performance governs:

- Share of rail-borne containers (% modal split)
- Yard congestion and dwell time
- Quay crane productivity (yard evacuation rate)
- Terminal throughputs without physical expansion

Fundamental Principle: A container terminal is rail-constrained long before it becomes quay- or yard-constrained.

ii) Planning Basis and Design Inputs

Core Planning Inputs

The following must be fixed before layout design:

- Annual throughput (TEU/year)
- Rail share (% of containers moved by rail)
- Train length standard (in flats/meters)
- Train frequency (trains/day)
- Container mix:
  - 20' / 40'
  - Empty/laden
- Terminal handling system:
  - RTG
  - RMG
  - ASC

Table- 3.12: Typical Rail Share Benchmarks

Terminal Type	Rail Share
Gateway port	25–40%
Transshipment hub	10–20%
Inland-focused terminal	35–50%

Planning Rule: Design rail facilities for the **ultimate target rail share**, not opening-day share.

Table- 3.13: Typical Container Train Configurations (India / Global)

Parameter	Typical Value
Wagon type	BLC / BLCS / DFC flats
Wagons per train	45–90
Operating length	700–1,500 m
Design future length	1,500 m
Containers/train	90–180 TEU

Table-3.14: Preferred Rail–Yard Interface Layouts

Yard Type	Suitability
End-loading rail yard	Small terminals
Side-by-side working	Medium throughput
On-the dock rail yard (RMG-served)	High throughput (best practice)

Table-3.15: Track Length Requirements

Item	Minimum	Recommended
Current trains	750–900 m	—
Future-proof	—	1,500 m

Non-negotiable: No mid-terminal train splitting.

Table-3.16: Number of Rail Lines (Guidelines)

Rail TEU/day	Arrival/Departure Tracks	Working Tracks
≤ 500	1	1
500–1,200	2	2
1,200–2,500	3	3–4
> 2,500	4+	4–6

Table-3.17: Alignment and Gradients Rail Yard

Parameter	Limit
Yard gradient	≤ 1 in 1,000
Working tracks	Level preferred
Crane zone	Strictly level

Table-3.18: Recommended Curvature of Rail Track

Location	Min Radius
Crane working zone	Straight only
Yard tracks	≥ 250 m
Lead/entry tracks	≥ 600 m

iii) Rail-Mounted Gantry (RMG) System (Container Handling Interface)

- Dedicated cranes over rail tracks
- Direct wagon-to-yard transfer
- Parallel handling on adjacent tracks

Productivity: 30–40 container moves/hour/crane

iv) Shunting & Train Handling Philosophy

Shunting Principles

- Through-running preferred
- Avoid direction reversal
- Dedicated rail yard locomotives
- No crossing of truck lanes by trains during crane working

Table-3.19: Typical Container Train Handling Cycle

Activity	Time (min)
Arrival & securing	15
Loading/unloading	90–180
Departure preparation	15
Shunting (internal)	30–60
Total terminal time	**2.5–4.5 hr**

v) Rail Yard Throughput Calculation

Track Capacity Formula

$$\text{Trains per track per day} = \frac{24}{T_{\text{occupancy}} + T_{\text{buffer}}}$$

Where: Occupancy = loading + shunting
Buffer = 0.5–1.0 hours

Typical value: **4–6 trains/track/day**

vi) Rail TEU Capacity

$$\text{Annual Rail TEU} = (\text{Trains/day}) \times (\text{TEU/train}) \times 365$$

vii) Safety, Clearance & Interface Requirements

Table-3.20: Typical Clearance Requirement of Track

Element	Requirement
Track centre to crane leg	≥ 2.5 m
Parallel rail tracks	≥ 6.5 m
Crane span	Track + safety margin

3.13 Manpower Planning for Marine & Port Terminals

Manpower planning in marine and port terminals involves determining the right number of skilled personnel needed to efficiently operate berth activities, cargo handling systems, yard operations, marine services, and administrative functions. The goal is to ensure safe, continuous, and cost-effective operations while matching workforce levels with cargo volume, equipment deployment, and vessel traffic.

3.13.1 Bulk Cargo Terminals : Operations Manpower

Table-3.21: General Manpower requirement for Bulk Terminal

Terminal Capacity (MTPA)	Typical Direct Manpower
5 MTPA	80–120 persons
10 MTPA	120–180 persons
20 MTPA	180–260 persons

***General Guidelines**: 0.20 – 0.35 manpower per 1,000 tonnes/day of throughput*

Typical breakdown of manpower for various works:

a. Berth operations (MHC/Shiploader/Unloader) : 8–15 per shift
b. Yard & conveyor operations : 10–20 per shift
c. Maintenance (mechanical + electrical) manpower :10–18% of total operations

In another way, it can be estimated as

a. 1 operator per equipment (reclaimer, stacker, shiploader, unloader) per shift
b. 1 supervisor per 8–12 operators
c. Maintenance ratio: 1 technician per 4–5 machines

3.13.2 Container Terminals: Operations Manpower

Table-3.22: General Manpower Requirement for Container Terminal

Equipment / Function	Thumb Rule
Quay Cranes (QC)	3 operators/ crane (to cover all shifts)
RTG / RMQC / ARMG	4–5 operators/crane
Reach Stackers	3 operators per machine
Trucks/Prime Movers	1 driver per truck (1:1 or 1:1.3 in some ports)
Supervision	1 per QC gang or 1 per 20–25 workers

***General Guidelines**: 1.8 – 2.5 persons per 1,000 TEUs annual throughput*

i. Total Direct Manpower Estimate

Table-3.23: General Manpower Requirement vs Container Terminal Capacity

Capacity (TEUs/year)	Direct Manpower
0.5 million TEU	250–350 persons
1.0 million TEU	400–600 persons

2.0 million TEU	700–900 persons

Productivity-based Rule - 35–40 direct manpower per quay crane deployed

3.13.3 Liquid Bulk Terminals (Petroleum, Chemical, LNG/LPG)

i. **Typical Staffing**

a. Marine & jetties: 6–12 per shift
b. Pipeline & pumping: 4–8 per shift
c. Tank farm management: 10–20 permanent staff

***General Rule**: 0.05–0.10 manpower per 1,000 tonnes of liquid handled annually*

ii. **Total Headcount**

Table-3.24: General Manpower Requirement vs. Liquid Terminal Capacity

Tank Farm Size	Total Manpower
Small (50–100 ML)	40–60
Medium (100–250 ML)	60–100
Large (>300 ML)	100–150

3.13.4 Ro-Ro / Vehicle Terminals

I. **Typical Terminal Staffing**

a. Small: 40–60
b. Medium: 80–120
c. Large OEM hub: 150–250
   - *1 person per 15–20 vehicles loaded/unloaded*
   - *One supervisor per 50–75 vehicles*

3.13.5 General Cargo Terminals

General Manpower Rule : 0.3–0.5 manpower per 1,000 tonnes per year with following breakdown:

a. Yard labour: 12–20 / shift
b. Crane & equipment ops: 8–12 / shift
c. Maintenance: 8–15% of total ops manpower

3.13.6 Marine Services (Common for All Terminals)

i. **Typical Staffing**

Table-3.25: General Manpower Requirement for Harbour Operation

Category	Manpower
Pilotage	1 pilot per 700 – 1,000 ship calls
Mooring crew	6 – 8 persons per shift (per berth cluster)
Tug crew	6 – 8 per tug × number of tugs
Harbour Master Office	12 – 20 staff

ii. **Administration, Safety & Support**

Approximate manpower requirements:

- Admin (HR, Finance, IT): 10–15% of operations strength
- HSE + Quality: 4–6% of total staff
- Security: 10–20 persons per gate + patrol teams

Key Trends in Maritime and Port Industries

Chapter Summary

Ports are essential connectors between land and sea, facilitating trade and contributing to economic growth. Their strategic positioning attracts a range of industries, generating regional value despite limited direct employment. Spillover effects often extend internationally, as demonstrated by the Port of Rotterdam's influence on Central European economies. The emergence of larger container ships has enhanced trade efficiency but necessitates substantial infrastructure upgrades, including deeper, stronger quay walls and advanced cargo handling systems. Leading global ports such as Suez, Singapore, Rotterdam, Hamburg, and Shanghai continue to modernise through infrastructure expansion, technological innovation, and strategic planning to accommodate larger vessels and improve global connectivity. Conversely, India's trade remains service-driven, with limited high-tech manufacturing exports. Port sector challenges—such as congestion, poor hinterland connectivity, and low containerization—continue to hinder competitiveness, despite tariff reductions and some infrastructure improvements.

Learning Objectives

- Understand the role of maritime transport in regional and national development
- Examine the evolution of ship sizes and current trends in the shipping sector
- Assess port adaptability to mega-ships and associated constraints
- Identify key parameters of efficient and competitive ports
- Review infrastructure upgrade initiatives by major maritime institutions
- Analyse the operational dynamics of Indian ports
- Evaluate India's port competitiveness and sectoral constraints.

4.1 Maritime Transport

Maritime transport is central to global trade and the economy. In today's interconnected world, no nation is entirely self-reliant; each depends on maritime trade to export its surplus and import necessities. Countries that have embraced maritime trade with open borders and markets have witnessed notable economic growth. This exchange of goods, services, resources, and knowledge across nations is termed "globalisation," defined as "the development of an increasingly integrated global economy characterised by free trade, free capital flow, and access to cost-effective foreign labour markets."

Most items we use daily—whether raw materials, components, or finished goods—are typically transported by sea. Shipping has long been a dominant mode of transport and a vital communication link between coastal cities, nations, and continents. Cost-effective and efficient maritime transport plays a crucial role in fostering growth and sustainable development, helping distribute the benefits of trade and commerce more evenly.

Since World War II, global trade has expanded rapidly, with commercial marine transport steadily increasing over the past five decades. Besides rail, water transport remains the most economical and environmentally efficient method of moving goods. Currently, about 90% of global trade is handled by the international shipping industry. In today's economic landscape, maritime transport and industrial growth are closely linked. The sector has expanded significantly alongside industrial production. Additionally, the cruise industry has gained global popularity, contributing to the economy as more recreational vessels visit diverse ports worldwide. Thus, the maritime industry holds immense value across trade, industry, science, and leisure.

Estimates suggest the shipping industry annually moves around 2 billion tons of crude oil, 1 billion tons of iron ore (essential for steel production), and 350 million tons of grain. These volumes are unfeasible for road, rail, or air transport. Such raw materials enable nations to build industries, develop infrastructure, support populations, and produce refined goods for export—driving prosperity and enabling growth in developing countries.

Other major bulk commodities like chemicals and refined fuels are also transported by sea, followed by the delivery of finished products to markets. Today, the maritime industry is a vital link in global trade, with ocean-going vessels being the most efficient—and often the only—means of transporting large quantities of essential goods. Ports serve as the key physical hubs supporting international trade flows.

Moreover, as a key driver of the global economy, the shipping industry is supported by advanced logistics networks that deliver just-in-time components and goods to manufacturers and consumers. Over the past five decades, seaborne trade has grown fourfold, offering worldwide benefits through competitive freight costs.

In summary, the advantages of the maritime industry include:

Shipping

- Maritime trade is vital to foreign trade and national economies, enabling access to essential raw materials for economic development.
- It facilitates the cost-effective manufacturing and export of goods and products. Consequently, economic pressures and technological progress have led to a trend toward larger vessels.
- Shipping has the lowest environmental impact per ton among all transport modes.

- This has, in turn, increased the need for channel deepening along key maritime routes.
- Port infrastructure and ongoing development are crucial to the health of maritime shipping, especially in light of the growing number and size of vessels.
- Port and land-side infrastructure should be on par with international standards.

4.2 Benefits of the Geographical Location of Port

Ports are critical components of the transportation system, serving as key interfaces between land and sea. Their economic advantages are numerous. Ports play a vital role in global supply chains, acting as facilitators of trade between port regions and countries.

Port cities—such as Rotterdam (Netherlands), Singapore, Shanghai (China), Los Angeles (USA), Mumbai (India), and Hamburg (Germany)—benefit from a portion of the economic impact generated by ports, with most direct port-related value addition still occurring within these cities.

The primary function of a port is to provide efficient and cost-effective intermodal and intramodal cargo transfer, inspection, storage, and control. To fulfil this role, ports must effectively accommodate ships, barges, and other transport vehicles. They form a vital link in the transport chain that moves cargo from origin to destination. Ideally, ports facilitate continuous cargo flow between land and water, and also between water-based transport modes.

Ports tend to attract businesses from various industries and services, including land transport, logistics, warehousing, and storage. Many ports also host resource-intensive industries such as refineries, chemicals, steel, coal, aerospace, and renewable energy production—including offshore wind and biomass. The presence and type of these industries are influenced by factors like available space, port strategies, and the regional economic structure. Strong connections often exist between ports and associated industries.

While port industries generate local employment, this is relatively limited compared to the broader regional economy. Even in major ports, direct port-related employment typically does not exceed a few thousand jobs.

Industries drawn to port areas can contribute significantly to employment and value creation in port regions. Resource-intensive industries continue to favour port locations due to reduced transportation costs. In larger, wealthier regions with strong service sectors, port traffic is more diversified and includes high-value goods like containers and consumer products. In contrast, agricultural and industrial regions often specialise in bulk cargo.

However, most indirect and catalytic effects of ports occur outside the port regions. Businesses in other areas also benefit from efficient ports through reduced transport costs and improved access to international markets.

In many cases, these spillover effects extend beyond national borders—for example, the Port of Rotterdam plays a key role in supporting German industries, and several landlocked Central European countries benefit from its operations.

4.3 Evolution of Ship Size

As global trade and economic activity have expanded, shipping demands have driven the development of larger and more specialised vessels.

In the 1980s, Handy-size carriers were the dominant dry bulk ships. By the late 1970s, Capesize vessels (150,000 DWT) began replacing Panamax vessels (70,000 DWT) for long-haul voyages. In the early 1990s, only about 30 ships exceeded 200,000 DWT. After 2000, Capesize vessels gradually became the primary ship type. By 2006, the classification threshold for Capesize ships increased from over 80,000 DWT to over 100,000 DWT, and their share in dry bulk capacity continued to rise. By the end of 2013, over 300 ships exceeded 200,000 DWT, and by 2020, this number reached approximately 526, with 86 vessels in the 200,000–300,000 DWT range—reflecting a 70% increase over seven years.

Since 2007, Very Large Ore Carriers (VLOCs) have been deployed for dedicated iron ore transport. These vessels exceed the size of standard ore carriers and are part of a strategic shift toward bulk transport efficiency. The economics of bulk shipping continue to shape trade routes and industrial locations. Simultaneously, intermodal transport has become a global standard, with containerization and mechanised handling reducing costs and improving efficiency across transport modes. Ports now function as key nodes in global commodity chains managed by logistics firms, shipping lines, freight forwarders, and transport operators. Technological advancements have significantly influenced the growth of bulk and containerised maritime transport.

In recent years, Valemax bulk carriers, ranging from 300,000 to 400,000 DWT, have entered production. These ships, designed for iron ore transport from Brazil to Asia, can carry up to 400,000 metric tons—2.3 times more than previous vessels—and reduce carbon emissions per ton by 35%. Valemax ships have set a new benchmark for large-scale express ore transport. As of [year], 526 Capesize bulkers were operating globally, connecting 121 ports across 31 countries, with approximately 84% of them in the 200,000–300,000 DWT range.

Beyond bulk shipping, containerization has revolutionised global trade and logistics, often referred to as the third transportation revolution. It has introduced faster transit, simplified packing, secure cargo handling, and specialised operations. Over the past 40 years, container ships have evolved from small feeder vessels to Panamax, Post-Panamax, Suezmax, and the latest Malaccamax class.

Container ships have grown rapidly in size—faster than any other vessel type. In just a decade, their average capacity has doubled. The largest container ships currently carry up to 20,000 TEU, with vessels exceeding 24,000 TEU already in operation or on order. Technical limits for container ship size have not yet been reached, with recent orders for 22,000 TEU ships. However, port infrastructure and shore-based cargo handling systems are under pressure, which may lead to a practical cap on ship size in the near future.

Out of approximately 55,000 merchant ships engaged in international trade, around 15,000 were general cargo vessels as of January 1, 2021.

In summary, the rise of larger container ships has led to cost efficiencies for carriers, reduced maritime transport costs, and supported global trade. However, these gains come with challenges, including the need for upgraded infrastructure and the management of traffic surges at ports—issues explored in subsequent sections

Global Shipping Industry

- Over 80% of global trade by volume is transported via maritime shipping.
- Dry cargo (e.g., containers, bulk goods) now dominates maritime trade, making up 62% of total volume, up from 46% in 2000.
- The global container fleet grew by 4.8% in 2024, showing resilience despite geopolitical challenges.
- Container port volumes are expected to decline by 1% in 2025, due to trade route disruptions and policy shifts.
- Developing countries now handle 54% of global maritime freight, up from 38% in 2000, with Asia leading the growth.

Indian Shipping Industry

- India's fleet includes 1,526 vessels totalling 13.75 million gross tonnage (GT) as of December 2023.
- The overseas fleet accounts for 12.02 million GT, with oil tankers making up 55.8% of total tonnage.
- Dry cargo liners increased their share from 14.1% in 2013 to 24.5% in 2023.
- 45.5% of Indian vessels are over 20 years old, indicating a need for fleet modernisation.
- India's Maritime Vision 2030 aims to expand fleet capacity, upgrade port infrastructure, and promote sustainable shipping.

4.4 Scenario of Current Fleet

The commercial marine transport industry is primarily categorised into three major cargo types: dry bulk, liquid bulk, and containerised cargo. Each of these broad categories includes several sub-segments.

- Dry bulk cargo mainly consists of iron ore, coal, and wheat.
- Liquid bulk cargo is divided into crude oil, gas (including liquefied natural gas (LNG) and liquefied petroleum gas (LPG)), and refined petroleum products.
- Containerised cargo typically includes manufactured and semi-manufactured goods, and is further split into transoceanic liner trade and regional feeder trade.

According to 2018 data:

- Dry bulk cargo represented 40% of global marine trade by tonnage.
- Oil and gas accounted for 35%.
- Containerised cargo made up 13%.
- The remaining 12% consisted of general cargo of various types.

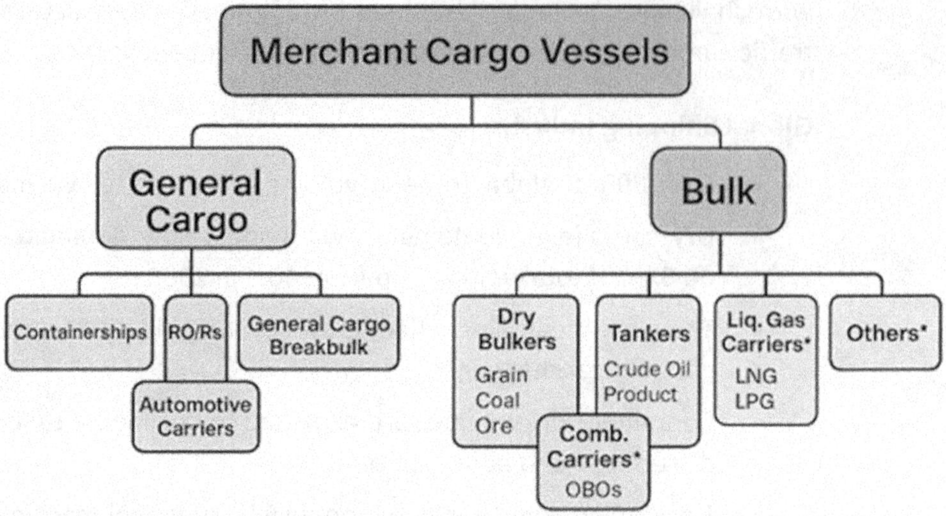

Figure 4-1. General Categorisation of the vessel

Maritime transport continues to be the leading mode for international trade, both for bulk commodities and containerised break-bulk cargo. Various vessel types are involved in this trade, typically illustrated in industry diagrams or fleet composition charts.

4.5 Current Trends in Different Shipping Sectors

Container ships are currently the longest vessels in the world in terms of overall length (LOA). The largest container ships measure up to 400 meters, which exceeds the maximum lengths of modern oil tankers (380 m), bulk carriers (362 m), and cruise ships (360 m). However, container ships typically have shallower drafts compared to tankers and bulk carriers, resulting in lower gross tonnage (GT) and deadweight tonnage (DWT). Historically, some oil tankers were longer than today's container ships—for example, vessels with LOA of 458 meters—but these are no longer in operation.

Table-4.1: Dimensions of Largest Ships by Type

Ship Type	Name	LOA (m)	Beam (m)	DWT	GT	Draft (m)	Year
Container	MSC Oscar	394	59	197,362	193,000	16	2015
Container	CSCL Globe	400	59	184,320	187,541	16	2014
Oil Tanker	TI Class	380	68	441,893	234,006	24.5	2002
Bulk Carrier	Valemax	362	65	400,000	200,000	23	2011
Cruise Ship	Oasis Class	360	60.5				

4.5.1 Growth of Containerisation

To ensure safe and efficient cargo handling and transportation, industries are increasingly adopting container shipping, known as containerisation. For instance, the paper industry has modified the dimensions of paper rolls to fit container sizes, enabling streamlined shipping. Today, a wide range of commodities—including paper, timber, and others—are shipped as containerised cargo for both import and export. This method allows goods to be transported directly from production sites to consignees or even end consumers, enhancing efficiency and minimising damage risks.

Containerised shipping is experiencing global growth, with some forecasts suggesting that up to 90% of general cargo may be shipped in containers within the next decade. Products such as malt, peat moss, fertilisers, timber, scrap, and waste materials are increasingly being containerised.

There is a mutually reinforcing relationship between maritime shipping and globalisation. For example, prior to the 1950s, crude oil was typically refined near its source and transported in small tankers (12,000–30,000 DWT). It was later recognised that locating refineries closer to consumer markets and transporting crude in larger volumes would be more economical. This led to the development of large crude carriers (exceeding 200,000 DWT), significantly reducing the per-unit cost of intercontinental energy transport.

A similar pattern applies to agricultural products. Today, it is common for such goods to be harvested on one continent, processed on another, and assembled or packaged on a third before reaching the final market. For example, cotton grown in North America may be sent to African mills, then to Asian factories, and finally returned to North America for retail. Products like orange juice and wine are also shipped across continents to meet seasonal demand or compete with higher-cost domestic production. In this context, containerisation supports integrated freight transport across multiple modes.

In the current global economy, key competitive factors include time, cost, and delivery reliability—all of which are supported by containerisation. Over the past decade, the doubling of maximum container ship size has reduced per-container vessel costs by roughly one-third, making maritime shipping more affordable. The trend toward larger container ships has been ongoing for decades, driven by the pursuit of economies of scale by shipping lines. This growth has been enabled by efficient container handling systems at ports, which allow operators to capitalise on the benefits of larger vessels.

However, the rise of mega-ships also necessitates further adjustments in port infrastructure and operations, as discussed in subsequent sections.

4.5.2 Adapting Ports to Mega-Ships

Existing ports must adapt to the ongoing increase in ship sizes. However, mega-ships place a significant strain on current port infrastructure, and many ports are struggling to keep up. Presently, next-generation container ships with capacities around 24,000 TEU are viable only at major ports that handle consistently high cargo volumes throughout the year.

Ports capable of accommodating these vessels must have advanced shipping systems suited to the layout of large container ships, and more importantly, sufficient infrastructure to ensure rapid cargo loading and unloading. This includes not only the port itself but also the berthing facilities. Mega-ships require larger equipment, expanded storage areas, and enhanced inland distribution networks. While some ports are investing billions in upgrades, others lack the space or resources to do so. Currently, only a limited number of ports can handle container ships of 20,000 TEU or bulk carriers of around 300,000 DWT. Although the number of mega-ships remains small, their high cargo capacity leads to increased peak traffic, placing additional pressure on port operations. Addressing this challenge requires more efficient infrastructure, adequate resources, and skilled manpower.

Ports must invest proactively, anticipating future shipping capacity rather than reacting to it. Smaller terminals face an uncertain future due to draft and length limitations, which prevent them from accommodating large vessels. Despite these challenges, the container shipping industry continues to show growth potential. While China remains a key market, other regions are also contributing to the rising demand for containerised goods.

To keep pace with the increasing scale of ship sizes, existing port infrastructure requires essential upgrades, as outlined in the following sections.

4.5.3 Requirement for Enhanced Quay Structures

The rapid increase in vessel size—particularly mega-carriers—has placed considerable strain on existing port terminals and shipping channels, many of which were not originally designed to accommodate such rapid growth. As a result, numerous ports now face the need to deepen, reinforce, and extend their quay walls to safely berth and moor the largest vessels and withstand the forces exerted during these operations.

Larger vessels typically have greater drafts, making quay wall height a critical design factor to prevent grounding. For instance, a 24,000 TEU container ship requires a retaining height of approximately 25.84 meters, which exceeds the structural capacity of many older port terminals.

In addition to the draft, quay length must also be increased to accommodate longer vessels, posing further challenges to quay wall design and load-bearing capacity. The overall increase in ship dimensions has significant implications for the strength and functionality of quay walls, which must support safe berthing, mooring, and cargo operations.

Furthermore, the deployment of larger cranes, container handling systems, and stacking equipment necessitates enhanced bearing strength of quay structures to support heavier loads and more intensive operations.

4.5.4 Constraint Due to Water Depth

The depth of approach channels and berths is a primary limiting factor for accommodating large vessels. As ship sizes increase, so do their fully loaded drafts, requiring greater water depth in navigation channels, berthing areas, and manoeuvring zones. For vessels exceeding 80,000 DWT, the typical design draft is around 14.5 meters, meaning port infrastructure must offer a minimum depth of 15 meters to ensure safe passage and berthing.

However, only a limited number of ports—such as Rotterdam and Singapore—currently meet these depth requirements. A growing concern is the reduction in Under Keel Clearance (UKC). As vessel dimensions expand, available UKC has diminished significantly, in some cases leaving less than 1 meter of clearance. This poses serious risks, including grounding and reduced manoeuvrability, especially in constrained or shallow waters.

4.5.5 Land Availability Constraints

Modern container port terminals require a substantial amount of land to function effectively. At a minimum, a well-equipped terminal must include multiple berths to handle several vessels simultaneously, a quay area for ship-to-shore operations, designated zones for container transfer, storage, and delivery/receiving, all integrated with road and rail connections. Additional facilities such as empty container depots, customs processing areas, and truck holding zones are also essential.

In many cases, logistics operations, including distribution centers, are situated within or adjacent to the port area. Furthermore, ports often host industries that benefit from proximity to maritime transport. Large, modern ports typically occupy thousands of hectares, depending on their specific roles and operational requirements.

This spatial requirement is closely tied to the port's organisational capabilities, which include planning, information systems, and coordination and competition among ports.

4.6 Expansion of Navigation Channels

Many ports are facing significant challenges, particularly related to navigational depth and the limited number of berths suitable for accommodating larger vessels. Additional concerns involve the landside operational capacity to efficiently manage the increased cargo volumes associated with these ships. To support the growing size and capacity of vessels, ports are upgrading their navigation and ship-handling infrastructure, a trend mirrored by the expansion of the two largest man-made shipping channels—the Suez Canal and the Panama Canal.

The Suez Canal, inaugurated in 1869, and the Panama Canal, opened in 1914, have both played pivotal roles in shaping global shipping routes and vessel design for over a century. In response to increasing vessel sizes, the Panama Canal underwent a major expansion, including the construction of larger locks to accommodate Post-Panamax ships with capacities up to 12,600 TEUs, compared to the previous limit of 4,800 TEUs for Panamax vessels. Corresponding upgrades were also made to adjacent ports to ensure they could handle these larger ships.

The new, wider Panama Canal is now operational, and the Suez Canal has been expanded to allow two-way traffic, creating opportunities for the development of new ship classes optimised for the increased channel capacity. To ensure full container ship accessibility, the Suez Canal Authority completed an expansion in January 2010, increasing the permissible draft to 22.12 meters. Further deepening of the western channels from 14.6 meters to 15.8 meters is underway to accommodate ultra-large container vessels and reduce transit times.

The Panama Canal expansion, a critical development in global trade, addressed the limitations imposed by its original locks, which had previously constrained vessel upsizing. Initiated in 2006 and completed in 2017, the project increased channel depth from 12.6 meters to 16.8–18.3 meters and widened the locks from 33.5 meters to 54.9 meters, enabling the passage of ships with capacities up to 13,500 TEUs.

These canal expansions are expected to significantly influence the composition and deployment of the global container fleet. According to Drewry's research, approximately 50% of existing Panamax vessels (4,000–5,000 TEU capacity) are likely to be phased out, as shipowners opt for larger vessels offering lower per-unit transport costs, contributing to overcapacity in the Panamax segment. Consequently, the trend toward larger containerships is being actively supported by these infrastructure developments.

4.7 Quay Cranes

The continual increase in vessel size necessitates ongoing upgrades to port infrastructure, particularly quay cranes. Larger ships present challenges in terms of crane outreach and lifting height. The latest generation of mega-container ships requires cranes capable of spanning 23 container rows, and some ports are already preparing for vessels even larger than those currently anticipated.

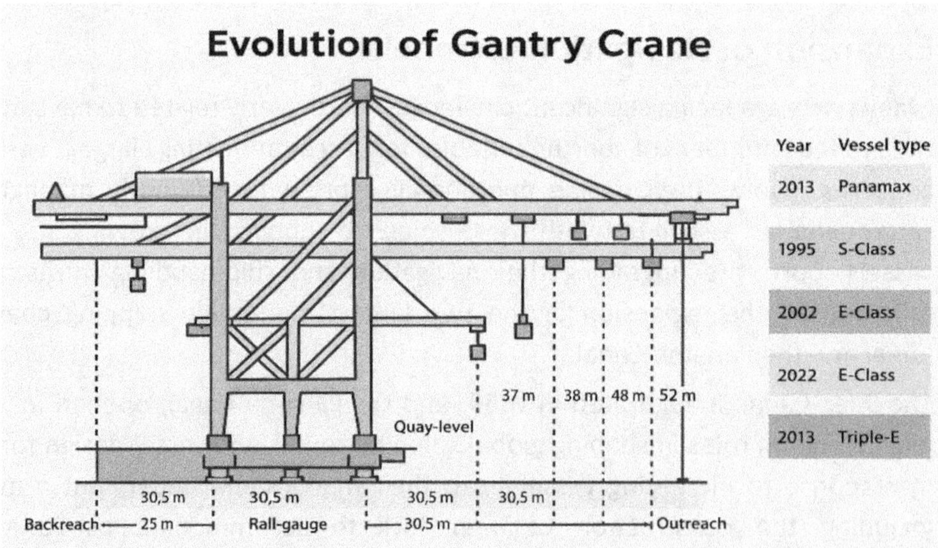

Figure 4-2: Evolution of Gantry Cranes

Super-Post-Panamax cranes, available at select major ports, are essential for handling these large vessels. Ships of this scale will not call at ports lacking the necessary equipment. If container ship capacities reach 20,000 to 25,000 TEUs, as projected by some experts, even the most advanced ports will need to undertake another round of crane upgrades. Similar infrastructure demands are emerging in other segments, such as bulk carriers and cruise ships, requiring enhancements to both terminal facilities and supporting structures.

Today's customers are increasingly sensitive to transport time, and fortunately, advancements in port infrastructure help mitigate delays. While the trend toward larger ships drives the development of port facilities, the reverse is also true—modern, efficient ports further enable the deployment of larger vessels. Over the past decades, ports have introduced automated systems, advanced cranes, and integrated loading/unloading controls, all of which contribute to operational efficiency for large ships.

For example, in 1970, container cranes could handle stacks up to 35 meters high. By 1995, this increased to 52 meters, and by 2002, to 58 meters. With the arrival of Triple-E class ships, cranes now manage stacks up to 69 meters (Preparing the Ports, 2013). Over time, not only have crane span and reach expanded, but handling capacity has also improved—from 15–18 containers per hour to 28–32 containers per hour, operated by a single gantry crane operator.

4.8 Future Shipping Trends

The rapid advancement in the development of mega-sized dry bulk carriers and container ships, including vessels exceeding 400 meters in overall length, suggests that even larger container ships may emerge in the near future. According to the Ministry of Oceans and Fisheries of Korea, the Jinhae New Port, part of Busan Port, is set to begin construction of berths capable of accommodating 30,000-TEU vessels.

When designing ports or maritime infrastructure—such as navigation channels or breakwaters—the projected size of target vessels is a critical consideration. In anticipation of future ship dimensions, several studies have analyzed the characteristics of mega-container ships using regression analysis.

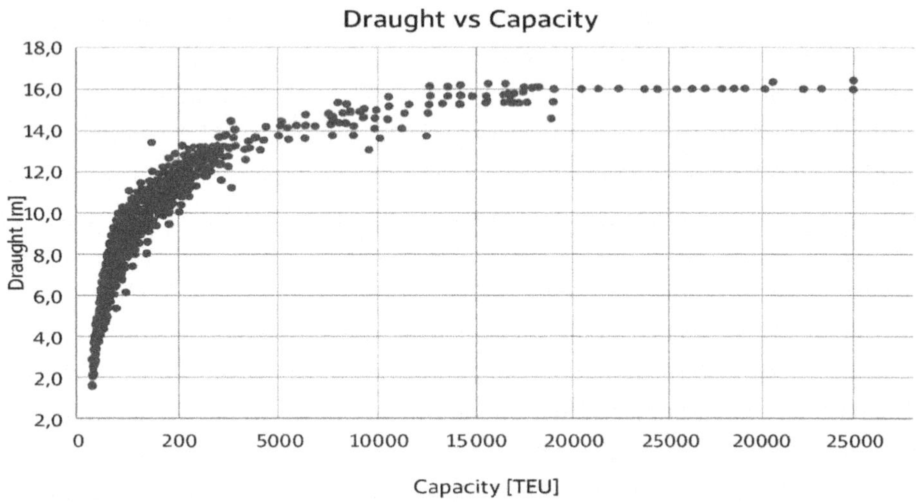

Figure 4-3: Relationship between draught and capacity (in TEUs) of container ships since 1964.Relationship between

For instance, predictions for the overall length of 30,000-TEU ships vary significantly, ranging from 453.0 meters to 536.9 meters, a deviation of 83.9 meters.

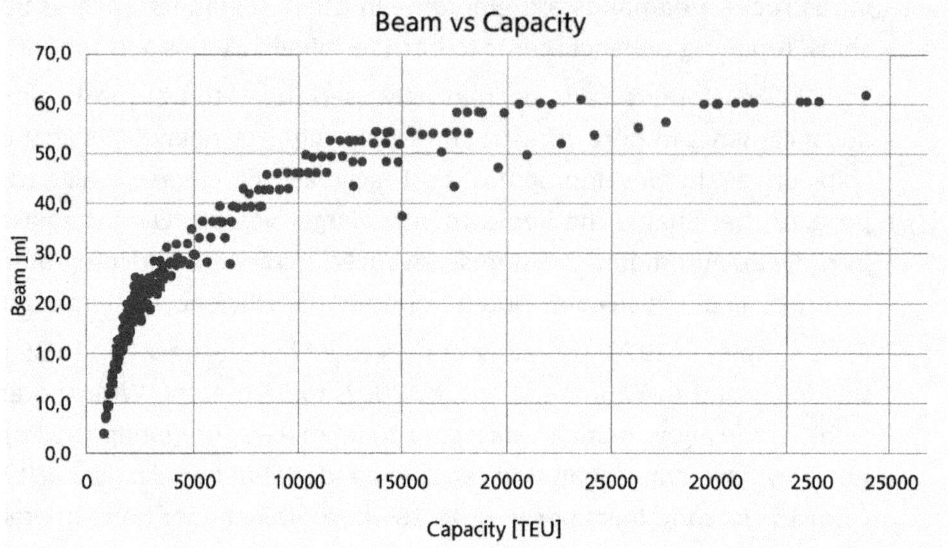

Figure 4-4: Relationship between beam and capacity (in TEUs) of container ships since 1964.

Institutions such as the Korea Maritime Institute have used simple linear regression to estimate the main dimensions of these vessels based on TEU capacity. Researchers like Cho et al. have modelled changes in ship size under different scenarios, while Park and Suh projected that a 30,000-TEU ship would be approximately 453 meters long, requiring a pier length of at least 500 meters. Current container ships in the 15,000–24,000 TEU range are typically around 400 meters in length and are being enlarged primarily through increased breadth.

Further studies, including those by Katayama et al., have examined the relationships between key ship dimensions—such as length overall (LOA), breadth (B), draft (d), and gross tonnage (GT)—to inform port design and planning. Akakura et al. conducted statistical analyses using ship specifications from Japan and Lloyd's Register, while Takahashi et al. utilised data from the Lloyd's Maritime Intelligence Unit.

These analyses reveal that draft tends to stabilise once container ships exceed 12,000 TEU, typically remaining within the 16–16.5 meter range. Similarly, for dry bulk carriers, draft stabilises around 22.0 meters.

4.8.1 Dimensional Trends and Key Influences in Mega Containerships

Among core ship dimensions, beam has experienced the most notable proportional increase. Adjustments in beam, length, and depth are made incrementally to accommodate added TEU or DWT capacity. For VLCS and ULCS, capacity growth is tied to all three dimensions, with beam being the most influential. Draught, however, remains the least modified aspect in these designs.

While vessel dimensions are expanding, the pace of growth is slowing, indicating that larger capacities are being achieved without major increases in overall size. Beam stands out as the most flexible dimension for adding TEU capacity, more so than length or draught.

Projections based on historical trends suggest that a 30,000-TEU ship would likely have a draught under 17 meters and a length below 425 meters, while beam could extend to around 65 meters. Despite rising capacities, physical dimensions are not expected to grow significantly.

However, technological shifts could change this outlook. Business strategies and canal constraints currently limit ship size more than technical capabilities. Environmental concerns are also shaping future designs—larger, more fuel-efficient ships may emerge to cut emissions per TEU, while port emission charges and urban pressures could encourage smaller, cleaner vessels.

4.9 Parameters of an Efficient Port

Port efficiency is a key factor influencing international transport costs. Among various determinants, the following are generally recognised as the most critical:

4.9.1 Port Planning

Effective port planning significantly enhances operational performance. This involves multiple layers of coordination. A port master plan outlines the core functions of the port, their spatial arrangement, and interconnections. It should specify the allocation of future traffic and the necessary actions to support it. This master plan serves as the foundation for more detailed planning activities, including:

- Berth allocation
- Backup Area and Yard layout
- Internal traffic flow
- Intermodal operations

Alignment across these elements is essential for smooth port operations. The benefits of sound planning are substantial; for example, yard planning quality alone can impact handling times by up to 30%, as noted by De Koster (2013).

Additionally, Many ports also carry out long-term planning, often described in public strategic vision documents. These plans, which can cover several decades, help identify future development needs, guide investment priorities, and highlight possible operational challenges. When done well, strategic planning encourages stakeholder involvement, improves client relationships, and builds local support. It works best when these visions inspire innovation and future-focused initiatives.

For instance, the Port Vision 2030 of Rotterdam, adopted in 2012, focused less on forecasting throughput and more on envisioning the future role of the port, based on a comprehensive analysis of emerging trends.

Ports are the key link between sea and land transport, supporting both national growth and global trade. But as ships get larger, ports need major infrastructure upgrades, which can lead to excessive spending and unnecessary competition. This is why port planning should focus on the right, well-targeted improvements and avoid building facilities that are inefficient or not needed.

4.9.2 Port infrastructure

Onshore infrastructure accounts for about 40% of coastal nations' transport costs, with strong evidence linking port quality to maritime expenses. Limao and Venables (2001) found that improving infrastructure from the 75th to the 25th percentile can cut transport costs by 30–50%.

Exporting countries' port infrastructure has a greater impact on costs than that of importers. Abe and Wilson (2009) showed that a 10% rise in port congestion increases maritime costs by 0.7%, largely due to inefficient logistics. Poor trucking services extend dock times, raise inventory costs, and reduce export volumes, leading to fewer shipping services.

Port depth is vital for connectivity, especially as modern vessels require drafts of around 20 meters—exceeding many ports' capacities. Ports like Hamburg rely on high tides to accommodate large ships. Consequently, many ports invest in dredging to deepen berths and channels, a trend accelerated by the Panama Canal expansion.

Port performance is assessed via operational metrics such as:

- **Utilisation**: TEUs/crane/year, vessels/berth/year, TEUs/hectare/ year, containers/lane/hour
- **Productivity**: crane moves/hour (25–40), vessel service time, truck turnaround, gang moves/man-hour
- **Dwell times**: 5–7 days for imports, 3–5 days for exports

While container terminals are widely researched, there is little benchmarking available for other types of terminals. In the past, cities developed around their ports, but today many urban ports face congestion and environmental challenges. As a result, some cities are moving port activities elsewhere to reuse central waterfront areas.

4.9.3 Maritime connectivity

Maritime connectivity plays a big role in how competitive a port is. The more connected a port is—with frequent and reliable shipping services—the more attractive it becomes to shippers. Ports that offer direct routes and faster delivery times tend to get more business. When there's a lot of trade between two ports, ships travel more often, which makes the service more dependable.

Even though having access to major sea routes (called maritime forelands) can give a port an edge, good connectivity also depends on how efficiently the port operates and how well it's linked to inland transport systems.

Sometimes, ports that are close to each other—like Hong Kong and Shenzhen—might share similar shipping routes, but they can also support each other by offering different strengths in connectivity.

Big ports often use transhipment (moving cargo from one ship to another) to grow their network. While gateway ports mainly serve nearby cities and regions, transhipment ports—

usually near busy sea lanes—focus on transferring cargo between ships, with little or no land transport involved.

4.9.4 Strong hinterland connections

Good hinterland connectivity depends a lot on how a port is built. Things like having rail tracks right next to the docks, smooth links to national rail networks, or canals that connect the port directly to rivers or inland waterways make a big difference. But not all ports have these features. In many cases, cargo has to go through several steps before it even reaches a train or barge, which makes the process slower and less competitive compared to using trucks.

Some ports don't have direct access to inland waterways, so barges have to come in through the sea, which isn't always practical—especially if the port layout or vessel size makes it difficult. These problems usually come from poor planning in the port's original design, and fixing them later is tough.

Ports that manage to balance different transport modes well—like rail, river, and short-sea shipping—usually have strong internal rail systems and special setups for water-based transport. For example, in Hong Kong, they use feeder barges to move cargo straight from ships to barges, skipping the usual dockside handling.

In some countries, they've built dedicated freight corridors to move cargo quickly without getting stuck in city traffic or interfering with passenger trains. India has developed Dedicated Freight Corridors (DFCs) like the Western DFC (connecting Delhi to Mumbai ports) and Eastern DFC (connecting Punjab to West Bengal). These are helping industries move goods faster and more reliably.

As ports start working more closely with dry ports and inland terminals, sharing data becomes really important. Expanding digital systems that connect shipping, port, and land transport operations can help move cargo more smoothly, cut costs, and make ports more competitive.

4.9.5 Making Ports More Competitive

A port is considered competitive when it consistently attracts more traffic than others, helping it grow its market share. Studies show that having multiple terminal operators within the same port actually improves performance. It keeps things fair, avoids monopolies, and encourages better, more flexible services.

To boost competitiveness, ports need to improve their shipping connections, run operations more efficiently, and have strong links to inland transport. It's also important to have local community support—especially in big cities where ports have been around for a long time.

Major cities like Guangzhou, Shenzhen, Tianjin, and Hong Kong in Asia, or New York and Los Angeles in the U.S., are good examples of this. Even cities without ports rely heavily on nearby port cities. The closer and smaller the port-city, the more closely it works with its surrounding region.

Port authorities often go for shorter lease periods (concessions) to stay flexible, attract more players, and allow for future upgrades. To stay competitive, it also helps to have a shared logistics platform where everyone—from producers to end users—can access clear, real-time information across the supply chain.

In developing world-class port infrastructure, gaining support from nearby communities, businesses, and local authorities is essential. When locals recognise the benefits—such as employment, trade, and improved infrastructure—they are more likely to support port growth. This backing helps avoid delays, speeds up project approvals, and promotes better coordination and planning that balances economic, environmental, and social needs.

4.9.6 Competing for Customers with New Technology

The logistics industry is going through a tech upgrade. Everyone—from manufacturers to end users—is investing in smarter systems to work faster and more efficiently. One big innovation is blockchain, which is even starting to show up in shipping.

Some terminals, such as those in Rotterdam and Hamburg, have adopted full automation, using Automated Guided Vehicles (AGVs) for container transport and Automated Stacking Cranes (ASCs) for yard operations. In contrast, ports in Korea and Taiwan have implemented semi-automated systems, where stacking is automated but container movement is handled by trailers. Although automated terminals require higher initial investment, they offer superior profitability, service quality, safety, and environmental performance (Yang & Shen, 2013). Automation also reshapes the workforce, replacing many low-skilled roles with fewer high-skilled supervisory positions. In high-wage regions, automation helps reduce labour costs. However, in low-wage countries, manual operations often remain more cost-effective and efficient (Drewry, 2010).

4.9.7 Making Ports Faster: Cranes, Yard Use, and Turnaround Time

Modern container cranes are super-efficient—they can lift two or more containers at once using special double or tandem lifts, which helps speed things up. The more cranes working on a ship, the faster it gets loaded or unloaded, though adding too many doesn't always make a big difference. The same goes for yard equipment—having more usually helps things run smoother.

Experts say a terminal works best when the yard is about 70% full. If it gets more crowded, operations slow down. For berths (where ships dock), around 65% occupancy is ideal—anymore, and ships start queuing up, and service quality drops. Some ports with tight schedules can handle more traffic, but others with random ship arrivals get jammed easily.

One of the best ways to check how well a port is doing is by looking at turnaround time—how long ships stay before leaving. Shorter times mean less congestion and more cargo moving. Data shows ports in East Asia, Europe, and the Caribbean are the fastest, while those in Africa and South Asia (including India) take longer. Interestingly, Chinese ports used to be slow in the '90s but became some of the fastest by 2006 and 2011.

4.9.8 Port Education and Skill Development

Many ports run learning programs—like school visits, boat tours, summer camps, and online resources—to teach people about port operations. Running a port well needs good equipment, land, and skilled workers. While ports used to rely more on manual labour, jobs have shifted toward tech-based roles over the years.

With containerization and digital tools changing how ports work, there's now a clear gap in skills across logistics and shipping. To fix this, many ports and terminal operators offer training for both new and current staff. For example, Euro-gate Terminal in Rotterdam moved from traditional gang-based teams to smaller, multi-skilled crews that keep automated cranes running smoothly.

Even with automation, skilled workers are still key to keeping ports efficient. Skilled labour is crucial in port operations because it ensures that complex equipment, systems, and processes run smoothly and safely. As ports become more automated and technology-driven, the need for workers who can operate, maintain, and manage these systems increases. Skilled workers help reduce errors, improve efficiency, and adapt to new technologies—making the entire port more productive and competitive. Even in highly automated terminals, human expertise is still essential for supervision, troubleshooting, and decision-making.

4.9.9 Environmental Regulations

The shipping industry is now putting more focus on being eco-friendly and cutting down its environmental impact. This shift is part of a global push to reduce emissions, and many companies are investing in cleaner tech and greener practices to stay in line with new expectations.

Even though shipping is already one of the most efficient transport modes in terms of emissions per ton-kilometre, stricter rules are being rolled out worldwide. These include limits on sulphur emissions, better fuel efficiency, and cleaner engines. Meeting these standards often means expensive upgrades, which can affect how shipping companies manage their costs.

Still, being sustainable is becoming a major factor in business decisions. More shippers, logistics providers, and customers now prefer working with companies that care about the environment and treat workers fairly. This trend is shaping how companies operate and compete—those that meet environmental standards are more likely to win long-term deals and stay ahead in the market.

Green transportation technologies are transforming logistics, with companies adopting electric vehicles, battery-powered locomotives, and sustainable aviation fuels to reduce emissions. Eco-friendly warehousing is on the rise, featuring solar panels, energy-efficient systems, and strategic placement to cut fuel use. Digital tools like AI, blockchain, and digital twins enhance sustainability by optimising routes, forecasting demand, and tracking emissions. Packaging is becoming more sustainable, with brands like Nike and GlaxoSmithKline leading circular supply chain efforts. Regulatory pressures and rising carbon pricing are driving compliance, while consumer demand for eco-friendly delivery is pushing businesses to prioritise environmental responsibility for long-term success.

4.10 Upgradation Effort by Key Maritime Institutions

4.10.1 Suez Canal: Ongoing Upgrades for Global Trade

The Suez Canal, a 193 km artificial waterway linking the Mediterranean Sea to the Indian Ocean via Port Said and Suez Port, is a vital global trade route that shortens shipping time between Europe and Asia by bypassing the Cape of Good Hope. It is the longest canal without locks, allowing uninterrupted transit in 13–15 hours, and handles around 50 ships daily. Since its inauguration on 17 November 1869, the canal has undergone multiple upgrades to accommodate larger vessels and increased traffic.

Enhancements over time include deepening, widening, and adding bypasses, enabling passage for ships up to 240,000 DWT with drafts of up to 66 feet, reflecting its strategic importance in transporting energy, goods, and components between Asia, the Middle East, and Europe.

Since opening in 1869, the Suez Canal has been steadily expanded to meet increasing maritime needs. Its total length grew from 164 km to 193.3 km by 2015, and its bypass channels increased from 27.7 km in 1956 to 113.3 km in 2015, improving traffic flow.

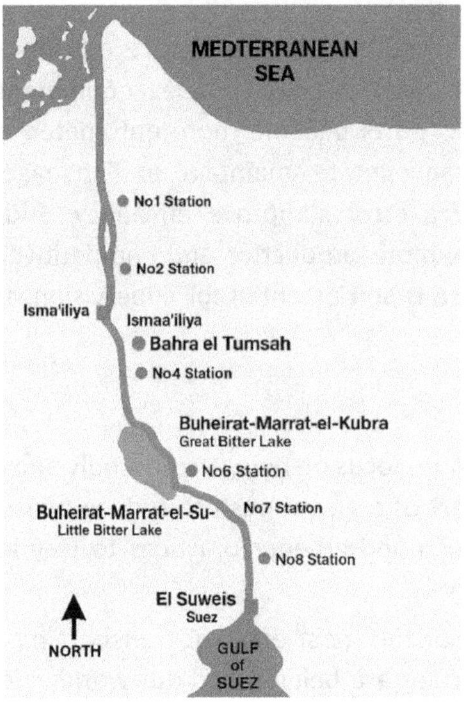

Figure 4-5: Navigational Route of Suez Canal

The canal's width at key sections expanded from 60 m in 1956 to as much as 225 m by 2015, allowing larger ships to pass. Its water depth increased from 8 m to 24 m, enabling vessels with drafts of up to about 20.1 m to transit. The cross-sectional area expanded from 304 m^2 to over 5,200 m^2, and the maximum deadweight tonnage (DWT) of ships increased from 5,000 tons to 240,000 tons, reflecting the canal's modernisation to support global trade.

Table-4.2: Chronology for the development of the Suez Canal is as follows;

Year	Overall Length (km)	Bypasses Length (km)	Width at Key Points (m)	Max Water Depth (m)	Cross-Sectional Area (m^2)	Max Ship Load (DWT)
1869	164.0			8.00	304	5,000
1956	175.0	27.7	60	14.00	1200	30,000
1962	175.0	27.7	89	15.50	1800	60,000
1980	189.8	77.0	160–175	19.50	3250–3600	1,50,000

1994	189.8	77.0	170–190	20.50	3600–4000	1,70,000
1996	189.8	77.0	180–200	21.00	3850–4300	1,85,000
2001	191.8	79.0	195–215	22.50	4350–4800	2,10,000
2010	193.3	80.5	205–225	24.00	4800–5200	2,40,000
2015	193.3	113.3	205–225	24.00	4800–5200	2,40,000

4.10.2 Expansion of Panama Canal

The Panama Canal connects the Pacific and Atlantic oceans for international shipping since its opening in August 1914. The size of the lock chambers and the depth of the water in the Canal dictate the size of ships that can pass through.

The original canal could allow vessels carrying up to 5,000 TEUs (Twenty-foot Equivalent Units), often referred to as "Panamax", to pass through; these ships were up to 106 feet wide, 965 feet long, and 39.5 feet of draft. Later on, with the development, the canal became too small for the most modern cargo ships to pass through. The old dimensions of the First Panama Canal could not suit the needs of the modern shipping industry.

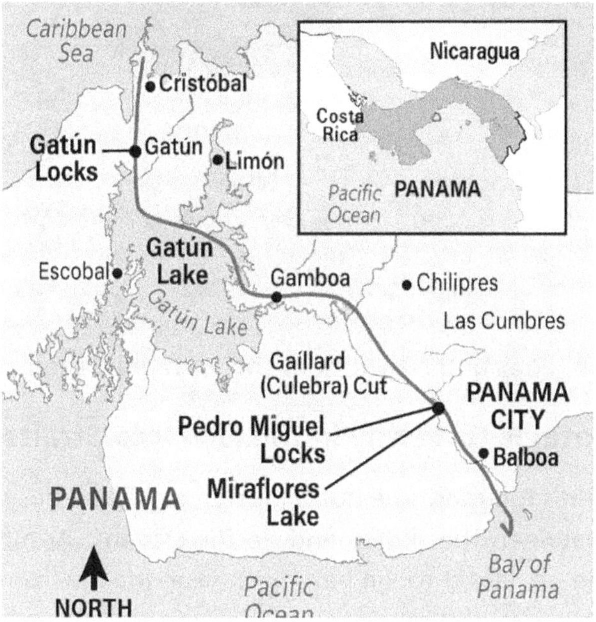

Figure 4-6: Navigational route of Panama Canal

The locks became too small for the new generation ships (Post-Panamax), and the global shipping industry was hindered due to the limited passage of ships through this important traffic route.

On account of this, the larger marine vessels either circumnavigate through South America by sailing an additional 12,875 kilometres or move through the Suez and Panama Canal, which was losing its traffic and revenue considerably.

Thus, the need for the expansion of the Canal has emerged to suit the new generation of ships called the Post-Panamax vessels. The maximum dimensions of these ships are 366 meters in length, 49 meters in beam, and 15.2 meters in draught. The problem of the Post-Panamax vessel transit was solved by the construction of the new expanded Panama Canal, which was opened in 2016.

With the completion of the expansion, post-Panamax vessels are now able to pass through the canal with up to 13,000 TEUs. Aside from increasing ship capacities (from 5000 to about 13,000 TEUs), the new locks also created a third lane of traffic for these new vessels, meaning that the other two lanes now have a greater capacity for smaller vessels. This expansion is significant because it allows greater sizes and frequencies of vessels to pass through that are heading to the US East and Gulf Coast ports, and now it can compete with the Suez, winning back what was lost to the Suez over the past years and significantly increasing toll revenues.

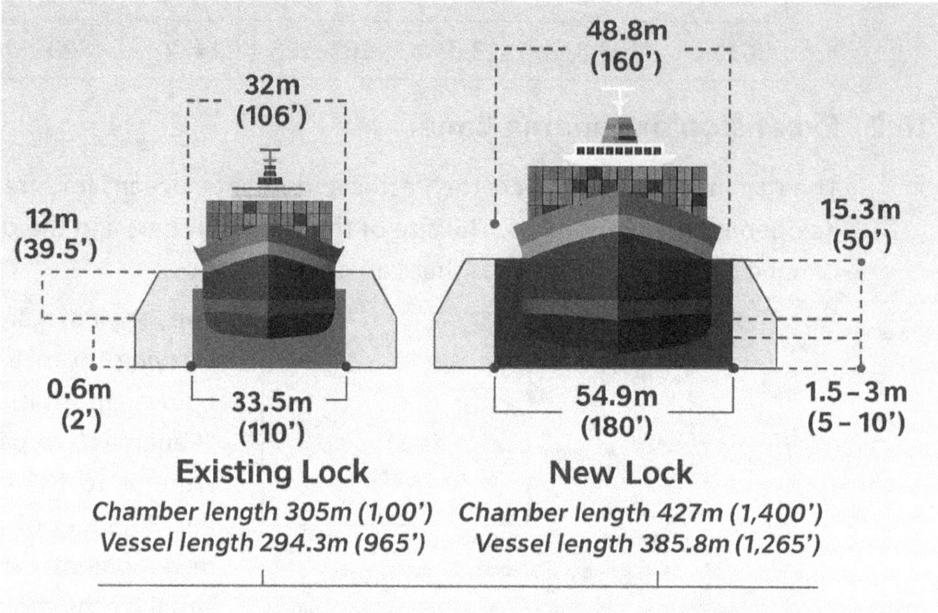

Figure 4-7 - Existing lock and new lock of the Panama Canal.

4.10.3 Port of Singapore: Strategic Gateway in the Malacca Straits

Singapore is one of the world's top maritime hubs, located along the busy Malacca Strait—the main shipping route between the Indian and Pacific Oceans. Around 25% of global traded goods and sea-borne oil pass through here each year, mostly from the Persian Gulf to markets in China, Japan, and South Korea.

The key players in this region are Indonesia, Malaysia, and Singapore. Together, their ports handle about 30% of global transhipment traffic. Major ports include Singapore, Port Klang and Tanjung Pelepas in Malaysia, and Tanjung Priok in Indonesia. In 2007, Singapore led globally with 27.9 million TEUs, followed by Port Klang (7.1M), Tanjung Pelepas (5.5M), and Tanjung Priok (3.9M). Even though Singapore is relatively small, it was the world's top container port in 2008 with 29.9 million TEUs, overtaking Hong Kong and closing in on Port Klang. Today, Singapore holds second place, reinforcing its strategic importance in global trade.

Table-4.3: Top 10 Container Ports in the World for 2024, ranked by TEUs handled:

Rank	Port	Country	TEUs (2024)	Growth Rate
1	Shanghai	China	51.51 million	+4.8%
2	Singapore	Singapore	41.12 million	+5.4%

Rank	Port	Country	TEUs (2024)	Growth Rate
3	Ningbo-Zhoushan	China	39.3 million	+11.3%
4	Shenzhen	China	33.4 million	+11.8%
5	Qingdao	China	30.87 million	+7.3%
6	Guangzhou	China	26.45 million	+4.1%
7	Busan	South Korea	24.4 million	+5.4%
8	Tianjin	China	23.29 million	+5.0%
9	Jebel Ali	UAE	15.53 million	+7.3%
10	Port Klang	Malaysia	14.64 million	+4.2%

4.10.4 Securing Singapore's Lead in Maritime Trade

Singapore's maritime growth over the past decade is the result of a clear strategy, solid execution, and strong collaboration between the government and the shipping industry. Even though Malaysian ports are well-positioned along the Malacca Strait, they still trail behind Singapore in the race for transhipment volumes.

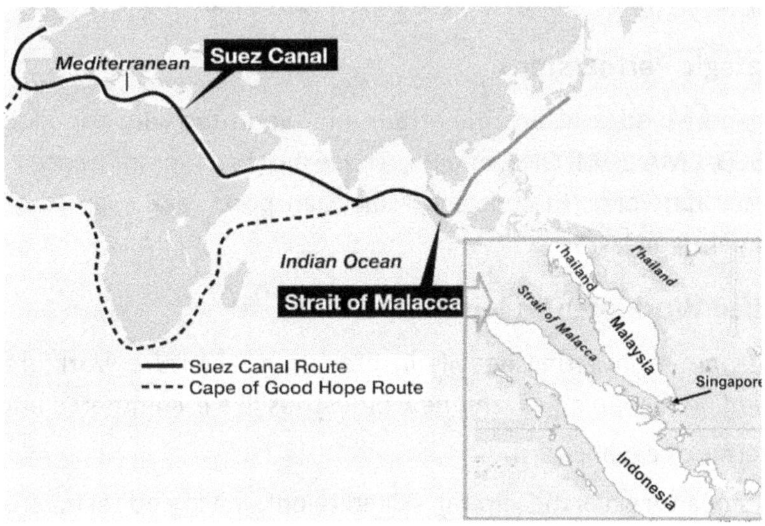

Figure 4-8: Port of Singapore in the Malacca Straits

One major reason is the absence of a coordinated national strategy in Malaysia, which has kept ports like Tanjung Pelepas (PTP) and Port Klang from competing effectively with Singapore. Transhipment—where goods are moved through an intermediate port before reaching their final destination—is a key area where Singapore excels.

Another issue is the slow pace of capacity upgrades at Malaysian ports. This has made it harder for major shipping lines to expand services there. In contrast, Singapore's ports offer greater scale and efficiency. Over the past few years, congestion at PTP and Port Klang has pushed more cargo traffic toward Singapore.

4.10.5 Applying Innovation and Automation for the shipping industry

Singapore's PSA Terminal is one of the most high-tech ports globally. It's invested in smart systems like CITOS and PORTNET, which help spot issues early, reduce theft and piracy, and save money for shipping companies. Singapore is also testing driverless vehicles to make port operations even smoother. As automation grows, Singapore's shipping rates are becoming more competitive.

The government is also working on the Tuas mega-port, set to be ready by 2040. It'll handle up to 65 million TEUs a year and even use drones to load ships. Innovation is a big part of Singapore's shipping strategy, and the International Maritime Centre (IMC) has laid out a 2030 vision to keep Singapore ahead in the region.

"Singapore has always been proactive when it comes to its national development."

Even though Malaysian ports have lower stevedoring costs, Singapore offers better network connectivity and cheaper bunker fuel, which helps offset the higher handling fees. One reason for this strong connectivity is Singapore's joint ventures with major global shipping lines like MSC, COSCO, CMA CGM, ONE, and PIL. These partnerships influence where carriers choose to trans-ship, giving Singapore an edge.

Malaysian ports, on the other hand, don't have similar JV setups, which is one reason they're still trailing behind Singapore in Southeast Asia.

4.10.5.1 Strategic Partnerships

Singapore's edge also comes from joint ventures with top global shipping lines like MSC, COSCO, CMA CGM, ONE, and PIL. These partnerships influence carrier decisions and expand service networks. In contrast, Malaysian ports lack such JV arrangements, limiting their competitiveness.

4.10.5.2 Skilled Workforce and Infrastructure

Singapore is investing heavily in preparing its future workforce—focusing on education, emerging technologies, and new industries like e-commerce and commodity trading.

In terms of capabilities:

- **Singapore** has the capital, skilled talent, advanced tech, strong business infrastructure, and global market access.
- **Malaysia** has a semi-skilled workforce, mid-level technology, basic infrastructure, and natural resources.
- **Indonesia** has mostly unskilled labour, basic technology, and plenty of raw materials and undeveloped land.

4.10.5.3 Free Trade Agreement and Regional Dynamics

Since 2007, the entire Riau archipelago near Singapore has operated as a free trade zone. Around 600 companies have set up in 13 industrial parks managed by Batam's industrial authority, including big electronics brands like Sanyo, Panasonic, Siemens, Sony, Toshiba, and Epson, along with naval suppliers. Batam has grown into a major industrial hub. Many businesses have moved just 20 km from Singapore to Batam due to lighter regulations and lower wages, while still benefiting from Singapore's free trade deals, especially with the United States.

Singapore currently ranks third globally in GDP per capita, behind only Qatar and Luxembourg. Malaysia trails but is still ahead of Indonesia.

Singapore's massive shipping industry keeps gaining momentum as more global players shift operations to the island. The Singapore Port Authority (PSA) has formed joint ventures with top global shipping lines like MSC, COSCO, CMA CGM, ONE, and PIL to run terminals in the city. These partnerships heavily influence which ports carriers choose for transhipment, expanding Singapore's service network and reinforcing its dominance.

4.10.6 Port of Rotterdam: Strategic Maritime Hub of Europe

The Port of Rotterdam stands as one of Europe's most vital maritime hubs. It held the title of the world's largest port for an impressive 42 years (1962–2004) before being overtaken by Singapore and later Shanghai. Despite this, Rotterdam remains Europe's largest port, spanning 41 square miles (105 sq. km) and handling an immense volume of shipping traffic.

4.10.6.1 Historical Growth and Infrastructure

Over the last 150 years, Rotterdam has consistently expanded, adapting its port infrastructure—including jetties, quay walls, and cargo handling systems—to meet evolving demands. Its focus on flexibility and continuous improvement has been key to maintaining its global relevance.

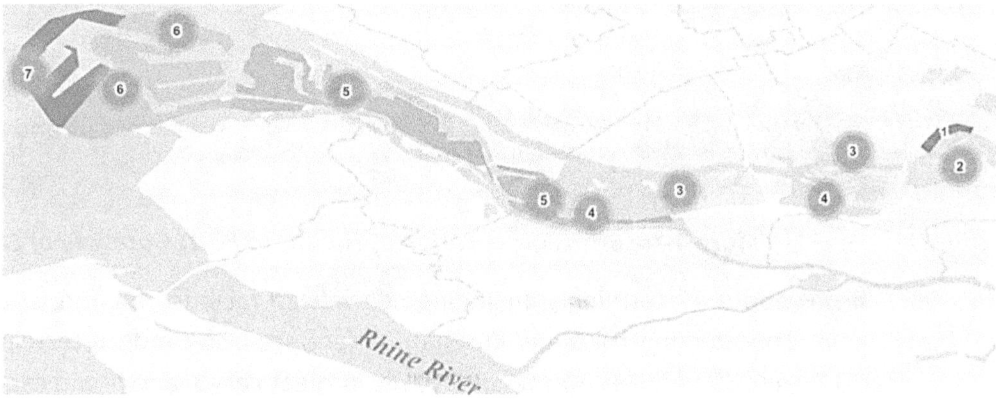

Figure 4-9: Historic Growth of Port of Rotterdam

The port's Port Vision 2030 emphasises its strengths: a strategic location, an entrepreneurial spirit, and a long-standing commitment to industry and logistics. Warehouses, railways, roads, and waterways have all been upgraded to accommodate larger ships, deeper drafts, and heavier loads.

4.10.6.2 Post-War Innovation and Global Leadership

Following the devastation of World War II, Rotterdam rebuilt with advanced terminals and modern cranes, making it a highly attractive port. In 1962, it surpassed New York to become the world's largest port, a position it held until 2004.

4.10.6.3 Adapting to Change

Rotterdam Port has consistently updated the design of its berthing facilities to keep up with changes in ship sizes and cargo handling methods. As vessels have grown larger and more complex, the port has responded by redesigning its infrastructure to meet new demands. It hasn't just kept pace—it's set the standard.

By using the latest construction materials and techniques, Rotterdam has built facilities that are not only durable but also highly efficient. These upgrades have helped the port handle deeper drafts, heavier loads, and faster turnaround times. Its commitment to innovation and adaptability has made it a benchmark for modern port development worldwide.

4.10.7 Port of Hamburg – Germany's Global Trade Gateway

The Port of Hamburg is Germany's top foreign trade hub, thanks to its prime location between the North Sea and the Baltic Sea and its strong port operations. As the country's largest multipurpose port, it plays a key role in serving European inland markets, reaching up to 450 million consumers. One of its biggest strengths is rail connectivity. Hamburg offers around 2,000 container train services every week to and from the European hinterland, with over 90% of cargo moving by rail, making it Europe's leading rail port.

The port can handle all types of vessels and cargo.

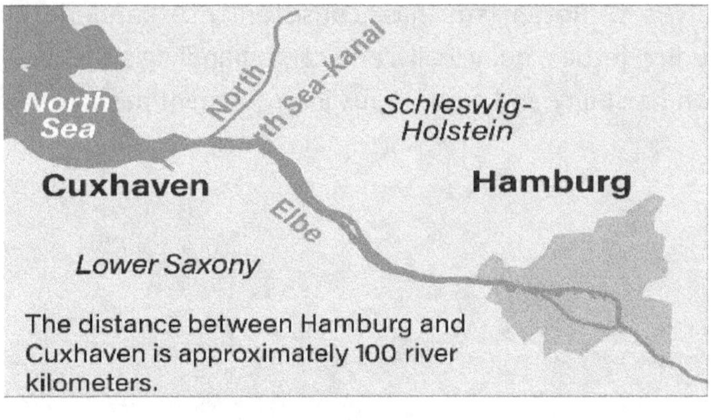

Figure 4-10: Strategic Location of Hamburg Port: Gateway of Europe

It has about 75 terminals, including specialised facilities for the world's largest container ships operated by HHLA and Eurogate. There are also modern terminals for general cargo, Ro-Ro goods, vehicles, timber, heavy lift, project cargo, and liquid cargo.

Hamburg's port activities generate around €20 billion in gross value annually and support over 260,000 jobs directly and indirectly. Its advanced handling technology, efficient transport systems, and strong feeder and hinterland links make it a key player in global trade. It's also vital for import-export logistics in Hamburg and the surrounding region.

Additionally, Hamburg plays a major role in the New Silk Road trade route to China, offering 200 scheduled rail connections per week. Since 2013, China has expanded its westward trade corridor along this historic route, strengthening Hamburg's position in Europe–Asia logistics.

4.10.7.1 Hamburg – Gateway to the New Silk Road

Hamburg isn't just Germany's largest seaport—it's also a key European gateway to the New Silk Road, China's ambitious trade initiative linking Asia, Europe, and Africa via land and sea. Originally called "One Belt, One Road," the project includes two major routes: the Silk Road Economic Belt (by land) and the 21st Century Maritime Silk Road (by sea). These routes form a vast network for moving goods across continents by rail, road, and ship.

Since the initiative launched in 2013, rail transport has become a major part of the Belt and Road strategy. Today, a web of rail corridors stretches from China's eastern coast all the way to Europe's western ports. According to the International Union of Railways, nearly 878,000 TEUs were moved along these Eurasian rail corridors in 2020.

Hamburg plays a central role in this network. In 2021, around 160,000 TEUs were transported by rail between Hamburg and China. The port handles about 2.6 million TEUs in China trade annually, with nearly one-third of all containers passing through Hamburg linked to Chinese imports or exports.

As costs rise in coastal Chinese cities like Shanghai and Shenzhen, China's "Go West" policy is encouraging industries to shift inland. Cities like Chongqing and Chengdu have become major manufacturing hubs. For example, Hewlett-Packard has shipped over four million laptops to Europe via the Chongqing–Duisburg rail line since 2011.

The Trans-Eurasia Railway, connecting Chongqing to Duisburg, is the most important link so far. Spanning 11,179 km across six countries—including Kazakhstan, Russia, Belarus, and Poland—it offers a faster alternative to sea freight. For instance, a car shipped from Europe to Chengdu by sea can take up to three months, while rail cuts that time to just 25 days.

With Chengdu and Chongqing now officially recognised as international land ports, competition among hub ports is intensifying. Ports like Colombo, Singapore, Port Klang, and Tanjung Pelepas are expanding their transhipment roles beyond regional boundaries. Despite Colombo's strategic location on the East–West sea route, it now faces stiff competition from these Southeast Asian ports, each offering alternative options for global shipping lines.

4.10.8 Port of Shanghai: Anchoring China's Trade Power

China's vision of the 21st century as a maritime era has driven major investments in marine infrastructure and resources. By 2005, the country had built a vast network of ports capable of handling growing volumes of shipping traffic. Today, China hosts more ports than any other nation, including seven of the world's ten busiest ports.

4.10.8.1 Global Leadership

At the forefront is the Port of Shanghai, which became the world's largest port in 2010, surpassing Singapore. It has maintained this position through strategic expansion and integration with nearby satellite ports, notably Yangshan Port—the largest automated container terminal in the world, responsible for handling the bulk of Shanghai's container traffic. Shanghai has earned global recognition, including the "Best Connected Port" award from UNCTAD, thanks to its extensive global shipping links and efficient operations.

4.10.8.2 Strategic Layout and Regional Integration

Shanghai's success is supported by a well-coordinated regional layout. It serves as an international shipping centre, while Ningbo-Zhoushan functions as a regional hub. Together with other feeder and extension ports in the Yangtze River Delta, this network ensures smooth cargo flow and high operational efficiency.

In addition to its role as a global container hub, the Port of Ningbo complements Shanghai by offering superior infrastructure for handling iron ore and crude oil shipments. To strengthen inland connectivity, several feeder ports have been developed upstream along the Yangtze River, serving as transit hubs that link Shanghai to the hinterland.

4.10.8.3 Infrastructure Capabilities

Shanghai's port layout is designed to accommodate the largest container ships in the world, including vessels with capacities of up to 24,000 TEUs. The port is equipped with over 100-ton lifts and a wide range of fixed, mobile, and floating cranes. It boasts 125 docks and 19 terminals, including:

- 3 break-bulk terminals
- 2 bulk cargo terminals located in Luojing, Wusong, and Longwu
- A cruise terminal capable of handling 1 million passengers annually

4.10.8.4 Operational Scale and Global Reach

Despite already being the world's largest port, Shanghai continues to expand and modernise, handling increasing volumes of cargo each year. It is now recognised as a key driver of the world's fastest-growing economy.

Each month, over 2,000 container ships depart from Shanghai, connecting to 281 shipping routes that span the globe. This extensive network and high-capacity infrastructure make the port highly attractive to global shippers and carriers.

4.10.9 Chinese Shipping Industry

China's vision of the 21st century as a maritime era has driven major investments in marine infrastructure and resources. By 2005, the country had built a vast network of ports capable of handling growing volumes of shipping traffic. Today, China hosts more ports than any other nation, including seven of the world's ten busiest ports.

4.10.9.1 Global Leadership

At the forefront is the Port of Shanghai, which became the world's largest port in 2010, surpassing Singapore. It has maintained this position through strategic expansion and integration with nearby satellite ports, notably Yangshan Port—the largest automated container terminal in the world, responsible for handling the bulk of Shanghai's container traffic. Shanghai has earned global recognition, including the "Best Connected Port" award from UNCTAD, thanks to its extensive global shipping links and efficient operations.

4.10.9.2 Strategic Layout and Regional Integration

Shanghai's success is supported by a well-coordinated regional layout. It serves as an international shipping centre, while Ningbo-Zhoushan functions as a regional hub. Together with other feeder and extension ports in the Yangtze River Delta, this network ensures smooth cargo flow and high operational efficiency.

In addition to its role as a global container hub, the Port of Ningbo complements Shanghai by offering superior infrastructure for handling iron ore and crude oil shipments. To strengthen inland connectivity, several feeder ports have been developed upstream along the Yangtze River, serving as transit hubs that link Shanghai to the hinterland.

4.10.9.3 Infrastructure Capabilities

Shanghai's port layout is designed to accommodate the largest container ships in the world, including vessels with capacities of up to 24,000 TEUs. The port is equipped with over 100-ton lifts and a wide range of fixed, mobile, and floating cranes. It boasts 125 docks and 19 terminals, including:

- 3 break-bulk terminals
- 2 bulk cargo terminals located in Luojing, Wusong, and Longwu
- A cruise terminal capable of handling 1 million passengers annually

4.10.9.4 Operational Scale and Global Reach

Despite already being the world's largest port, Shanghai continues to expand and modernise, handling increasing volumes of cargo each year. It is now recognised as a key driver of the world's fastest-growing economy.

Each month, over 2,000 container ships depart from Shanghai, connecting to 281 shipping routes that span the globe. This extensive network and high-capacity infrastructure make the port highly attractive to global shippers and carriers.

4.10.9.5 Local Impacts in Port Cities

Port cities in China experience direct economic benefits from shipping activities. Most of the value added from port operations is generated within these cities, especially due to the clustering of industries around port areas. This clustering leads to economies of scale and knowledge sharing, attracting resource-heavy industries that benefit from reduced transportation costs due to proximity to the port.

The nature of the local economy also influences port traffic. In wealthier regions with strong service sectors, ports tend to handle a diverse mix of high-value goods, such as containers and consumer products. In contrast, agricultural and industrial regions are more focused on bulk cargo, including raw materials and commodities.

4.10.9.6 Wider Regional and International Spillovers

While direct impacts are concentrated in port cities, indirect and catalytic effects extend far beyond. Businesses in other regions benefit from efficient port operations, which help lower transport costs and facilitate both imports and exports.

These spillover effects are not limited to domestic regions. In many cases, they extend across borders. For example, the Port of Rotterdam plays a crucial role in supporting German industries, and several European ports serve as gateways for landlocked Central European countries.

4.10.9.7 China's Shipping Industry – Growing Strong

Over the past decade, China's merchant shipping sector has made huge strides. It's now a major player in global trade and a clear sign of China's rising maritime influence. With over 400 domestic and international shipping companies involved in sea trade, China has earned A-class membership in the United Nations Maritime Organisation—and has held that status for five years straight. On top of that, China has a seat on the IMO Steering Committee, showing its growing leadership in global maritime affairs.

Today, over 90% of China's foreign trade moves by sea, making shipping a key driver of the country's economic growth. The Chinese-flagged fleet includes a wide range of vessels—bulk carriers, tankers, LNG and chemical ships, container vessels, Ro/Ro ships, and even passenger liners.

Foreign investment in China's shipping sector has surged since 1984, when the first foreign ship docked at a Chinese port. But while the maritime side has grown rapidly, inland transport—especially rail and road—is still underdeveloped, putting extra pressure on sea routes. China's main goal is to keep energy supplies flowing and maintain easy access to raw materials and global markets. As its economy continues to grow, China has become more focused on protecting its sea lines of communication (SLOCs)—the key shipping routes that keep trade moving.

In the last decade, China's maritime identity has shifted dramatically. With booming trade, fast economic growth, rising defense spending, and a modernising navy, China is clearly building up its maritime power to play a bigger role in the region—and possibly beyond. China's rise as a sea power has seriously boosted its industries and overall development. Here's how its shipping fleet helps drive that change:

4.10.9.8 Fuelling Export-Led Growth

China's massive shipping network is the backbone of its global manufacturing success. Since over 95% of its international trade moves by sea, having a strong fleet means it can ship huge amounts of products quickly and cheaply. This includes everything from smartphones, laptops, and home appliances to industrial machinery, electric vehicles, solar panels, and clothing. Thanks to this setup, China can keep pumping out goods and sending them all over the world without delays or high costs.

4.10.9.9 Leading the Shipbuilding and Logistics Game

China builds more than 1,700 commercial ships every year—over half of all ships made globally. That's a huge number, and it's possible because the government backs its shipyards with funding and smart industrial policies. This means Chinese shipbuilders can offer lower prices and faster delivery than many competitors. But it's not just about ships—this success boosts other industries too. For example, it drives demand in steel production, heavy engineering, and machinery manufacturing. So when China builds a cargo ship, it's also creating jobs and growth in places like steel plants and equipment factories. It's a ripple effect that strengthens the whole economy.

4.10.9.10 Smart Investment in Ports, at Home and Abroad

China has poured a ton of money into building and upgrading ports—not just within the country, but all over the world. Big domestic ports like Shanghai, Ningbo-Zhoushan, and Guangzhou are super-efficient and located close to major factories, which helps cut down transport time and costs. On the global front, Chinese companies like COSCO and China Merchants Port have bought stakes in key ports across Europe, Asia, Africa, and Latin America. This gives China more influence over important trade routes and helps ensure smoother, cheaper movement of goods like electronics, machinery, textiles, and even vehicles. It's a smart strategy that strengthens China's grip on global trade.

4.10.9.11 Smart Tech and Automation in Port Operations

China is leading the way in making its ports smarter and more efficient through cutting-edge technology. Many of its major ports now use AI-powered cranes, autonomous trucks, and real-time cargo tracking systems to speed up operations. This high level of automation helps reduce labour costs, cut down on delays, and handle massive volumes of cargo with precision. For example, ports like Shanghai Yangshan use driverless vehicles to move containers, while smart logistics platforms help track shipments instantly from the dock to the destination. These innovations are crucial for keeping up with global demand and ensuring China's industrial output flows smoothly across international markets.

4.10.9.12 Shaping Global Trade Through Maritime Influence

China's maritime strength gives it a powerful edge in steering global supply chains. Its state-owned enterprises (SOEs)—like COSCO Shipping and China Merchants Group—don't just operate efficiently; they also align with national strategic goals. This unique mix of business savvy and government backing allows China to negotiate favourable trade deals and invest in critical infrastructure worldwide.

By securing stakes in ports and logistics hubs across continents, China ensures reliable access to essential resources like iron ore, crude oil, natural gas, and other raw materials needed to keep its manufacturing sector running. This influence helps stabilise supply routes, reduce dependency on external players, and maintain steady production of goods ranging from electronics and vehicles to construction materials and industrial equipment.

4.10.9.13 Expanding Global Reach Through Port Ownership

China's maritime strategy goes far beyond its borders. With ownership or stakes in over 129 ports worldwide, China has positioned itself to influence major shipping routes and global trade flows. This expansive network not only boosts its economic footprint but also strengthens its geopolitical leverage, allowing it to play a more active role in shaping international commerce.

By controlling or partnering in key ports across Europe, Asia, Africa, and Latin America, China can ensure smoother logistics, secure access to vital resources, and respond quickly to shifts in global demand. These ports act as strategic gateways for importing raw materials like oil, gas, and minerals, and exporting finished goods such as electronics, vehicles, and industrial equipment.

Thus, China's maritime fleet is far more than a tool for shipping—it's a strategic engine driving the country's industrial expansion, powering its manufacturing sector, and reinforcing its position as a global economic leader. Through its dominance in shipbuilding, smart investments in port infrastructure, and cutting-edge logistics technology, China has built a comprehensive maritime ecosystem. This system not only supports its domestic growth but also extends its influence across global supply chains, making maritime logistics a cornerstone of its long-term development strategy.

4.10.10 Port of Colombo: Strategic Transhipment Hub in South Asia

The Port of Colombo, situated on the southwestern coast of Sri Lanka along the Kelani River, stands as the country's largest and most active maritime hub. It is also recognised as the busiest port in South Asia. Its strategic position in the Indian Ocean has made it a vital transhipment point within the Asian shipping network. In the early 1980s, the port underwent significant modernisation to accommodate containerised cargo, including the installation of cranes and gantry systems typical of modern container terminals. Further enhancements in the 1990s, such as deepening the access channel, boosted its capacity and efficiency. Today, Colombo ranks among the top 30 container ports globally. Notably, it handles approximately 2 million TEUs annually that are either originating from or destined for India, underscoring its critical role in regional trade logistics.

The Port of Colombo has emerged as a preferred transhipment hub in South Asia due to its strategic location, high cargo volumes, advanced infrastructure, and operational efficiency.

4.10.10.1 Strategic Location

Colombo is positioned directly on the East–West shipping route, one of the busiest maritime corridors globally. This makes it a natural stop for vessels connecting Asia, the Middle East, and Europe. Its proximity to India's southern coast allows it to efficiently handle Indian cargo that many Indian ports cannot accommodate due to depth limitations.

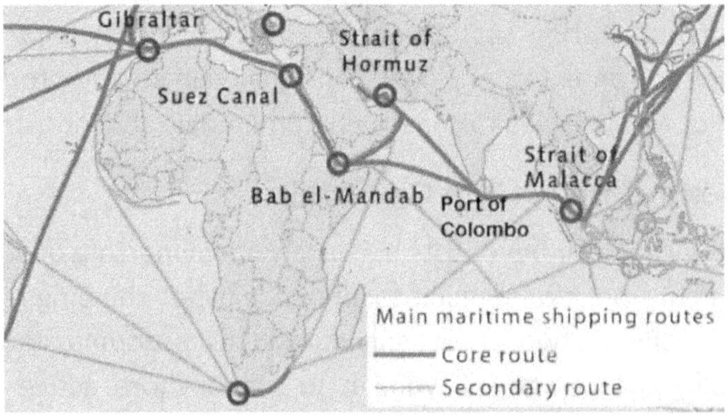

Figure 4-11: Strategic Location of Port of Colombo

4.10.10.2 High Transhipment Volume

In 2024, transhipment cargo accounted for 81% of Colombo's container throughput, totalling 6.31 million TEUs—a 9.7% increase year-on-year. Over 1.2 million TEUs of Indian cargo are rerouted through Colombo annually, highlighting its importance in regional logistics.

4.10.10.3 Advanced Infrastructure

Colombo's terminals—CICT, SAGT, JCT, and ECT—feature deep-water berths and modern equipment capable of handling the largest container ships. Expansion projects like the Colombo West International Terminal (CWIT) and ECT Phase 2 are underway, expected to push the port's capacity beyond 10 million TEUs by 2025.

4.10.10.4 Operational Efficiency

Despite global shipping disruptions, Colombo has maintained high throughput and fast vessel turnaround times. Continued investments in automation and digital systems are enhancing cargo handling speed and reducing delays.

4.10.10.5 Regional Standing: South Asia

- **Efficiency**: Ranked #1 in South Asia and the Indian Subcontinent for port efficiency by the World Bank and S&P Global.
- **Global Ranking**: Positioned 22nd globally among 370 ports in the Container Port Performance Index (CPPI).
- **Transhipment Role**: Serves as the primary transhipment hub in South Asia, handling over 2 million TEUs annually for Indian cargo.
- **Connectivity**: Recognised as South Asia's best-connected port, facilitating robust trade between India and global markets.

Though not among the highest in volume, the Port of Colombo excels in efficiency, strategic relevance, and transhipment specialisation. Ongoing infrastructure upgrades and international partnerships position it as a key player in global maritime logistics.

4.10.10.6 Can Indian Transhipment Ports Compete with Colombo?

Transhipment terminals work like hubs—small ships bring in containers, which are then loaded onto bigger vessels heading to final destinations. But many Indian ports near the key east-west shipping route aren't deep enough for large ships. This route handles about 80% of India's transhipment cargo and most of East Asia's trade with Europe.

On India's west coast, ports mostly send containers directly to their destinations. In contrast, East Coast ports depend heavily on nearby transhipment hubs. Right now, around 2.8 million TEUs, or 25% of India's container traffic, are trans-shipped outside the country—mainly through Colombo, Singapore, Port Klang, Salalah, and Jebel Ali. Colombo alone handles 1.2 million TEUs of Indian cargo.

In fact, 78% of container traffic from India's east coast is transhipped. Only a few southern Indian ports have the deep water needed to handle large ships efficiently and act as proper transhipment hubs.

Because of this, exporters and importers in South India pay an extra ₹5,000–₹6,000 per container, which makes them less competitive and adds a high cost to the economy. Indian ports lose about ₹1,500 crore in business every year. According to the shipping ministry, India is also missing out on becoming a major hub for container trade between Asia, Africa, the U.S., and Europe.

4.10.11 Economic Benefits of Ports – An Overview

Ports bring a wide range of economic benefits, starting with their key role in global supply chains. They act as vital connectors between regions and countries, helping goods move efficiently across borders. Since maritime transport costs make up a big part of the total value of traded goods, ports help reduce these costs—especially for containerised cargo, which is generally cheaper to ship per tonne than non-containerised goods, particularly on major trade routes.

Beyond lowering trade costs, ports also create jobs, add economic value, and attract industries. In fact, improving port efficiency has helped many countries boost their trade volumes. On the flip side, higher maritime transport costs tend to reduce trade. For example, if shipping costs between two countries double, import values can drop by 66–80%, and overall trade volume can fall by 26–28%.

Among various port features, efficiency stands out as the biggest factor influencing international transport costs. It even outweighs other elements like infrastructure quality, private sector involvement, and connectivity between ports.

Other characteristics—such as terminal capacity, technology, and logistics integration—also play a role in shaping maritime transport costs and overall port performance.

4.10.11.1 Port Infrastructure and Its Economic Impact

Port infrastructure plays a major role in shaping transport costs, especially for coastal countries. Studies show that onshore infrastructure can account for up to 40% of total transport costs. For example, if a country with poor infrastructure improves to a higher standard, it could cut transport costs by 30–50% (Limao & Venables, 2001).

Interestingly, the exporter's port infrastructure has a bigger impact on transport costs than the importer's. Also, port congestion matters—a 10% rise in congestion can lead to a 0.7% increase in maritime transport costs (Abe & Wilson, 2009). This is often tied to the quality of logistics services. Poor trucking operations can cause delays, increase inventory costs, and reduce export volumes, which in turn leads to fewer shipping services.

Better trade performance, supported by efficient ports, can boost economic growth. Research consistently shows that trade has a strong positive effect on national output and development.

Ports also attract a wide range of industries. These include logistics, warehousing, and storage, as well as heavy industries like refineries, chemicals, steel, coal, aerospace, and renewable energy. However, the type of industries varies depending on available space, port strategies, and the local economy. Ports often have strong links with these industries, creating regional economic ecosystems.

While ports do create local jobs, direct port-related employment is usually modest—even in large ports, it rarely exceeds a few thousand jobs. However, port-attracted industries can contribute significantly to regional employment and economic value.

4.10.11.2 Facilitating Competitive Ports: Key Considerations

To ensure cities benefit from their ports, it is essential that ports remain competitive. A competitive port is one that is chosen more frequently than others, thereby expanding its market share. Evidence suggests that intra-port competition—where multiple terminal operators compete within the same port—can enhance performance by preventing monopolistic practices and promoting flexible, multi-service operations (De Langen & Pallis, 2005).

Competitiveness can be enhanced through stronger maritime connectivity, efficient port operations, and well-developed hinterland links. Local support for port activities is equally important and can be built through inclusive urban–port development strategies. Historically, ports have played a major role in shaping cities, with many large urban centres—particularly in Asia and North America—developing around their ports. Examples include Shanghai, Osaka–Kobe, Guangzhou, Shenzhen, Tianjin, Hong Kong, New York, and Los Angeles. In India as well, cities such as Mumbai, Kolkata, Chennai, and Kochi have grown around their port infrastructure, demonstrating the close relationship between ports and urban development.

However, urban congestion and environmental concerns now require better transport policies around ports. In some cases, port relocations have helped free up valuable urban land for other uses. Even cities without ports depend heavily on their connectivity to nearby port cities, especially when the port is small and closely integrated with the metropolis.

Port authorities often prefer shorter concession periods to maximise revenue, lower entry barriers, and allow for redevelopment flexibility. A competitive port is shaped by choices made by shippers, followed by forwarders, shipping lines, and terminal operators.

4.10.11.3 Maritime Connectivity and Operational Performance

Maritime connectivity is vital for port competitiveness, influencing the frequency and reliability of shipping services. Ports with strong maritime links attract more cargo due to faster and more direct delivery options. Transhipment also boosts connectivity, especially for ports located on major shipping routes. While most large port cities operate as gateway ports, some strategically located ports function primarily as transhipment hubs, with minimal land-side interaction.

Nautical access is another critical factor. As container ships grow larger and deeper—some requiring drafts of up to 20 meters—ports must invest in dredging to maintain competitiveness. For example, Hamburg can only accommodate the largest ships during high tide, prompting widespread dredging efforts across estuary ports and along rivers, especially in response to infrastructure upgrades like the Panama Canal expansion.

4.10.11.4 Global Cooperation and Performance Metrics

Some ports are enhancing competitiveness through international cooperation. Rotterdam, for instance, has investments in Sohar (Oman) and Suape (Brazil), while Antwerp is involved in Duqm (Saudi Arabia) and India. These partnerships often include consultancy and training services.

Port performance is measured using various indicators across cranes, berths, yards, gates, and labour teams. Metrics include TEUs per crane per year, vessel service time, truck turnaround time, and crane moves per hour (typically 25–40). Container dwell times range from 5–7 days for imports and 3–5 days for exports.

A key indicator of port efficiency is vessel turnaround time—the average duration a ship stays in port. Faster turnaround reduces congestion and increases throughput. Data from Lloyd's Marine Intelligence Unit shows that the most time-efficient ports are in East Asia, Europe, and the Caribbean, while ports in Africa and South Asia (including India) lag behind. Notably, Chinese ports have dramatically improved their turnaround times since 1996, becoming among the most efficient globally by 2011.

4.10.11.5 The Trend of Consolidation among Shipping Lines

Recently, the highly competitive market has been pushing ocean carriers—especially container carriers—to reduce costs and improve service quality.

Shipping has historically been a vital human activity, especially where prosperity relied on international and inter-regional trade. Maritime transport is a key driver of globalisation. Technological, economic, and socio-cultural changes have made it nearly impossible for any country to remain fully isolated from global economic activities. Over the decades, the maritime industry has adapted its technologies, national registries, and labour resources to meet the demands of globalisation.

There is a mutually beneficial relationship between maritime shipping and globalisation. For instance, until the 1950s, crude oil was typically refined at its source and transported in small tankers ranging from 12,000 to 30,000 DWT. It was later recognised that placing refineries closer to consumer markets and transporting crude in larger volumes would be more cost-effective. This led to the development of large crude carriers (over 200,000 DWT), significantly reducing the per-unit cost of intercontinental energy transport.

A similar pattern is seen in agricultural products. Today, it is common for such goods to be harvested on one continent, processed on another, assembled on a third, and finally sold in a fourth. For example, cotton grown in North America may be sent to African mills, then to Asian factories, and finally returned to North America for retail. Products like orange juice and wine are also shipped globally to meet seasonal or climatic demands or to compete with higher-cost domestic production.

In a globalised economy, containerization enables integrated freight transport across all modes. Currently, time, cost, and delivery reliability are key competitive factors, and containerization supports a balanced approach to these demands.

4.10.12 Dynamics in Indian Ports

4.10.12.1 Scenario of Indian Ports

India has a long coastline—about 7,500 km—with nearly 200 ports. But none of them makes it to the list of the world's top 20. Port productivity is usually measured by how much cargo (in tonnes or TEUs) is handled per hour. Better productivity means faster ship turnaround times. Unfortunately, most major Indian ports still lag behind global standards.

To fix this, government-run ports started privatising terminals and outsourcing cargo handling back in 1991. But one big issue is how port pricing works. In many Indian ports, ships pay based on how long they stay at the berth. So, if a port is inefficient and ships take longer to load or unload, the port actually earns more. This discourages quick improvements, and ports often act only after ships start choosing other destinations. Plus, port projects take a long time to complete.

Compared to ports like Singapore and Hong Kong, Indian ports handle much less cargo. Studies point to poor infrastructure and slow procedures as key reasons why Indian ports struggle to stay competitive. These inefficiencies lead to higher transport costs, so international shipping companies often send smaller feeder vessels to India. On top of that, port charges in India are much higher than in other Asian countries.

Looking at the overall trend from 2015 to 2024, India's container traffic has grown at a faster average rate than China's. China already operates at a much larger scale, but its year-to-year growth has stayed relatively moderate—typically between 4% and 9%. India, meanwhile, shows stronger percentage growth, rising from lower base volumes. For example, India's container throughput increased by about 12.6% in 2023, compared with China's 4.9% in the same year, and Indian ports continued to post around 10% growth in FY 2024–25. Together, these trends indicate that, on average, India's growth rate over the past decade has been higher than China's, even though China still handles far more total containers.

Two key indicators of port performance are turnaround time (TAT)—how long a ship stays in port—and dwell time—how long cargo sits at the port. Both are crucial for measuring how efficient a port really is.

4.10.12.2 How Port Systems Work and Why They Matter

Ports work best when they're located near factories or facilities that handle bulk or liquid cargo—this setup allows for smooth, direct cargo transfers. Modern ports are built to handle specific types of goods and ships, which helps speed up loading and unloading, cutting down the time ships spend docked. Older, general-purpose ports tend to be slower because they try to serve all kinds of vessels.

In global trade, transport costs make up a big chunk of the total cost of goods. That includes port charges, freight, insurance, and inland transport. Since port-related costs can be higher than ocean or inland transport, it's important to improve cargo handling, port tech, and operations. Ports should be able to keep cargo moving smoothly between land and sea transport.

If ports don't keep up, they risk losing business. For example, Haldia Dock in India couldn't provide enough draft for crude oil carriers, so the cargo shifted to Paradip Port, which now sends the oil by pipeline to Haldia refineries.

Studies show that turnaround time (TRT or TAT)—how long a ship stays in port—is a key measure of port efficiency. Shorter turnaround means better performance. But when ports are slow to adapt, it makes planning difficult. They need to benchmark their performance against industry standards and plan ahead by adding infrastructure when demand grows.

4.10.12.3 Shipping Improvements and Containerization

Ship design has come a long way. Large container ships now travel faster—up from 23 knots to around 25.5 knots—and use more efficient engines and propellers. They also make better use of space, with wider and deeper hatches and improved systems for securing containers. Ships can now stack up to 10 tiers of containers on deck, making them more cost-effective and profitable.

After World War II, shipping saw two major trends: specialisation and gigantism. Containerization played a big role in this. Though it required huge investments, it revolutionised transport by making it faster, cheaper, and more standardised. It works for almost any type of goods and allows for automated handling. As container use grew, so did the size of ships—each trend pushing the other forward.

As one expert put it, containerization opened the floodgates for global trade, and it's still driving demand for bigger, more efficient ships.

4.10.12.4 Upgrading Port Facilities

Some of the world's biggest ports now use Super-Post-Panamax cranes, which are needed to handle the largest container ships. But not many ports have them yet. If ship sizes grow to 20,000 or even 25,000 TEUs—as some experts predict—then even the best-equipped ports will need to upgrade their cranes again. Similar upgrades are happening in other areas, too, like bulk cargo and cruise terminals, to keep up with bigger ships.

A few ports, like Rotterdam and Hamburg, have gone a step further and automated parts of their terminals. These use automated guided vehicles (AGVs) to move containers and automated stacking cranes (ASCs) to organise them. In places like Korea and Taiwan, some terminals are semi-automated—they automate stacking but still use trailers to move containers between the dock and the yard. Automated terminals cost more to build, but they're seen as better in terms of profitability, safety, quality, and environmental impact. Automation also changes the workforce—replacing many low-skilled jobs with fewer high-skilled supervisory roles. In high-wage countries, automation helps cut labour costs. But in countries with lower wages, sticking to manual operations is often cheaper and more efficient.

4.10.12.5 Making Ports Ready for Bigger Ships

Ports are a key link between land and sea transport. But not all ports can handle the biggest ships. For example, Maersk's Triple-E ships can carry over 18,000 containers (TEUs), but they usually load only around 15,000–16,000 because only two ports—Yantian (China) and Tanjung Pelepas (Malaysia)—can fully accommodate them. This means the size of ships is often limited by what ports are capable of handling.

As ships get even bigger—possibly up to 20,000 or 25,000 TEUs—ports will need to upgrade their container cranes and other infrastructure, even the ones that are already well-equipped. This trend isn't just limited to container ships; bulk carriers and cruise ships are also getting larger, which means terminals need to keep upgrading.

Some ports, like Rotterdam and Hamburg, have started using automated terminals. These use automated guided vehicles (AGVs) to move containers and automated stacking cranes (ASCs) to organise them. In places like Korea and Taiwan, some terminals are semi-automated—they automate stacking but still use trailers to move containers around.

Automated terminals cost more to build, but they're considered better in terms of efficiency, safety, quality, and environmental impact. They also change the kind of jobs available—replacing many low-skilled roles with fewer high-skilled supervisory positions. In countries with high labour costs, automation helps save money. But in places where wages are low, manual operations are often still more cost-effective.

4.10.12.6 Port Planning – The Basics and Beyond

Good port planning can greatly improve how efficiently a port operates. Everything begins with a master plan—a big-picture blueprint that explains the port's purpose, how its facilities are arranged, and how different areas work together. This plan also identifies where new cargo will be handled and what upgrades or changes are needed as the port grows. Once the master plan is in place, it becomes the base for more detailed planning—such as deciding where ships will berth, how containers will be stored, how trucks and cargo will move through the port, and how road and rail connections will link to the terminal. When all these parts fit well together, the port functions smoothly, and delays are minimised.

Even something very specific, like how the yard is planned—meaning how containers are positioned, stacked, and moved—can make a major difference. Studies show that efficient yard planning alone can cut cargo handling time by as much as 30%, improving overall productivity and reducing congestion.

But it's not just about day-to-day operations. Many ports also look far ahead and create strategic visions that guide development over decades. These long-term plans help ports decide where to invest, identify future challenges, and stay competitive. They also help build trust with customers, local communities, and other stakeholders. When done right, these visions spark innovation and open up new possibilities. A great example is Port Vision 2030 from Rotterdam. Instead of just predicting how much cargo the port would handle in the future, it took a creative approach—imagining what the port of the future could look like based on global trends that might reshape the role of ports entirely.

4.10.13 Who Drives Port System Implementation?

Putting port systems into action usually involves three main players: the national government, the port community, and the port authorities. Each plays a different role. The national government often leads projects that are part of broader digital or customs initiatives. These are usually tied to efforts like National Single Window systems, which aim to streamline trade processes and support eGovernment goals. In these cases, Port Community Systems (PCS) are seen as a key tool for improving how different agencies and stakeholders share information and manage cargo flows.

Thus, the roles of key actors involved in implementing port community systems are;

- **National Government**: Sets the overall direction and policies. The main focus often leads to projects that are part of national digital initiatives, like customs modernisation or eGovernment programs. For example, PCS may be introduced as part of a National Single Window system to streamline trade and improve coordination between different government agencies.
- **Port Authorities**: Manage and oversee port operations, ensuring that PCS aligns with port infrastructure, logistics, and operational needs. For example, they help integrate PCS with terminal operations, cargo handling, and vessel scheduling to improve efficiency.
- **Port Community**: It includes shipping lines, freight forwarders, customs brokers, terminal operators, and logistics providers. These are the daily users of PCS, so their input is crucial for designing systems that are practical and user-friendly. For example, they provide feedback on system functionality and help ensure smooth data exchange across the supply chain.

Together, these three actors ensure that PCS is not only technically sound but also meets the needs of all stakeholders—from government regulators to private logistics companies.

4.10.14 Why Strong Hinterland Connections Matter

For a port to work efficiently, it needs solid connections to the inland areas—this is what we call hinterland connections. That means having things like rail tracks that go right up to the docks, smooth links to the national railway network, and canals that connect the port directly to inland waterways. But not all ports have these features.

In many cases, containers have to be moved multiple times before they even reach a train or barge. The more steps involved, the less attractive rail or water transport becomes compared to trucks. Some ports don't even have direct access to inland waterways, so barges have to come in by sea, which isn't allowed for many of them and would require special ships or design changes. These issues often come down to poor port design, and fixing them isn't always easy.

Ports that do it right—those with dedicated rail tracks, river terminals, and short-sea shipping facilities—can move cargo more sustainably and efficiently. For example, in Hong Kong, feeder barges transfer cargo directly from large ships to smaller barges without needing to unload at the quay.

Some countries are building dedicated freight corridors, which are special lanes or railways just for cargo. These help speed up freight movement by keeping it separate from passenger traffic. India's Delhi–Mumbai freight corridor, which stretches about 1,500 km, is a good example. While these corridors aren't the only way to boost rail transport, they're especially useful in places where passenger and freight trains share the same tracks.

As ports become more connected to dry ports and inland terminals, it's also important to share data across the entire network. A more integrated Port Community System (PCS) that links sea, port, and land operations can make cargo movement smoother, cut costs, and make the port more attractive to users.

4.10.15 Issues and Challenges with Indian Ports

4.10.15.1 High turnaround times

Ports in India suffer from high turnaround times for ships. For example, in Singapore, the average ship turnaround time is less than a day. However, in India, it is over two days. Port congestion: Port congestion due to container volume, shortage of handling equipment and inefficient operations is a major concern. Example: In the Nhava Sheva port, Sub-optimal Transport Modal Mix: Lack of requisite infrastructure for evacuation from major and non-major ports leads to sub-optimal transport modal mix. Limited Hinterland Linkages: There is inefficiency due to poor hinterland connectivity through rail, road, highways, coastal shipping and inland waterways. This, in turn, increases the cost of transportation and cargo movement

4.10.15.2 Inadequate Infrastructure and Technology Issues

Lack of adequate berthing facility, number of berths, sufficient length for proper berthing of the vessels at the Non-Major Ports is another problem. Most Non-Major Ports do not have proper material handling equipment in place which could facilitate a quick turnaround.

4.10.15.3 Capability to Handle Large Vessels

In principle, the positive impact of port infrastructure investments on increasing the trade volume is emphasised. For example, the research studies carried out for Brazilian ports have shown that an increase in port infrastructure is associated with large increases in Brazilian exports. At the same time, it is possible to identify some negative effects of the spatial expansion of port infrastructure and conflicts of interest between the port and the port city, as well as uneven shares in the benefits ensuing from the port development, perceived by various stakeholder groups.

Draft is also a major limitation in India as terminals and ports are unable to cater to vessels beyond Panamax (Draft over 15 meters) size that are increasingly dominating global trade. Most Indian ports lack equipment for handling large volumes.

Vessel drafts that can be accommodated by India's Major Ports are generally limited; there are very few ports that can presently receive fully laden Capesize dry bulk ships (the largest ore and coal carriers) of more than 180,000 dwt, large tankers, or container ships of over 8,000 TEU carrying capacity. The limitation of the draft restricts the opportunities for India's shippers to gain the transport cost advantages of direct services by the largest vessels. Partly for this reason, India's ports are facing increased competition for direct shipping calls from major transhipment ports in the region, which can berth larger vessels. Trans-shipped Indian traffic is then consigned to/from Indian ports by smaller vessels, adding a transport cost penalty.

The draft issue is particularly serious for container trades. No major port in India is presently capable of receiving container ships capable of carrying more than 8,000 TEU.

For example, the new South Container Terminal in Colombo (a day closer than Mumbai or Chennai to international shipping routes) is scheduled to start operating in 2014 with 2.4 million TEUs of container throughput capacity and the ability to handle both New Panamax and Triple E super container vessels. Not only in Sri Lanka but also in ports in the United Arab Emirates (UAE) have deeper drafts for container ships and will compete for the direct container service traffic at India's ports, which may add both time and international transport costs for India's trading enterprises.

Further, many ports also lack adequate navigational aids, facilities and IT systems

4.10.15.4 Issues with Regulations

The major problem with regulation is that major and non-major ports fall under different jurisdictions. Further, the regulatory framework is rigid. Cabotage laws in India continued to remain restrictive. Foreign-flagged vessels are not allowed to ship cargo from one Indian port to another, as that remains a protected turf for domestic shippers. Land acquisition and environmental clearances are some specific challenges for non-major ports.

Most port PPPs impose strict limits on what private operators are allowed to do, usually in terms of the types of cargo they are allowed to handle. Until recently, Other problems were related to tariff regulation and the absence of a dispute resolution mechanism. Discriminatory provisions for Indian vessels: Foreign vessels are exempt from duty on bunker fuel, while Indian vessels have to pay this duty. Seafarers aboard Indian flag vessels are subject to Indian income tax, while those working aboard foreign vessels are not subject to this tax. Port management should integrate the process of developing new resources into its strategic planning.

4.10.15.5 Collection of physical and intangible assets and capabilities

The effectiveness of a port is determined by these assets and capabilities. In order to become or remain competitive, port management has to identify, cultivate and exploit its core competencies. Core competencies are the collective learning in the organisation, especially how to coordinate diverse skills and integrate multiple streams of technology.

4.10.16 Port-Driven Growth in India and China

Seaports play a vital role in regional economies by generating positive economic impacts through their operations. The adoption of modern technologies and innovations enhances port efficiency, reduces operational costs, improves transparency, and attracts new business opportunities. As critical nodes in international trade, ports serve as essential economic infrastructure, driving regional development by facilitating transport and enabling global commerce.

Port infrastructure often becomes a catalyst for regional growth, stimulating logistics activities such as warehousing, distribution, and containerization. These activities not only add economic value but also create employment and attract ancillary industries, thereby fostering broader economic development around port areas.

4.10.16.1 Comparative Trade Integration: India Vs. China

A comparison of trade integration and economic outcomes in China and India reveals that while both nations have made substantial progress, China's reforms—particularly in manufacturing—have been more extensive. This deeper integration is a key factor behind China's stronger economic performance.

China began its transition from a centrally planned to a market-oriented economy over 25 years ago. Integration into the global trading system provided access to foreign markets, capital, and advanced technologies, which accelerated its growth. Since the onset of reforms, China's merchandise trade has shifted significantly, with a surge in manufacturing exports and increased imports of fuel, energy, and capital goods.

Trade and investment liberalisation in China has created a highly attractive business environment, leading to substantial foreign direct investment (FDI) inflows. Since 1993, China has been the leading FDI recipient among developing countries and ranked first globally in 2002. It currently stands as the third-largest FDI destination after the US and UK.

4.10.16.2 Trade as a Driver of Growth

The parallel rise in GDP and trade in both China and India raises the classic question: Did trade liberalisation drive economic growth, or did growth naturally lead to increased trade? In China's case, trade expansion is partly attributed to its specialisation within Asia's production networks, where it serves as a hub for final processing and assembly of goods destined for global maritime.

4.10.16.3 India's Trade Competitiveness and Port Sector Constraints

India has significantly reduced tariffs on non-agricultural goods and some non-tariff barriers. Though it has strengths in certain service sectors, its services trade policy remains more restrictive than China's.

India's population and labour force—especially skilled labour—are growing faster than China's. Yet, China's economy is nearly three times larger, with double the GDP per capita (PPP).

India's trade growth has been led by services, not manufacturing. Despite its skilled labour and capital, India's manufacturing exports remain focused on low-tech goods. High-tech exports have stayed below 5% since the mid-1990s, compared to China's 30%. The export mix is still dominated by petroleum, jewellery, furniture, chemicals, and textiles—similar to China's early 1990s profile.

While some advanced items like auto parts have emerged, traditional exports like gems, apparel, and food products have grown, indicating limited integration into global high-tech production networks.

4.10.16.4 Port Sector Challenges

Despite progress, India's ports face congestion and lack supporting infrastructure. The absence of a unified plan for Inland Container Depots, CFSs, and dry ports is a concern. Hinterland connectivity remains weak, and containerization is just 25%, far below the global average of 60–70%.

Part-II

Marine, Structural and Geotechnical Studies for Port Development

Coastal Geomorphic Process: A Practical Understanding

Chapter Summary

Coastal geomorphology examines how natural forces shape the Earth's surface, especially along coastlines where metocean forces—the combined effects of weather and ocean conditions—are most active. Understanding tidal cycles, longshore currents, and wave dynamics is essential, as these continuously act on coastal landforms, driving erosion, sediment transport, and deposition. These processes shape diverse features such as beaches, dunes, mudflats, barrier islands, lagoons, estuaries, cliffs, and coral atolls. A natural balance exists in sediment movement along shorelines, but disruptions—whether from sea level rise or human activities like land reclamation—can shift erosion and deposition patterns to new locations.

Learning Objectives

- Understand how natural forces shape coastlines through erosion, transport, and deposition.
- Learn the origin and role of geomorphic agents like tides, waves, and currents.
- Explore how metocean forces (weather + ocean) drive coastal changes.
- Identify key coastal landforms: beaches, dunes, mudflats, estuaries, etc.
- Grasp sediment transport mechanisms and shoreline balance.
- Recognise impacts of disruptions like sea level rise and land reclamation.

5.1 Introduction

Coastal geomorphology is shaped by various natural agents and processes that interact with the earth's materials along shorelines. These processes involve the removal, transportation, and deposition of sediments. The primary forces driving these changes are metocean forces, which are the combined effects of meteorological (weather-related) and oceanographic (sea-related) factors. Metocean, a term derived from the fusion of 'meteorological' and 'oceanographic', encompasses the study of atmospheric and sea conditions that affect coastal and marine environments. Meteorological conditions include elements like wind patterns, air temperature, atmospheric pressure, humidity, precipitation, and the occurrence of severe weather events like storms and cyclones. These factors play a crucial role in determining the energy and direction of waves and the movement of ocean currents, which in turn influence coastal morphology.

Oceanographic conditions focus on the sea and include water levels caused by storm surges and tidal changes, variations in water temperature, salinity, bathymetry (underwater depth profiles), and the layering of water masses with different densities. This data is crucial for various projects, from coastal engineering to marine conservation, as it helps in predicting changes and planning for resilience against extreme events. Metocean studies are crucial for understanding and predicting coastal changes. They require careful data collection, analysis, and validation to establish what is considered normal or extreme in a given area. This information is vital for various projects, from coastal management to offshore construction, ensuring safety and sustainability in the face of dynamic natural forces.

5.2 Processes of Wind-Generated Waves

Ocean waves are mainly generated by the action of wind on water. Wind blowing over the surface of water transfers energy to the water through friction, and the energy transferred from wind to water causes waves to form. Waves move as individual oscillating particles of water in the form of a crest and trough, which is briefly discussed in successive sections.

5.2.1 Wave Theory

The propagation of waves has a complex process of resonance and shearing action, in which waves of differing wave height, length, and period are produced and travel in various directions. Once waves are formed, ocean waves can travel for vast distances. When it is spreading in a larger area, wave heights are reducing, but wavelength and period of waves are maintained, as shown in Figure 5.1. In a storm zone, waves are generated with high frequency and energy, often characterised by a short period. As these waves move away from the storm zone, their energy is not only dissipated but also redistributed from higher to lower frequencies. This phenomenon occurs because waves of different frequencies travel at varying speeds. Consequently, outside the storm generation area, the sea state undergoes a transformation due to this dispersion of frequencies. Waves of differing frequencies travel at different speeds, and therefore, outside the storm generation area, the sea state is modified by the various frequency components. The low-frequency waves travel faster than the high-frequency waves, resulting in a swell sea condition as opposed to a storm sea condition. Swell is typically smoother and more uniform compared to the chaotic waves of a storm sea condition.

This process, known as dispersion, is fundamental in understanding and predicting sea state conditions, which are crucial for navigation, coastal management, and marine engineering projects. Thus, wind waves and swell waves are two types of surface waves that occur in the oceans. Wind waves, also known as sea waves, are generated by the immediate local wind. They tend to be more irregular, with a shorter wavelength and a steeper crest, reflecting a more chaotic state influenced by the variability of the wind. These waves contain a broad spectrum of frequencies and directions due to the changing wind conditions. In contrast, swell waves are longer-lasting waves that have travelled out of their generating area. They are more uniform and regular, with longer wavelengths and less steep crests. Swell waves are formed by distant weather systems and maintain their energy over great distances.

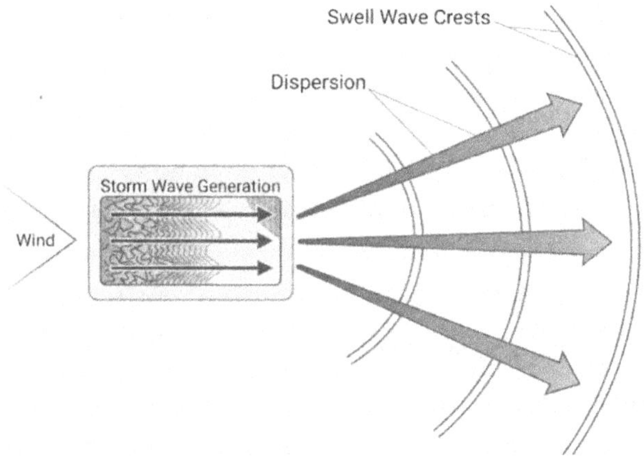

Figure 5.1: Schematic diagram of Wind-Driven Surface Wave Generation

Their frequencies and directions are more narrow and consistent, as they are the result of a more stable and sustained wind source, as shown in Fig. 5 .2.

Furthermore, the following Important terms are added here in brief to understand the function of sea waves:

- Wave crest is the highest point of the wave; the trough is the lowest point of the wave.
- Wave height—vertical distance between the crest and trough of a wave.
- Amplitude—is the maximum displacement of vibrating particles from their mean position.
- Wavelength is the horizontal distance between consecutive wave crests.
- Wave velocity is the speed at which a wave crest moves forward and is related to the wave's energy.
- Wave period is the time interval it takes for adjacent wave crests to pass a given point.

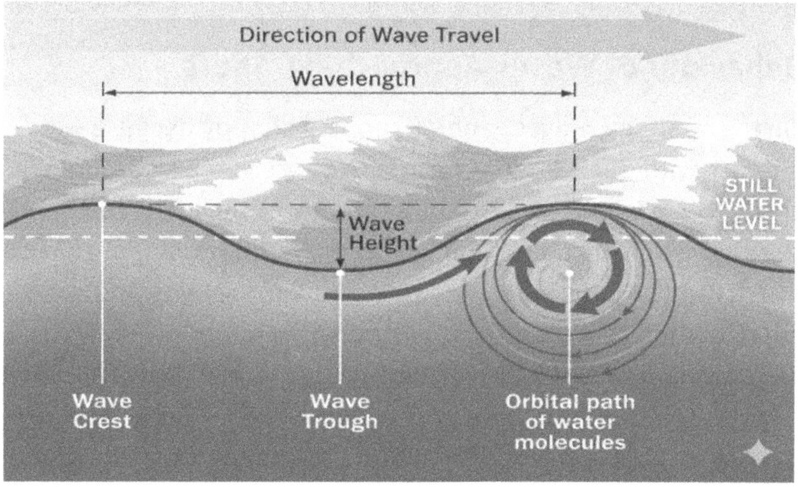

Figure 5.2 ***Diagram*** *describing wave base* Details

Thus, small magnitude waves are primarily generated by the wind's energy transferring to the water. This interaction creates a circular orbital motion in the water particles, leading to the characteristic undulating movement of waves. The variations in atmospheric pressure can also influence wave formation. As the wind continues to apply force, it propels the waves across the water's surface, allowing them to travel vast distances.

The crest is the wave water moving forward, and the trough is the water moving backwards. The circular motion of water particles diminishes with depth and is negligible at about one-half wavelength.

Wave base is the vertical depth at which water is disturbed by waves, above which waves will ripple the shore sand.

Wave base is measured at about one-half wavelength deep, where the water particles' circular motion diminishes. If waves approach a beach crest at about 6 m intervals, this wave motion disturbs the water to about 3 m deep.

This motion is known as a f**air-weather wave base**. In strong storms such as hurricanes, both wavelength and wave base increase dramatically to a depth known as **storm wave base**, which is approximately 90 m.

Waves are generated by wind blowing across the ocean surface. The amount of energy imparted to the water depends on wind velocity and the distance across which the wind is blowing. This distance is called **fetch**. Waves striking a shore are typically generated by storms hundreds of miles from the coast and have been travelling across the ocean for days.

Winds blowing in a relatively constant direction generate waves moving in that direction. Such a group of approximately parallel waves travelling together is called a **wave train**. Because wave trains are generated from different areas, they can spread in different directions and carry different amounts of energy. When different wave trains interact, they produce the choppy sea surface seen in the open ocean. Also, a wave train coming from one fetch can produce various wavelengths.

Longer wavelengths travel at a faster velocity than shorter wavelengths, so they arrive first at a distant shore. Thus, there is a wavelength sorting process that takes place during the wave train's travel. This sorting process is called the **picking process.**

5.2.2 Behaviour of Waves Approaching Shore

On the open sea, waves generally appear choppy because wave trains from many directions interact with each other. Constructive interference occurs where crests align with other crests. The aligned wave height is the sum of the individual wave heights, a process referred to as wave amplification. Hollows occur where troughs align with other troughs. Destructive interference occurs where crests align with troughs and cancel each other out. As waves approach the shore and begin to make frictional contact with the sea floor at about one-half wavelength or less, they begin to slow down. However, the energy carried by the wave remains the same, so the waves build up higher. Remember that water moves in a circular motion as a wave passes, and each circle is fed from the trough in front of the advancing wave. As the wave encounters shallower water at the shore, there is eventually insufficient water in the trough in front of the wave to supply a complete circle, so the crest pours over, creating a **breaker**.

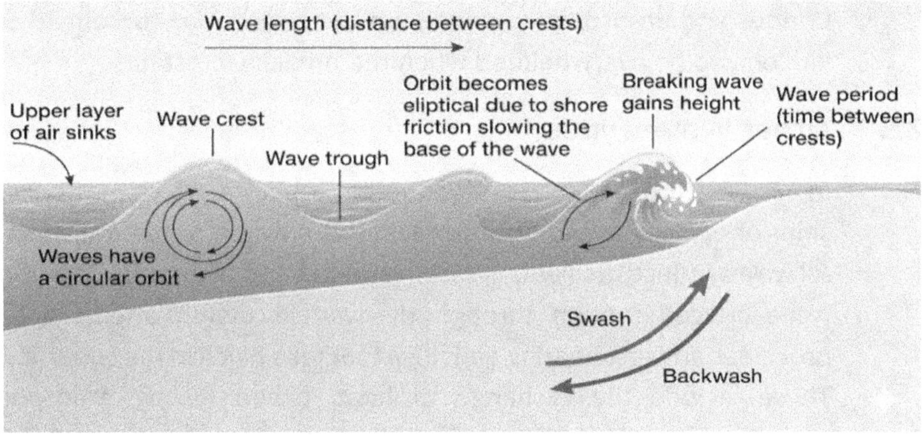

Figure 5.3: Figure of Wave Breaking

5.2.3 Storms

Waves generated by storms when they reach the shore surge onto the land, known as storm surges. Also, a tidal wave is a regular recurring wave caused by the effects of gravitational interaction between the sun, the moon, and the Earth.

The pounding by wind or storm-generated waves has a recurrence time interval of a few *seconds*, while tidal waves have a few hours, and cyclonic storm surges vary in between but are seasonal. The storm surges, when clubbed with tidal waves, can have a height of several meters and can be very devastating. When the bottommost part of these waves hits the seabed, it breaks, causing the development of the surf zone, the most active part of the foreshore, and resulting in spillage over the shore as forward, ebb, and backward rip currents. The extent to which they go up-shore is termed the fetch of waves. When mighty storms or tidal waves break, they acquire high elevation and result in inundation beyond the shoreline areas.

5.2.4 Tsunami

A tsunami is a special type of wave generated by energetic events affecting the sea floor, such as earthquakes, submarine landslides, and volcanic eruptions. During *earthquakes*, tsunamis can occur when the moving crustal rocks below the sea abruptly elevate a portion of the seafloor. Water is suddenly lifted, and a wave train spreads out in all directions, travelling over 322 kph (200 mph) and carrying enormous energy. Tsunamis may pass unnoticed in the open ocean because the wavelength is very long, and the wave height is very low. But, as the wave train approaches shore and each wave makes contact with the shallow seafloor, friction increases, and the wave slows down. Next, wave height builds up, and the wave strikes the shore as a wall of water over 30 m (100 ft) high. The massive wave, called a tsunami runup, may sweep inland well beyond the beach, destroying structures far inland. *All waves, like tsunamis, slow down as they reach shallow water. This causes the wave to increase in height.* Tsunamis deliver a catastrophic blow to people at the beach. As the trough of water in front of the tsunami wave is drawn back, the seafloor is exposed.

Curious and unsuspecting people on the beach may run out to see exposed offshore sea life, only to be overwhelmed when the breaking crest hits.

5.2.5 Other Wave Types

In nature, waves are the result of multiple factors, which may be: seismic, the movement of ships or explosions near the surface of the water, or natural, such as earthquakes or tides. All waves induce oscillations within a wide band of periods and associated wavelengths. The wave-induced motion through the water column affects a large number of marine processes and engineering activities that take place in the superficial sublayers of the ocean. These include the exchange of heat, ocean mixing, transport of sediment, coastal morphology, and offshore engineering, among many others. A description of the different types of ocean surface waves and their concurrent classification is presented in Table 5-1.

Table:5-1: Ocean Wave Classification

Classification	Period Band	Generating forces	Restoring Forces
Capillary waves	< 0.1 s	Wind	Surface tension
Ultra-gravity waves	0.1 – 1 s	Wind	Surface tension and gravity
Gravity waves	1 – 20 s	Wind	Gravity
Infra-gravity waves	20s to 5 min.	Wind and atmospheric pressure gradients	Gravity
Long Period Waves	5 min to 12 h	Atmospheric pressure gradients and earthquakes	Gravity
Ordinary Tidal Waves	12 – 24 h	Gravitational attraction	Gravity and Coriolis force
Trans-tidal Waves	>24 h	Storm and gravitational attraction	Gravity and Coriolis force

For planning and design purposes of a marine facility, their study becomes necessary.

i) Assessment of wave force on marine structures is very important during the design phase.

ii) If the waves formed within the marine area may cause troubles in the loading and unloading operation of ships, as well as many problems, accordingly, studying the waves and their refraction and spread is very important.

iii) Planning and designing the breakwater, the entrance, and shipping lanes.

5.2.6 Wave Statistics and Their Critical Role in Port Infrastructure Design

Wave data is one of the most critical inputs in the planning, layout, and engineering design of any port. Waves directly influence the safety, structural reliability, and operational efficiency of harbour infrastructure. Understanding the wave climate—its height, direction, period, frequency, and extreme behaviour—is essential for designing breakwaters, navigation channels, quay walls, harbour basins, and berth orientations.

Wave data informs decisions related to:

- Breakwater alignment and crest height
- Navigation channel alignment
- Harbour tranquillity
- Berth placement and orientation
- Quay deck elevation
- Vessel operations and safety
- Downtime and operational limits

A reliable wave climate assessment reduces design uncertainties, enhances port safety, and ensures long-term operational resilience.

5.2.6.1 Importance of Wave Data in Port Infrastructure Design

Wave data influences almost every major component of port development:

A. Breakwater Alignment and Crest Height

Wave data determines:

- Optimum alignment of the breakwater to block dominant wave directions
- Required crest elevation to prevent overtopping
- Armour layer sizing based on wave energy
- Stability of the toe and foundation under varying tide and wave conditions

Without accurate wave information, breakwaters may become under-designed (causing failures) or over-designed (leading to unnecessary cost).

B. Navigation Channel Alignment

Wave direction and height influence:

- Safe vessel approach and departure routes
- Wave-induced vessel motions affecting steering control
- Under-keel clearance due to vessel heave, pitch, and roll
- Cross-wave impacts on manoeuvrability

Wave-aware alignment ensures safe and efficient navigation for large vessels.

C. Harbour Tranquillity

Residual waves inside the harbour affect:

- Berthing safety
- Cargo handling efficiency

- Vessel motions at berth (surge, sway, heave)
- Downtime during rough weather

A carefully designed breakwater system and harbour layout ensure tranquillity under most environmental conditions.

D. Berth Placement and Orientation

Wave climate dictates:

- How berths should be oriented relative to prevailing waves
- Which locations inside the harbour will be naturally calmer
- How vessel motions will affect loading/unloading equipment

Different cargo types (container, dry bulk, liquid bulk, Ro-Ro, passenger) require different motion tolerances.

E. Quay Wall and Jetty Deck Levels

Wave run-up and splash affect the safe elevation of quay decks. Deck levels must remain safe and dry during:

- High waves
- Seasonal storms
- Extreme events and cyclones

Wave data, combined with tide and storm surge information, is essential for fixing non-flooding deck elevation.

5.2.7 Wave Data Sources and Measurement Methodology

A robust wave climate assessment requires the systematic integration of multiple complementary data sources, each addressing different spatial and temporal scales of wave processes. Generally, it combines remote sensing, numerical hindcast modelling, and direct in situ measurements, ensuring both long-term statistical reliability and site-specific accuracy for engineering design.

5.2.7.1 Satellite-Based Wave Measurements

Satellite altimeters and synthetic aperture radar (SAR) missions provide extensive, long-term datasets of significant wave height (Hs), wave period, and wave direction over global oceans. Altimeter systems derive wave height by analysing the radar return signal footprint, while SAR instruments infer wave direction and wavelength through spectral analysis of sea surface roughness patterns.

Satellite-derived wave data are particularly valuable for:

- Establishing offshore wave climatology
- Identifying seasonal and interannual variability

- Supplementing areas where long-term in situ measurements are sparse or absent

However, satellite observations are subject to limitations in nearshore zones (<10–20 m water depth) due to land contamination, coastal geometry, and complex bathymetric effects. Consequently, satellite data are primarily applied in deep-water or open offshore conditions and are used in conjunction with numerical model outputs rather than as standalone inputs.

5.2.7.2 Numerical Wave Hindcast Datasets

Long-term, continuous wave information is obtained from global and regional numerical hindcast models, including:

- ERA5 reanalysis
- WAVEWATCH III (WW3)
- INCOIS regional wave datasets

These models generate 20–40+ years of hourly to 3-hourly wave data, based on historical wind forcing and atmospheric reanalyses. The methodological strength of hindcast models lies in their:

- Temporal consistency
- Capability to provide return period statistics (e.g., 1-in-50, 1-in-100-year waves)
- Suitability for extreme value analysis and climate variability assessment
- Wave parameters extracted typically include:
- Significant wave height (Hs)
- Peak and mean wave period (Tp, Tm)
- Mean and peak wave direction
- Directional wave spectra

The accuracy of hindcast datasets depends on:

- Resolution and quality of wind forcing
- Model physics
- Grid resolution and bathymetry representation

Therefore, hindcast outputs are treated as baseline offshore boundary conditions, requiring validation against measured data before engineering application.

5.2.7.3 In Situ Wave Measurements

In situ wave measurements provide the highest level of accuracy and form the cornerstone of model calibration and validation. The following instrumentation is commonly deployed:

A. Directional Waverider Buoys

- Measure wave height, period, and direction using accelerometers and gyroscopes
- Provide high-resolution time series (typically 30-minute records)

- Well-suited for offshore and nearshore deployments (water depths >15–20 m)

B. ADCPs with Wave Measurement Modules

- Deployed on the seabed and measures wave orbital velocities and pressure fluctuations
- Suitable for intermediate depths
- Capable of simultaneous current and wave measurements

C. Nearshore Pressure Sensors

- Installed in shallow water zones
- Capture pressure variations, converted to surface wave characteristics using linear/nonlinear wave theory corrections
- Particularly useful for surf zone and shallow coastal wave transformation analysis

Wave measurements are typically recorded over 12–24 months to capture:

- Seasonal monsoon/non-monsoon conditions
- Storm events
- Directional variability

Quality control procedures include:

- Spike removal
- Sensor drift correction
- Cross-comparison between instruments
- Elimination of records affected by biofouling or signal loss

5.2.7.4 Nearshore Wave Transformation Modelling

Offshore wave data alone are insufficient for port and harbour design. Therefore, project-specific nearshore numerical wave modelling is undertaken using advanced spectral models such as:

- SWAN
- MIKE 21 Spectral Wave
- Delft3D-WAVE

The adopted methodology involves:

- Extraction of offshore boundary wave spectra from validated hindcast datasets
- High-resolution nearshore bathymetry incorporation
- Simulation of wave transformation from deep water to the coast, including:
    - Refraction
    - Shoaling
    - Diffraction

  - Wave breaking
  - Bottom friction

A. Assessment of wave penetration into harbour basins and along berths

These models form the basis for:

- Harbour tranquillity assessment
- Breakwater design wave estimation
- Berth orientation optimization
- Quay wall crest level and freeboard determination
- Vessel operability and downtime analysis

Nearshore modelling accuracy is particularly sensitive to:

- Bathymetric resolution
- Coastal features such as headlands, reefs, and channels
- Wave approach angle and multi-directional spreading

5.2.7.5 Model Calibration and Validation

Calibration and validation are critical steps to ensure confidence in numerical model results.

- **Calibration** involves adjustment of model parameters (e.g., bottom friction coefficients, white-capping terms, breaking indices) to match observed wave parameters during measured periods.
- **Validation** is performed using independent datasets, including storm or high-energy events not used during calibration.

Model performance is evaluated using standard statistical indicators:

- Root Mean Square Error (RMSE)
- Mean Bias
- Scatter Index
- Correlation Coefficient (R)

Acceptance criteria are defined based on international coastal engineering practice, ensuring the model reliably reproduces both average conditions and extreme events.

5.2.7.6 Integrated Wave Data Framework for Port Design

The final wave climate dataset represents a synthesised, validated product, combining:

- Long-term offshore wave climatology
- High-accuracy field measurements
- High-resolution nearshore wave transformation models

This integrated framework provides dependable wave inputs for:

- Breakwater armour design and stability analysis
- Navigation channel operability

- Harbour basin tranquillity criteria
- Berth downtime and vessel motion analysis
- Coastal protection and shoreline response studies

By adopting a multi-source, calibrated, and validated methodology, the wave assessment ensures that port infrastructure is designed to be safe, operable, resilient to extremes, and adaptable to future climate variability.

5.2.8 Acceptance Criteria for Wave Measurement Data Quality

This section defines the acceptance criteria to be adopted for wave measurements, numerical model performance, and design application

Table:5-2: In Situ Wave Measurement Acceptance Criteria

Parameter	Acceptance Criteria	Reference / Basis
Measurement duration	Minimum 12 months (preferred 18–24 months)	PIANC WG-52; BS 6349-1
Sampling interval	≤30 minutes	BS 6349
Data recovery	≥90% over the deployment period	PIANC best practice
Directional accuracy	±10° for dominant wave direction	PIANC WG-52
Significant wave height accuracy	±5–10%	BS 6349
Instrument drift	No systematic bias over deployment	PIANC
Storm event capture	At least one representative severe event	BS 6349

Notes:

- PIANC recommends longer deployments where monsoonal or seasonal variability is dominant.
- Shorter datasets may only be used in combination with validated hindcast data.

Table 5-3: Offshore Wave Climate Dataset Acceptance (for Satellite and Hindcast Data Use)

Aspect	Acceptance Requirement	Standard Alignment
Length of hindcast	≥30 years (preferred ≥40 years)	BS 6349
Wind forcing	Reanalysis-quality (e.g., ERA-Interim / ERA5)	BS 6349
Nearshore use	Not permitted as sole design input	PIANC

Satellite data role	Supplementary, offshore only	PIANC WG-52
Directional spread	Must include full directional spectrum	BS 6349

Key Requirement:

Satellite and hindcast data shall not replace in situ measurements for final nearshore or harbour design unless fully validated.

Table 5-4: Model Calibration Performance Criteria (Acceptance Criteria for Numerical Wave Model Calibration)

Parameter	**Typical Acceptance Range**	**Basis**
Bias (Hs)	±5–10%	BS 6349
RMSE (Hs)	≤20% of the mean measured Hs	PIANC
Scatter Index	≤0.25	BS 6349
Correlation coefficient (R)	≥0.85	PIANC
Directional agreement	±10–15°	PIANC WG-52
Period agreement	±10%	BS 6349

Calibration Requirement:

- Calibration must include both average and storm conditions
- Parameters adjusted shall be justified and documented

Table 5-5: Model Validation Requirements (Criteria for Model Validation)

Aspect	**Acceptance Criteria**	**Reference**
Dataset independence	Validation data not used in calibration	BS 6349
Seasonal coverage	Validation during different seasons	PIANC
Extreme event testing	At least one storm event	BS 6349
Performance consistency	Similar or marginally degraded statistics vs calibration	PIANC
Physical realism	Correct refraction and sheltering behaviour	PIANC WG-52

It is important to note that BS 6349 explicitly requires validation to be independent of calibration data for design confidence.

Table 5-6: Acceptance Criteria for Harbour Tranquillity (Typical PIANC Harbour Tranquillity Limits)

Facility Type	**Significant Wave Height Inside Harbour**
Container berths	≤0.3 m
Bulk carrier berths	≤0.3–0.5 m
Liquid bulk berths	≤0.2–0.3 m
Ro-Ro / passenger berths	≤0.2 m
Anchorage areas	≤0.5–0.8 m

Notes: Above data is based on PIANC WG-52 guidance, and final limits depend on vessel size, mooring system, and operational criteria

Table 5-7: Design Wave Selection/Acceptance Criteria for Maritime Structures (Based on BS 6349)

Design Condition	**Typical Return Period**	**Application**
Operational wave	1–5 years	Berthing, cargo handling
Serviceability limit state	10–25 years	Overtopping checks
Ultimate limit state	50–100 years	Breakwater stability, quay wall design
Accidental/ resilience check	>100 years (where required)	Critical infrastructure

BS 6349 Guidance: Different limit states must be assessed separately and Operational and extreme conditions must not be conflated

Table 5-8: Numerical Model Resolution Requirements (Acceptance Criteria for Nearshore Model)

Model Domain	**Typical Grid Resolution**	**Reference**
Offshore	250–1000 m	PIANC
Nearshore	20–50 m	BS 6349
Harbour basin	5–20 m	PIANC WG-52
Around structures	≤5 m	BS 6349

5.3 Tides

5.3.1 Tides Theory

Gravity is one major force that creates tides. Tidal forces are based on the gravitational attractive force. Regarding tidal forces on the Earth, the distance between two objects is usually more critical than their masses. Tidal-generating forces vary inversely as the cube of the distance from the tide-generating object. Gravitational attractive forces only vary inversely with the square of the distance between the objects. The effect of distance on tidal forces is seen in the relationship between the sun, the moon, and the Earth's waters.

Tides are larger oscillations of water around the Earth, caused by the gravitational pull of the sun and moon, aided by the Earth's rotation.

Spring Tide, when the sun, moon, and earth are all aligned, and ***Neap Tide***, when the sun and moon are at a right angle to each other, result in the equatorial and polar bulge of water whose fluctuation results in the creation of waves of much higher magnitude as compared to waves created by wind

5.3.2 Types of Tidal Cycles

The three basic tidal patterns that occur along the Earth's major shorelines of continents are known as Diurnal, Semidiurnal, and Mixed Semidiurnal.

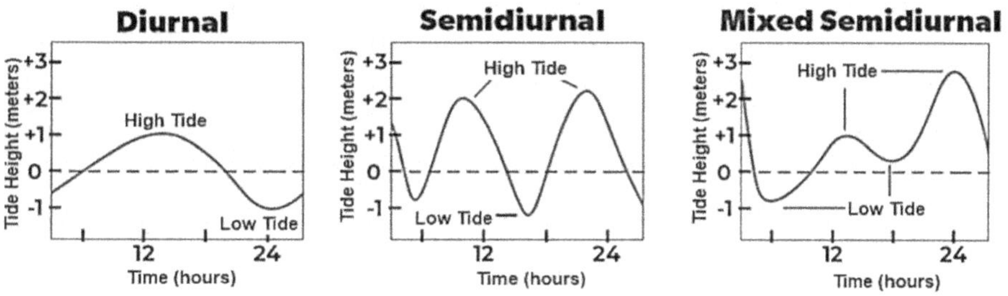

Figure 5.4: Types of Tidal Cycles

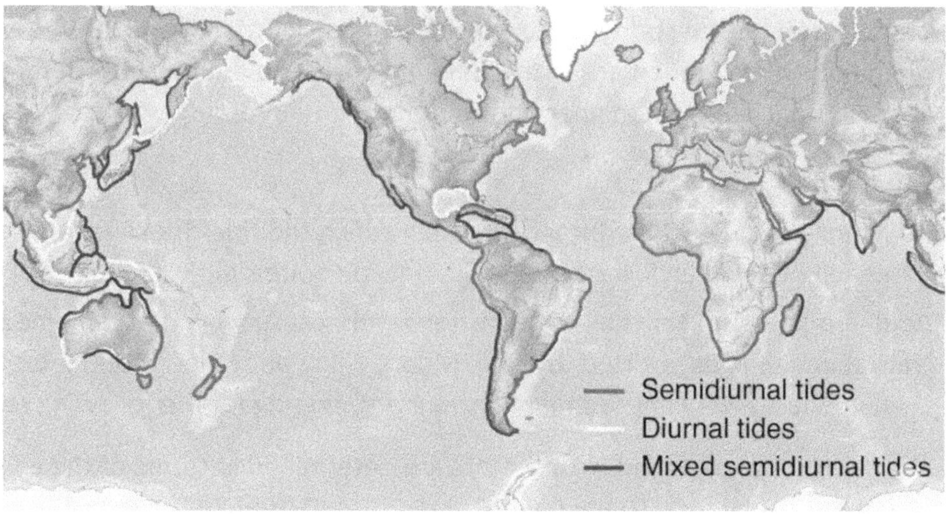

Figure 5.5: Tides types on the world map

In general, most areas have two high tides and two low tides each day. When the two highs and the two lows are about the same height, the pattern is called a semidiurnal tide.

If the high and low tides differ in height, the pattern is called a mixed semidiurnal tide. Some areas, such as the Gulf of Mexico, have only one high and one low tide each day. This is called a diurnal tide, as shown in **Figure 5.4** and **Figure 5.5.**

5.3.3 Tidal Variations - The Influence of Position and Distance

The moon is a major influence on the Earth's tides, but the sun also generates considerable tidal forces. Solar tides are about half as large as lunar tides and are expressed as a variation of lunar tidal patterns, not as a separate set of tides.

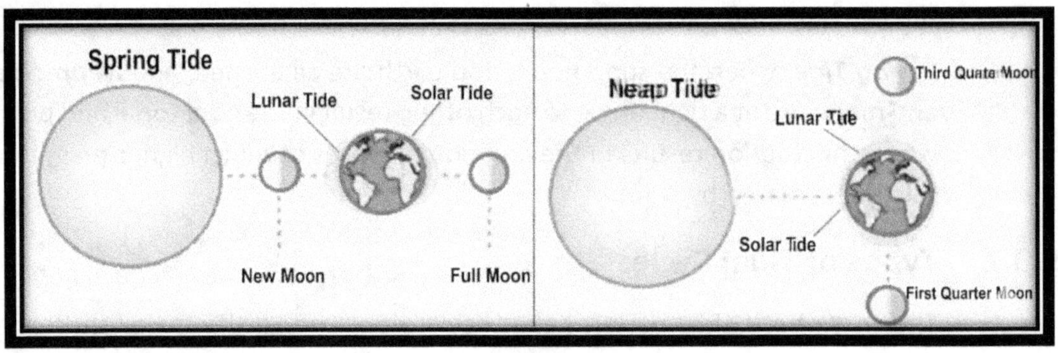

Figure 5.6 shows the Spring tide & Neap tide position

When the sun, moon, and Earth are in alignment (at the time of the new or full moon), the solar tide has an additive effect on the lunar tide, creating extra-high tides and very low tides—both commonly called Spring Tides. One week later, when the sun and moon are at right angles to each other, the solar tide partially cancels out the lunar tide and produces moderate tides known as Neap Tides. During each lunar month, two sets of spring tides and two sets of neap tides occur, as shown in Figure 5.6.

5.3.3.1 Equinoctial Tide

It is a special case of spring tides, which occurs on 21 March and 23 December every year, when the sun and moon are in the same line over the equator, which means the shortest distance, which causes a higher tide.

5.3.3.2 Tidal Range

Tidal range is the vertical distance through which the tide rises and falls, the difference in water height between low tide and high tide, or, quite simply, the "size" of the tide.

Tides are the rise and fall of seawater levels caused by the combined effects of the gravitational forces exerted by the Moon, Sun, and the rotation of the Earth. Tide phenomena are very important in harbour engineering for the following reasons:

1- The heights of piers and breakwaters are determined in terms of the change in sea level from the lowest level to the highest level.

2- The change in sea level is taken into consideration as a factor that determines the length and height of the side walls of the dry basins.

3- These phenomena give rise to what is known as tidal currents, and it is an important factor that must be studied when determining port entrances.

4- The type of port may depend entirely on the amount of change that occurs in the sea level in the region. If this change exceeds a certain limit, the port may be established in the form of closed basins.

The highest tide level in the world can be found in Canada's Bay of Fundy (17.0m), and second only in India at Bhavnagar, where the maximum spring tide recorded is 12.5 m.

5.3.3.3 Tide Data To be Considered

Design levels are finalised with respect to standard tidal datums (as applicable to the project location):

- **HAT** – Highest Astronomical Tide
- **MHWS** – Mean High Water Springs
- **MHHW** – Mean Higher High Water
- **MSL** – Mean Sea Level
- **MLWS** – Mean Low Water Springs
- **LAT** – Lowest Astronomical Tide
- **Storm** Surge / **Cyclone** Surge Levels
- **Sea-level rise allowances** (as per national/international guidelines)

5.3.4 Importance of Tide Levels in Port Infrastructure Design

Tide levels are a key design parameter in port planning. They directly influence the quay deck height for non-flooding and safe operations, the breakwater crest level for wave protection, and the dredged depth needed for safe navigation. Their importance for each component is summarised below:

5.3.4.1 Tidal Considerations for Quay Wall Deck Elevation

- To ensure non-flooding and structural safety, the deck level must be set above the Highest Astronomical Tide (HAT) + storm surge + wave run-up/splash so that overtopping and flooding do not occur during extreme conditions. Incorrect deck height can cause operational delays, equipment damage, and unsafe working areas.
- Tidal variations affect vessel freeboard, which influences mooring arrangements and overall operational safety at the berth.
- Differences in tidal range and extreme high tides also change hydrostatic pressures acting on the quay wall, which must be considered in structural design.

5.3.4.2 Tidal Levels for Breakwater Design

The crest height and armour sizing of a breakwater depend heavily on tidal conditions:

Breakwater Crest Level

The water depth at the breakwater toe changes with the tide.

At high tide, waves carry more energy, increasing overtopping risk and often requiring a higher crest level or berm adjustments.

At low tide, the toe armour and foundation may become exposed, making stability more critical.

Wave Transformation and Harbour Calmness

Tide levels influence wave shoaling, wave breaking, and the wave climate inside the harbour. These factors determine whether the harbour remains calm enough for safe berthing and cargo operations.

Structural Safety in Extreme Events

The combined effects of HAT + storm surge + wave setup define the design water level, which governs armour size, crest height, overtopping limits, and overall structural stability.

5.3.4.3 Tide Levels for Finalising Dredge Depth of Navigation Channels

Tide levels are essential in determining the **safe navigation depth**, based on:

Under-Keel Clearance (UKC)

Channels and turning basins must maintain safe depth at all tide levels. The minimum dredged depth is usually based on Lowest Astronomical Tide (LAT), considering vessel draft, squat, wave motions, and sediment allowance.

Using LAT sets the maximum dredge requirement, while using Mean Low Water (MLW) can reduce dredging if operational windows allow.

Tidal Windows

For deep-draft vessels, higher tides can temporarily increase available depth, reducing dredging needs through tidal-assisted navigation.

Sedimentation and Maintenance

Tidal currents strongly affect siltation in approach channels, turning circles, and berthing areas. Accurate tidal analysis helps predict long-term maintenance dredging requirements

5.3.5 Tidal Measurement Methodology

Tidal measurements are carried out to understand sea level variations at the port site and to define design water levels for navigation, dredging, quay heights, reclamation works, and numerical modelling. Accurate tidal data are essential to ensure safe vessel movement and reliable port infrastructure design.

i) Tide Gauge Location

Tide gauges are installed at stable locations that represent true sea level conditions. The main gauge is usually fixed to a permanent structure such as a jetty pile, quay wall, or dedicated gauge pillar, with free connection to open water but protected from waves and vessel wakes. For larger or complex ports, additional check gauges may be installed near channels or harbour basins to confirm consistency across the site.

ii) **Instruments Applicable**

Non-contact acoustic or radar tide gauges are preferred because they provide reliable measurements and are not affected by sediment buildup or marine growth. Pressure-based gauges may be used as backup. A staff gauge is always installed nearby for visual checks. Water levels are typically recorded every 1–5 minutes with an accuracy of about ±5–10 mm.

iii) **Vertical Datum and Benchmarks**

Permanent benchmarks are established near the tide gauge and tied to the national vertical datum using precise levelling or GPS surveying. This allows measured water levels to be referenced to standard levels such as Mean Sea Level, Lowest Astronomical Tide (Chart Datum), and Highest Astronomical Tide. Using a common datum ensures consistency across all design and modelling work.

iv) **Measurement Duration**

Tidal data are recorded continuously for at least 30 days to cover a full tidal cycle. For port design, a longer measurement period of around 12 months is preferred to capture seasonal variation and meteorological effects. If only short-term data are available, they are analysed and correlated with long-term nearby tide station records.

v) **Data Checks and Analysis**

Recorded data are checked for gaps, spikes, drift, and timing errors. Readings are cross-checked with staff gauge observations and predicted tides. Harmonic analysis is then carried out to identify tidal constituents and determine key tidal levels such as high tide, low tide, and mean sea level.

5.4 Currents

The *currents* in the sea and oceans are the unidirectional movement of water due to the result of differential heating of waters and salinity. The excessive heating of waters at the equator makes it light. These light waters rise up wells and move towards Polar Regions, while cold, heavy waters at the poles sink and move towards equatorial regions. These unidirectional movements of water are deflected as they hit different coasts, for example, the *Arabian Current and* the *Bengal Gyre*, along the west and east coasts of India. Similarly, regions with high sediment influx from rivers have high salinity and become heavy. These dense currents sink and move towards areas of low salinity and vice versa, causing uninterrupted unidirectional circulation of waters.

Such circulations are necessary not only for the much-needed distribution of nutrients for marine life. These currents transport millions of tons of sediments, choking harbours, a concern for civil engineers and engineering geologists. These currents also play a key role in the development of *El Niño* and *La Niña*, which may cause abnormal weather and cyclonic storms, of interest to climatologists, civil engineers, and engineering geologists alike.

5.4.1 Study of Sea and Ocean Currents for Port Planning

In the Indian context, the study of sea and ocean currents—closely linked with monsoon systems and tidal behaviour—is indispensable for safe navigation, optimal port layout, sediment control, structural stability, and environmental sustainability. Integrating current and tidal studies into port planning ensures that Indian ports are operationally efficient, economically viable, and resilient to natural and climatic variability.

Tidal currents are periodic horizontal water movements generated by the rise and fall of tides under the gravitational influence of the Moon and the Sun. They reverse direction between flood (incoming) and ebb (outgoing) phases and vary in magnitude with tidal range, seabed topography, and coastal geometry, playing a key role in navigation, sediment transport, and port design.

- Tidal current studies are essential to design safe approach channels and turning basins, determining berth alignment and mooring forces, assessing tidal windows for large vessel movement

Ports such as Kandla, Mumbai, and Kolkata require detailed tidal and current analysis due to strong tidal flows and reversing currents.

- **Sediment Transport and River Influence:** Indian ports are highly affected by littoral drift and river-borne sediments, especially along the east coast, where large rivers discharge into the sea. Current studies help in predicting siltation in channels and harbour basins, planning dredging frequency and sediment management systems, selecting suitable locations for spoil disposal
- **Structural Design and Coastal Stability:** Currents in coastal waters exert significant hydrodynamic forces on maritime structures. Appropriate information of current velocity and direction is essential for designing quay walls, piles, jetties, and breakwaters. It is important for preventing scour around foundations and ensuring long-term structural stability, especially during monsoon conditions.
- **Environmental and Regulatory Considerations:** Currents govern the dispersion of pollutants, dredged material, and thermal discharges, which is critical under India's coastal regulation and environmental clearance framework. Current studies support environmental Impact Assessments, selection of safe disposal sites of dredged materials and protection of sensitive coastal and marine ecosystems.

5.4.2 Hydrodynamic Study of Sea and Ocean Currents

A sea and ocean currents study is carried out to understand how water flows in coastal and offshore areas, including its speed, direction, depth-wise variation, and how it changes over time. The study examines different types of currents, such as tidal currents, wind-driven currents, residual currents, and seasonal circulation patterns, as these directly affect ship navigation, sediment movement, port structures, and environmental conditions.

Currents are measured in terms of velocity and direction, both at the surface and at different depths. The study takes into account time-based changes such as flood and ebb tides, spring–neap tidal cycles, monsoon-related seasonal reversals, and extreme events. It also considers how currents vary from place to place, including nearshore areas, approach channels, harbour basins, estuaries, and offshore regions.

Field measurements are made using modern instruments, mainly Acoustic Doppler Current Profilers (ADCPs), which record current speed and direction at multiple depth layers. These instruments are deployed for sufficient periods to cover complete tidal cycles and seasonal conditions, and the data is collected along with tide levels, wind information, and river flow to identify the factors controlling the currents.

The recorded data is then analysed to determine average and maximum current speeds, dominant flow directions, vertical variation, and residual circulation. This information is used in hydrodynamic models to simulate existing and future current conditions and to assess the impact of planned port developments such as navigation channels, breakwaters, dredging, and reclamation works.

5.4.3 Methodology for Current Measurement

i) Measurement Locations

Current measurement instruments are installed at representative locations covering the offshore approach area, navigation channel, harbour basin, and near-berth zones. Locations are selected to capture dominant tidal and residual current patterns while avoiding excessive disturbance from vessel movements, structures, or very shallow water effects.

ii) Instruments Used

Acoustic Doppler Current Profilers (ADCPs) are commonly used for sea current measurements at port sites as they provide current speed and direction over the full water depth. ADCPs are typically seabed-mounted and record current profiles in multiple vertical layers. In shallow or constrained areas, point current meters may be used. Measurements are usually recorded at intervals of 10 to 30 minutes with an accuracy of about ±1–2 cm/s.

iii) Deployment and Setup

The ADCP is mounted on a stable seabed frame or tripod and deployed such that it remains vertical and secure throughout the measurement period. Care is taken to protect the instrument from anchors, fishing gear, and strong wave action. Water depth, seabed condition, and shipping activity are considered during deployment planning to ensure reliable data collection.

iv) Measurement Duration

Current measurements are carried out continuously for at least 30 days to capture full tidal cycles and spring–neap variation. For port planning and design, a longer deployment of around 3 to 6 months is preferred to capture seasonal or monsoonal current behaviour. Short-term datasets are later correlated with tidal data and regional current information, where required.

v) Data Checks and Analysis

Recorded current data are checked for gaps, noise, and errors caused by instrument tilt, vessel passage, or signal interference. After quality control, current speed and direction are analysed by depth and tidal phase to identify peak currents, average conditions, near-bed currents, and any residual circulation patterns.

vi) Acceptance Criteria

Current measurements are considered acceptable when data availability exceeds 90%, and model-simulated current speed and direction match measured values within ±15% in magnitude and ±10° in direction at key locations.

5.5 The Coastal Region

The coastal zone occurs at the interface between the three major natural systems at *the* Earth's surface: atmosphere, ocean, and land surface. Processes interacting in all three of these systems are responsible for shaping the coastal zone, and this interaction makes the coastal zone extremely dynamic. The interactions of both subaerial and marine processes on materials along the coast, its lithology, and resistance to erosion are major defining factors for the evolution of a coastline, and the most prominent forces involved in this process are waves, currents, tides, along with the existing geological and lithological conditions of the sea coast.

Therefore, the coastal zone is also a zone of transfer of material from the land surface to the ocean, being moved to the beach and nearshore. In some areas, the accumulation of sediments may add to the land mass. Here, the physical processes that act to shape the sea coast and the landforms are mainly discussed, known as coastal geomorphology.

For the development of marine and coastal areas, understanding coastal geomorphology is also very important, and it is not merely limited to a narrow belt developing under present conditions, but also the much wider zone over which the sea has migrated in the recent geological past, which is discussed here in brief.

5.5.1 Coastal Geomorphology

Geomorphology comprehends the study of the landforms of the earth and the processes that form them, especially those that lead to the erosion, transport, and deposition of sediments. The focus of coastal geomorphology is therefore on the morphology of the coastal zone and on processes such as waves, tides, and currents that act to shape features as disparate as high rock cliffs, low coral atolls, and sandy beach and dune systems. The study of coastal geomorphology is greatly influenced by other fields of geomorphology, especially fluvial and aeolian geomorphology, and there is a common set of paradigms, instrumentation, and methodology.

As is the case in most areas of geomorphology, there are many terms used to describe coastal features

- **Shoreline** can be defined instantaneously as the intercept of the mean water level along the beach, but it is often used loosely as the swash limit or the landward edge of the backshore. However, in a broad sense, it can be divided into five primary zones— offshore, nearshore, surf, foreshore, and backshore.

- The **offshore zone** portion of the profile, where there is no significant transport of sediment by wave action, is the outer limit of the breaker zone.
- The **nearshore zone** is the area of the shore affected by the waves where the water depth is one-half wavelength or less. The width of this zone depends on the maximum wavelength of the approaching wave train and the slope of the seafloor.
- The **nearshore zone** includes the shoreface, which is where rocks are deposited. The shoreface is broken into two segments: the **upper and lower shoreface**. The upper shoreface is affected by everyday wave action and consists of finely laminated and cross-bedded sand. The lower shoreface is the only area moved by storm waves and consists of hummocky cross-stratified sand.

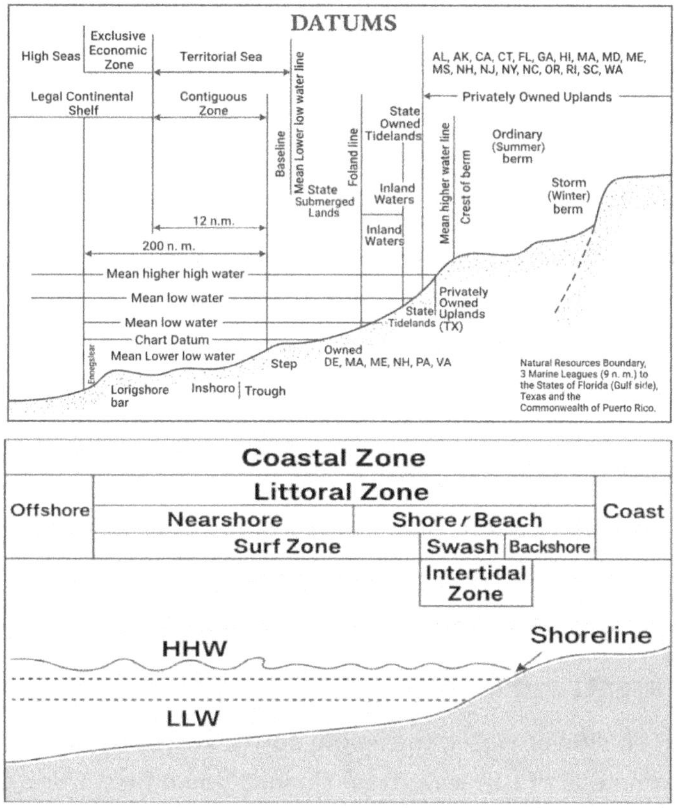

Figure 5.7: Different elements of the shore with respect to the sea water level.

- The **surf zone** is where the waves break. The foreshore zone overlaps the surf zone and is periodically wet and dry due to waves and tides. The foreshore zone is made of planar-laminated, well-sorted sand.
- The **beach face** is the part of the foreshore zone where the breaking waves swash up, and the backwash flows back down.
- **The foreshore** portion of the beach is subject to wave action during non-storm conditions.

- **The backshore** portion of the beach is subject to wave action only during storms. During non-storm conditions, sand may be reworked by aeolian action. This is the beach for recreational purposes.
- **Coastal zone: a** broad term for the area influenced by proximity to the coast; both the onshore and offshore limits are (deliberately) imprecisely defined.
- **The littoral zone** portion of the coastal profile, where sediment can be transported by wave action.
- The **seaward limit** is defined by the water depth at which significant sediment transport by wave action ceases (it can be equated with the boundary between the offshore and the nearshore), and the landward limit is the limit of (normal) wave action.
- **Nearshore zone** portion of the profile extending from the limit of significant sediment transport by waves to the low tide line. Note that where the offshore is defined as beginning at the seaward edge of the breaker line, then this will also mark the seaward limit of the nearshore.
- **The shoreface zone** dominated by wave action from the offshore landward to the low tide line is synonymous with the definition of nearshore used in this book.
- The **seaward limit** is defined by the spring low tide line and the landward limit by the limit of swash uprush during normal storms. The landward limit is often taken as the vegetation line.
- **Surf zone** of broken waves (surf bores) extending from the breaker zone to the foreshore. Varies with beach slope and wave conditions and will only be present when waves break some distance offshore.
- **Swash zone** of wave run-up on the beach and return of water in the form of backwash. It varies with wave and slope conditions.
- **Intertidal zone** between high and low water (tide).

5.5.1.1 Longshore Currents and Littoral Drift

As waves enter shallower water, they slow down. Waves usually approach the shoreline at an angle, with one end of the wave train slowing down first. This angle causes the waves to bend toward the beach. This bending action is called wave refraction. From the beach face, this bending looks like waves are approaching straight on, parallel to the beach. However, as refracted waves approach the shoreline at an angle, they create a slight difference between the swash as it moves up the beach face and the backwash as it flows back down. The swash and backwash along the beach create a current called the longshore current. Waves stir up sand in the surf zone and move it along the shore. This movement is called longshore drift. Longshore drift along the west and east coasts of India moves sand north to south on average. Longshore currents can carry longshore drift down a coast until it reaches a bay or inlet, where it will deposit sand in the quieter water. As the spit grows, it may extend across the mouth of the bay, forming a barrier called a baymouth bar. Where the bay or inlet serves as a boat anchorage, spits and baymouth bars are a severe inconvenience. Often, inconvenienced communities create methods to keep their bays and harbours.

5.5.1.2 Emergent and Submergent Coasts

Emergent coasts occur where sea levels fall. Submergent coasts occur where sea levels rise. Tectonic shifts and sea level changes cause the long-term rise and fall of sea levels. Some features associated with emergent coasts include high cliffs, headlands, exposed bedrock, steep slopes, rocky shores, arches, stacks, wave platforms, and wave notches.

In emergent coasts, wave energy, wind, and gravity erode the coastline. The erosional features are elevated relative to the wave zone. Sea cliffs are Wave notches carved. Refracted waves that attack bedrock at the base of headlands may erode or carve out a sea arch, which can extend below sea level. When a sea arch collapses, it leaves a rock column called a stack. When rock behind the stack erodes, sand can erode from the headland. Eroded sand that collects from the headland to the stack forms a tombolo: a sand strip that connects the stack to the shoreline. Where sand supply is low, wave energy may erode a wave-cut platform across the surf zone, exposed as a bare rock with tidal pools at low tide. This bench-like terrace extends to the cliff's base. When wave energy cuts into the base of a sea cliff, it creates a wave notch. Submergent coasts occur where sea levels rise due to tectonic subsidence—when the Earth's crust sinks—or when sea levels rise due to glacier melt. Features associated with submergent coasts include river mouths, fjords, barrier islands, lagoons, estuaries, bays, tidal flats, and tidal currents. In submerged coastlines, river mouths are flooded by the rising water. Fjords are glacial valleys flooded by post-ice age sea level rise.

Barrier islands are elongated bodies of sand that formed from the old beach sands that used to parallel the shoreline. Often, lagoons isolate barrier islands from the mainland. Barrier island formation is controversial: some scientists believe that they formed when ice sheets melted after the last ice age, raising sea levels. Another hypothesis is that barrier islands formed from spits and bars accumulating far offshore.

5.5.1.3 Tidal flats—or mudflats

Tidal flats—or mudflats- form where tides alternately flood and expose low areas along the coast. Tidal currents create combinations of symmetrical and asymmetrical ripple marks on mudflats, and drying mud creates mud cracks. In the central Wasatch Mountains of Utah, ancient tidal flat deposits are exposed in the Precambrian strata of the Big Cottonwood Formation. These ancient deposits provide an example of applying Hutton's principle of uniformitarianism (see Chapter 1). Features common on modern tidal flats indicate that these ancient deposits were formed in a similar environment: there were shorelines, tides, and shoreline processes acting at that time, yet the ancient age indicates that there were no land plants to hold products of mechanical weathering in place, so erosion rates would have been different. Geologically, tidal flats are broken into three different sections: barren zones, marshes, and salt pans. These zones may be abundant or absent in each individual tidal flat. Barren zones are areas with strong flowing water, coarser sediment, with ripples and cross-bedding common. Marshes are vegetated with common sand and mud. Salt pans or flats, less often submerged than the other zones, are the finest-grained parts of tidal flats, with silty sediment and mud cracks. Lagoons are locations where spits, barrier islands, or other features partially cut off a body of water from the ocean. Estuaries are a vegetated type of lagoon where fresh water flows into the area, making the water brackish —a salinity between salt and fresh. Groins gather sediment from longshore drift. Kara-Bogaz Gol lagoon, Turkmenistan. water. However, terms like lagoon, estuary, and even bay are often loosely used in place of one another. Lagoons and estuaries are certainly transitional between land and water environments where littoral, shallow shorelines, lacustrine, lakes or lagoons, and fluvial, rivers or currents can overlap. For more information on lagoons and estuaries.

As geomorphological agents, sea waves and currents erode the marine features along the sea coast, forming **significant landforms**. The Coastal Landforms are shaped by the constant motion of the waves and currents. **Currents** are created when waves are refracted while entering shallow water. These longshore currents carry along with themselves large quantities of material and sediments.

Waves and currents also erode sediments from cliffs and shorelines and transport them onto beaches.

When waves break in the surf zone, the water moves onto the shore with considerable force and turbulence, and sediments on the seabed are scoured at the same time. Further, storm and tsunami waves, unlike conventional breaking waves, can have far-reaching effects in a short period of time. On calm days with mild winds, waves cause little damage to the shoreline and can even assist in the formation of beaches and other depositional **features**.

5.5.2 Landforms Due to Waves

As geomorphic agents, waves and currents are responsible for the formation of the following features;

5.5.2.1 Depositional Landforms Due to Waves

- Beaches are the features of shorelines where deposition is the dominant factor. Beaches are made out of sand-sized components that are only present for a short time.
- Shingle Beaches are defined as beaches with an abundance of small pebbles and even cobbles.
- Sand dunes form as long ridges parallel to the coastline right behind the beaches.

5.5.2.2 Bars, Spits, and Lagoons

- Waves and currents deposit sand and gravel on the beach, creating bars that separate the shoreline from the sea. Between the mainland and the sea, they serve as a barrier.

- A **spit** is formed when one end of such a **bar** is attached to the coast and the other stretches into the sea.
- A **tombolo** is a type of deposition landform in which an island is connected to the mainland by a small spit or bar.

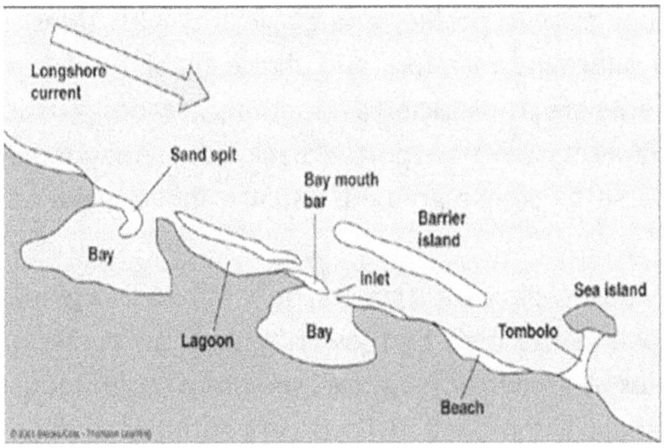

Figure 5.8: Depositional Landforms due to Waves

Both ends of the bar occasionally combine to enclose a section of seawater between the coast and the bar due to wave and current deposition. The confined area of the sea becomes a saline lake known as the **Lagoon**. In most cases, a lagoon is connected to the sea via a short strait. Lagoon Lakes include **Chilka** and **Pulicat** Lakes.

5.5.2.3 Landforms due to Currents

- Tides and currents make very little direct attack on the coastline when they come into contact with it.
- Tides primarily affect maritime erosion by extending an erosion line into an **erosion zone**. This zone corresponds to the area between the low water level and the high water level.
- Currents help to move eroded debris and deposit it as silt, sand, and gravel along the coasts.

5.5.3 Shoreline Processes and Products

The continuous combined effect of waves, tides, surges, and currents results in the erosion of coasts, transport, and deposition of sediments, resulting in the development of varied topographic features. Along any shoreline, erosion, production, or availability of sediment, their transportation and deposition tend to achieve balance. If this balance is disturbed due to natural or anthropogenic reasons, such as sea level changes or land reclamation, respectively, it results in dynamics of erosion or deposition at newer places.

5.5.4 Shoreline Erosion

The process of erosion is carried out by the processes of direct hydraulic action, milling, and solution. This results in severe coastal erosion, especially above the wave base, wherein sediment-filled coasts are more readily eroded as compared to rocky ones, forming sea cliffs, wave-cut benches, notches, sea arches, bays, stacks, etc. Erosion of sediment and rocks at and below the waterline undercuts the cliffs, leading to falls, slumps, and sliding, causing landward shifting of the shoreline. Many houses built on shoreline cliffs and other infrastructural features are always under the threat of vanishing into the sea. This needs proper engineering geological considerations before taking up such constructions, as post-construction engineering solutions may be very costly.

5.5.5 Shoreline Sediment Transport

Depending upon the topographic profile, energy vested with tides, surges, waves, and currents at a particular shoreline location, and the availability of sediments in terms of amount and size, sediments are transported as traction, saltation, suspension, and solution load. Normally, fine sediments are transported backshore in lagoons, while the onshore may vary from very coarse to coarse-grained, usually above wave base, and again fine-grained offshore.

Some shorelines can have wind-deposited coastal dunes as well as gravels termed as shingle deposits. The zone between the high and low tide sees an active process of sediment transport. Coastal waves and currents move sediment continuously, especially from promontories or headlands. The main activity is concentrated along the surf zone where waves are broken, resulting in swash and back swash of water, moving sediment upward and downward. The refraction of waves after hitting the coasts at oblique angles results in the movement of water slightly at a small angle to the beach, termed long-shore currents. These longshore currents lead to the deposition of spit beaches, also termed ***littoral drift***.

5.5.6 Shoreline Deposition

Sediments are deposited usually below the wave base as a normal process. Sediments are also deposited in protected areas between seaward-projected promontories, where the wave's energy is dissipated and sediments are deposited. Most of the wide beaches are formed in these areas. The most important depositional feature on the shores is beach bars, in the form of low-lying linear mounds parallel to the coast. Barrier Islands are low-height, long, narrow islands paralleling the coastline but cut off from the shore.

Theories suggest that they may have formed by the action of longshore currents on delta sands present at the river mouth. These barrier islands face relative erosion at their fronts with the sea, while receiving sediment deposition at their fronts with land. These islands are often so low in height that they are completely drowned and sometimes obliterated during severe high storm tides. But on the other hand, they provide the first line of defence against the fury of routine waves. Some coasts have carbonate deposition as coral reefs, precipitation of microcrystalline calcite, and broken shells from dead organisms.

5.5.7 Seawater Composition

Seawater Composition Seawater is a complex solution of inorganic, organic, and biological components. These can interact with materials to cause corrosion and to degrade their properties. Chemical Composition of Seawater (19 ppt Chlorinity)

Salinity (ppt); $S = 0.03 + 1.805 \times Cl^{-}$

i) **Anions**

Table:5-9 Ion Concentration in seawater (Anion)

Component	Concentration (g/kg of water)
Chloride	19.35
Sulphate	2.70
Bicarbonate	0.14
Bromide	0.067
Borate	0.0044
Fluoride	0.0014

ii) **Cations**

Table:5-10: Ion Concentration in seawater (Cation)

Component	Concentration (g/kg of water)
Sodium	10.76
Magnesium	1.29
Calcium	0.41
Potassium	0.39
Strontium	0.0079

The average chemical composition of seawater is (35 g/kg) different from one region to another. Specific weight for seawater is 1.025 tons/m^3.

5.5.7.1 Considerations of Environmental Salinity in Design

From the above details, it is evident that seawater is a complex chemical medium comprising dissolved salts, predominantly chlorides, sulphates, sodium, magnesium, and other minor constituents. The chemical composition of seawater presented above represents an average global value and is intended only as a reference baseline. In reality, the salinity profile of seawater is **site-specific** and exhibits significant variation due to geographical location, bathymetry, prevailing meteorological and oceanographic (met-ocean) conditions such as temperature, evaporation rate, tidal interaction, and circulation patterns, as well as the magnitude of freshwater inflow from rivers, estuaries, and land-based drainage systems. Therefore, reliance solely on average seawater composition without consideration of site-specific factors may lead to an underestimation or overestimation of durability demands.

A representative salinity profile for the Indian Ocean is presented in Section 6.6. Similar information is essential for assessing the durability performance of concrete and steel materials. Maritime structures located in environments with high salinity exposure require special consideration during structural design and detailing, including adequate provision of protective systems such as coatings and corrosion mitigation measures.

In addition to direct marine exposure, **atmospheric salinity** plays a significant role in the deterioration of structures located in coastal zones. Sea breeze action transports saline aerosols inland, with the intensity of deposition being influenced by seawater salinity, wind velocity, topography, and distance from the shoreline. Even structures not in direct contact with seawater may therefore be subjected to aggressive saline environments, resulting in accelerated ageing of exposed concrete and steel surfaces.

Protective painting and coating systems for exposed steelwork and reinforced concrete surfaces shall be selected based on the assessed atmospheric corrosivity category and expected salinity deposition rates. Furthermore, inspection regimes and maintenance cycles shall be clearly defined at the design stage, ensuring periodic condition assessment, timely repair, and renewal of protective systems. Such proactive planning is essential to achieve the intended design service life, minimise life-cycle costs, and maintain structural safety and functionality over the operational life of the asset.

5.6 Strategies for Marine and Coastal Investigations

Building a port needs a well-planned, step-by-step approach that brings together all marine, coastal, and navigation studies. These studies are the technical backbone for safely and efficiently planning, designing, and running a port. A strong investigation plan helps gather the right data, reduces design risks, and supports good engineering decisions.

An effective site investigation plan for port development begins with broad desktop studies that consolidate remote-sensing data, public-domain information, and previous investigation reports. This foundation is progressively strengthened through targeted, multidisciplinary data collection to close knowledge gaps and reduce early-stage risks.

Once preliminary insights are established, detailed investigations are undertaken, including on-site surveys, numerical modelling, physical modelling, and navigation simulations. Together, these studies build a comprehensive understanding of the site's hydrographic, geophysical, hydrodynamic, geomorphological, and navigational characteristics.

All numerical models—hydrodynamic, wave, and sediment—are calibrated and validated using field measurements, while physical models and navigation simulators confirm hydraulic behaviour and vessel-manoeuvring feasibility under realistic environmental conditions.

The entire investigation process is guided by international standards such as IHO S-44, PIANC guidelines, BS 6349, and CIRIA/CUR recommendations to ensure technical robustness and regulatory compliance.

A risk-based approach prioritises critical zones, including breakwater heads, entrance channels, seabed obstructions, soft soil pockets, and unstable shorelines. Continuous engagement with key stakeholders—port users, pilots, engineers, contractors, and regulators—helps align the design and operational assumptions with practical and real-world requirements.

5.6.1 Phases of Investigations for Port Development

The investigation process is executed in distinct phases, each building on the previous one.

Phase 1 (Desktop Studies & Preliminary Assessment)

All existing and readily available information is compiled to build an initial understanding of the site's physical and environmental setting. This includes analysing satellite imagery to interpret coastal features, reviewing historical shoreline changes, studying available hydrographic charts, and assessing metocean datasets such as waves, tides, currents, and wind conditions. Past records on sediment transport behaviour and erosion–accretion trends are also examined, along with information on environmentally sensitive areas, regulatory restrictions, and nearby infrastructure. This phase helps identify early opportunities and constraints, screen potential risks, and develop initial port layout options that are broadly aligned with site conditions and project objectives.

Phase 2 (Reconnaissance & Baseline Field Surveys)

Reconnaissance & Baseline Field Surveys focus on validating the assumptions made during the desktop study through quick on-ground checks. This phase involves carrying out preliminary bathymetric surveys to understand water depths, inspecting the shoreline to assess erosion, access, and physical features, collecting basic sediment samples to understand seabed conditions, and recording simple tide and current measurements to confirm metocean behaviour. These early field activities help verify whether the site is broadly suitable for port development, identify any constraints not visible in desktop studies, and define the exact scope and methodology needed for the more detailed investigations that follow.

Phase 3 (Detailed Marine Investigations)

Detailed Marine Investigations focus on collecting high-accuracy data required for engineering design. Hydrographic surveys using MBES capture precise seabed levels and morphology, while geophysical tools such as SSS, SBP, and magnetometers map subsurface layers, detect boulders, obstructions, and hard strata. Geomorphological assessments analyse shoreline evolution, sediment pathways, and coastal stability, and ADCP deployments record detailed tide and current patterns for hydrodynamic and navigational modelling. Together, these datasets provide the definitive baseline needed for accurate numerical modelling and design development.

Phase 4 (Numerical / Mathematical Modelling)

Numerical / Mathematical Modelling) applies the collected field data to develop detailed hydrodynamic, wave, sediment-transport, morphological, and flushing models. These models assess wave transformation, harbour tranquillity, current behaviour, erosion and siltation trends, and long-term coastal changes, helping optimise breakwater design, entrance channel alignment, and dredging strategies.

Phase 5 (Physical Modelling)

Physical Modelling involves creating scaled models in specialised wave basins or flumes to verify complex hydraulic behaviour that numerical models may not fully capture. These physical tests examine breakwater stability, wave penetration and overtopping, scour and sediment movement, and the response of moored vessels under varying conditions. The results provide a high-confidence check on the hydraulic design, ensuring that critical structures and layouts can withstand extreme wave and storm events.

Phase 6 (Navigation Simulation Study)

The Navigation Simulation Study assesses the operational safety and practicality of vessel movements using real-time or fast-time simulators that replicate site conditions. In this phase, virtual ship models—equipped with accurate hull, propulsion, environmental, and manoeuvring characteristics—are used to test approach paths, turning basins, channel alignment, and berthing operations. The simulations also evaluate tug assistance requirements, emergency manoeuvres, and under-keel clearance under different tide, wind, wave, and current scenarios. These exercises help confirm whether the proposed port layout, channel dimensions, and operational procedures can support safe and efficient navigation for all target vessel classes in both normal and challenging conditions

Finally, Report - Integrated Interpretation & Final Engineering Recommendations brings together all findings from the surveys, numerical models, physical tests, and navigation simulations to create a complete engineering picture of the site. In this phase, experts review every dataset together—bathymetry, geophysics, hydrodynamics, sediment behaviour, vessel manoeuvring, and structural performance—to ensure that all results align and support each other. Based on this integrated understanding, the final engineering recommendations are developed. These typically include the breakwater layout and crest levels, entrance channel size and orientation, turning basin dimensions, berth alignment, dredging and disposal strategy, shoreline protection works, navigation rules and operational limits, and long-term maintenance plans. The goal of this phase is to make sure the final port design is technically reliable, safe for operations, cost-effective, and environmentally sustainable, with all decisions backed by solid data and analysis.

Thus, a well-planned, multi-phase investigation approach ensures that port development is based on reliable data, thoroughly validated models, and realistic operational simulations. By combining detailed surveys, numerical modelling, physical hydraulic testing, and navigation simulation, engineers gain a complete understanding of marine conditions and how the port will function in real life. This integrated method helps reduce uncertainties, lowers construction and operational risks, and supports the creation of a port that remains efficient, robust, and environmentally sustainable throughout its entire life cycle.

The following section provides a brief outline of the methodology and approach used for the detailed site investigations.

5.6.1.1 Hydrographic & Geophysical Survey

For developing a port, it's important to clearly understand what the seabed looks like and what lies underneath it. A combined hydrographic survey and geophysical survey helps engineers get this full picture, so they can plan navigation channels, decide dredging needs, and design safe marine structures.

Hydrographic Survey mainly deals with measuring water depths and mapping the seabed. Modern tools like Multi-Beam and Single-Beam Echo Sounders, along with GPS systems and tide gauges, provide accurate underwater maps. The output includes depth charts, 3D seabed models, and information about slopes, shoals, and underwater features. This helps in designing approach channels, turning circles, berth pockets, and estimating how much dredging will be required.

Geophysical Survey goes a layer deeper—quite literally. It looks at the seabed surface and the soil layers beneath. Tools like Side Scan Sonar show images of the seabed, helping detect debris, rocks, or obstacles. Sub-Bottom Profilers show how thick the sediment is and what kind of layers are underneath. Magnetometers detect buried metal objects, while seismic tools like Boomers or Sparkers help identify deeper geological layers. These results help engineers understand what type of soil or rock is present and what challenges might arise during construction.

Both surveys follow well-known international standards (like IHO S-44 and PIANC guidelines), which ensures that the results are reliable and accurate for engineering decisions. When the results of both surveys are combined, they give a strong understanding of dredging conditions, possible risks, and the best foundation options for marine structures. Engineers can identify hard layers, soft mud, buried obstacles, and bedrock levels. This makes planning much safer and helps avoid unexpected issues during construction.

The final report usually includes depth maps, seabed images, seismic profiles, soil layer sections, anomaly maps, and engineering recommendations. These findings guide dredging plans, structure foundations, and overall port layout. In short, combining hydrographic and geophysical surveys gives a complete and dependable understanding of the seabed and what lies below it. This leads to safer, cost-effective, and well-planned port development.

5.6.1.2 Hydrodynamic Study

A hydrodynamic study helps us understand how the sea behaves around a proposed port site. It covers waves, currents, movement of sand along the coast, and how water circulates inside the harbour. These insights are essential for designing breakwaters, optimising port layout, planning dredging, and ensuring smooth ship movement.

1. Waves

The wave study looks at how offshore waves behave as they approach the coast. It checks their height, period, and direction, and how they transform due to refraction, shoaling, and breaking. Using long-term wave data and models like SWAN or MIKE21, engineers estimate design wave conditions and wave behaviour inside the harbour. This helps determine breakwater alignment, port entrance conditions, and wave agitation inside the basin.

2. Currents

The current study focuses on how water flows due to tides, winds, and density differences. Field measurements using ADCPs and numerical modelling show how fast currents move, in which direction, and how they influence navigation, berthing, and sediment movement.

Current maps from these models help assess scour near structures and guide the design of safe and efficient port geometry.

3. Littoral Drift (Movement of Sand Along the Coast)

Littoral drift describes how sand naturally moves along the shoreline under wave action. Understanding this helps predict whether the port might cause erosion nearby or if siltation may block the approach channel. By analysing shoreline changes and modelling sediment transport, engineers estimate net and gross drift rates. This helps determine siltation levels near breakwaters and the need for maintenance dredging or shoreline protection works.

4. Recirculation Inside Harbour

Recirculation study identifies zones inside the harbour where water movement is very slow or stagnant. These dead zones can lead to heavy siltation or poor water quality. Using 2D/3D hydrodynamic models, engineers map areas with long water-retention times and suggest layout improvements, such as changing breakwater gaps or increasing basin flushing.

5. Integrated Findings

When wave, current, and sediment results are combined, they give a complete picture of site conditions. This helps in:

- Choosing the best breakwater alignment
- Positioning the port entrance for safe navigation
- Predicting siltation hotspots and planning dredging
- Checking erosion risks along the coastline
- Improving circulation inside the harbour

In short, a hydrodynamic study gives engineers the full picture of how the sea behaves around a port. It helps them design a safer, more efficient, and low-maintenance port that works well with natural coastal processes rather than fighting against them.

5.6.1.3 Geomorphological Study

A geomorphological study provides an engineering-grade understanding of coastal and seabed landforms, their evolution, and the physical processes that control shoreline stability at a port site. It supports decisions on port layout, breakwater orientation, dredging zones, and long-term coastal management.

A. Study Objectives

- Map coastal landforms (beaches, dunes, tidal flats, estuaries, rocky outcrops).
- Identify seabed morphology relevant to navigation and dredging (shoals, bars, ridges, channels).
- Assess long-term shoreline movement (erosion/accretion trends) and characterise sediment types and transport pathways.
- Determine geomorphic stability and implications for port structures.
- Geomorphic studies typically incorporate topographic and bathymetric datasets, multi-decadal satellite imagery for shoreline change analysis, beach profile

measurements and sediment grain-size data, hydrographic and geophysical survey outputs (MBES, SSS, SBP), and local metocean records—including waves, currents, and tides—for process interpretation.

- Finally, a geomorphological study offers a technically robust understanding of how the coastline and seabed have formed, how they are changing, and how they will react to new port structures. This information is essential for site selection, layout optimisation, dredging strategy, and ensuring that the port operates safely and sustainably over the long term.

5.6.1.4 Integrated Mathematical Model Study

A mathematical model study is basically a way to "test" how the sea will behave around a new port—without waiting years to see it in real life. Using computer simulations, engineers check how waves, tides, currents, and sediments will interact with the proposed breakwaters, channels, and port layout. This helps catch potential problems early and design a port that stays safe, efficient, and low-maintenance.

A. Aim of Study

The main idea is to predict the hydraulic behaviour around the port before construction actually starts. The model helps assess wave conditions, harbour agitation, tidal currents, siltation patterns, shoreline changes, and water circulation inside the harbour. By comparing different layout options, engineers can select the most stable and cost-effective arrangement for breakwaters, entrance channels, and berths.

B. Dataset Required for the Model

To make the model realistic, it uses measured wave data, tide levels, current profiles, detailed bathymetry, and sediment test results. It also relies on shoreline geometry, wind data, and long-term coastal trends. All these inputs help create a virtual environment that behaves much like the actual project site.

C. Tools Used for Modelling

The simulations are run using well-known coastal modelling software like MIKE21, Delft3D, SWAN, or TELEMAC. These tools allow engineers to run 2D or 3D simulations and see how the port performs under various conditions—normal seasons, monsoons, and even rare storm scenarios. They're flexible enough to test multiple layout configurations.

D. Simulation of Combined Model

The wave model shows how waves bend, break, reflect, and enter the harbour, helping check if the basin will stay calm. Hydrodynamic modelling predicts tidal flows and currents that influence navigation and may cause scour. Sediment transport modelling estimates where sand or silt will accumulate, which helps plan dredging needs and predict shoreline impacts. Recirculation modelling highlights where water may stagnate inside the harbour and whether flushing is adequate. Together, these models give a complete hydraulic picture of the proposed port.

E. Calibration of Model

To make sure the model behaves realistically, it is calibrated using actual field measurements—wave records, tide data, and current profiles. Then the model is validated by checking if it can reproduce known historical events. This step ensures the predictions are reliable for engineering decisions.

F. Output of Integrated Mathematical Model Study

The final outputs include wave height maps, harbour agitation levels, current vectors, siltation maps, and shoreline change predictions. Engineers also get estimates of longshore drift, erosion and deposition zones, and flushing-time results. Based on these, practical recommendations are made for breakwater alignment, entrance orientation, basin layout, dredging levels, and coastal protection measures.

This integrated modelling approach is crucial because it helps ensure the port will work well under real conditions—especially during monsoons or storms. It minimises risks of excessive siltation, erosion, and poor navigation conditions. Most importantly, it helps design a port that is safer, more efficient, and more economical to maintain over its operational life.

Coastal Geoscience of India: Framework and Insights

Chapter Summary

India's coastline is largely smooth due to ancient tectonic shifts, with certain areas gradually rising or sinking over time. The coastline varies by region and is divided into three main segments: the west coast, the east coast, and the southern tip known as Cape Comorin. The west coast, shaped by submergence and faulting, forms a narrow strip suitable for major ports such as Kandla, Mumbai Port, Jawaharlal Nehru Port, and Cochin. However, India's relatively smooth coastline offers fewer natural harbours compared to Europe's deeply indented shores. Coastal landscapes are continuously reshaped by the dynamic interaction between land and sea. These changes are often unpredictable and vary by location. To fully understand coastal evolution, it is essential to study weather patterns, ocean currents, landforms, and geological processes that collectively influence shoreline transformation over time.

Learning Objectives

- Understand the overview of the Indian coastline
- Explore the forces shaping India's coastal landscape
- Examine the geology of the East Coast, West Coast, and Cape Comorin
- Study the impact of monsoons, winds, waves, and currents on coastal regions
- Analyse the salinity distribution pattern in the Indian Ocean

6.1 Overview of the Indian Coastline

The Indian subcontinent is geographically bounded by three marginal seas and one ocean: the Arabian Sea to the west, the Bay of Bengal and the Andaman Sea to the east, and the Indian Ocean to the south. The total length of India's coastline is approximately 7,516.6 km, comprising both passive continental margins and active tectonic margins, indicative of varied geodynamic settings. Of this coastline, approximately 6,100 km is associated with the mainland, while 1,197 km pertains to the island systems, including the Andaman & Nicobar Islands and the Lakshadweep archipelago, which are part of distinct tectonic and coral atoll formations, respectively. The eastern coastal plains, adjacent to the Bay of Bengal, are characterised by deltaic formations and low-gradient sedimentary basins, whereas the western coastal plains, bordering the Arabian Sea, exhibit narrow, fault-controlled coastal strips. India's Exclusive Economic Zone (EEZ) extends up to 200 nautical miles from the baseline, as defined under the United Nations Convention on the Law of the Sea (UNCLOS, 1982). The EEZ confers sovereign rights to the coastal state for the exploration, exploitation, conservation, and management of natural resources, including offshore hydrocarbons, marine minerals, and renewable energy sources such as wind and wave energy.

The baseline, from which this zone is measured, typically follows the low-water line along the coast, as delineated by hydrographic surveys.

It is essential to distinguish between territorial waters and the EEZ. Territorial waters extend up to 12 nautical miles from the baseline and grant the coastal state full sovereignty over the water column, seabed, subsoil, and airspace. In contrast, the EEZ provides functional jurisdiction primarily over subsurface and seabed resources, without full sovereignty over the water column or airspace.
The relatively linear and tectonically stable nature of India's coastline is attributed to plate tectonic reconfigurations during the Late Mesozoic Era, particularly the Cretaceous period, when the Gondwanaland supercontinent underwent fragmentation. The Coromandel Coast in Tamil Nadu and the Malabar Coast in Kerala exemplify emergent coastlines, shaped by isostatic rebound and epeirogenic uplift. Conversely, the western coast displays a complex interplay of tectonic uplift and subsidence, with normal faulting contributing to the formation of submerged coastal plains. A notable example is the hypothesised submergence of the ancient city of Dwaraka, which may represent a paleo-shoreline or archaeological marine site.

This tectonic subsidence has resulted in a narrow coastal corridor along the western margin, conducive to the development of deep-water ports and natural harbours. Major ports such as Kandla, Mazagaon, Jawaharlal Nehru Port (Nava Sheva), Marmagao, Mangalore, and Cochin are strategically located along this stretch. However, India's coastline lacks the highly indented, ria-type coastlines found in regions like Western Europe, which naturally facilitate harbour formation. The trailing-edge coasts of India and Africa, being relatively smooth, offer limited natural harbour potential.

Furthermore, the coastal geomorphology of India is subject to continuous modification due to marine transgression and regression, longshore drift, wave refraction, sediment transport, and coastal erosion. These processes are influenced by monsoonal dynamics, sea-level fluctuations, lithological variability, and tectonic activity. A comprehensive understanding of these interactions necessitates an integrated approach involving geomorphological mapping, remote sensing, geophysical surveys, and sedimentological analysis to assess the evolving nature of India's coastal landscapes.

6.2 Shaping Forces of the Indian Coastline

India's coastline is subject to continuous and dynamic transformation, primarily governed by a combination of natural forces. These include:

- Seasonal rainfall patterns
- Prevailing winds
- Wave dynamics

- Tidal fluctuations
- Oceanic currents
- Extreme weather events, notably cyclones

These interrelated elements collectively influence the geomorphology of the coastal zone, reshaping shorelines and significantly impacting the coastal ecosystem. Beyond meteorological influences, sediment transport and subsurface geological structures play a vital role in shaping and stabilising India's coastal zones. Comprising sand, silt, clay, and gravel, these materials are continuously transported and redistributed by natural forces such as waves, tides, currents, and wind. Their movement governs the formation, erosion, and accretion of coastal landforms. Sediments are derived from multiple sources, including fluvial systems such as the Ganges and Brahmaputra, marine depositional processes, aeolian transport in arid zones, and anthropogenic activities like dredging, land reclamation, and coastal infrastructure development. Regional variations in sediment composition influence the resilience of coastal features—sandy shores are more dynamic and erosion-prone, while muddy coasts offer greater stability but are sensitive to tidal changes. Transport mechanisms like longshore drift, tidal currents, monsoonal flows, and cyclonic events contribute to seasonal and spatial changes in coastal morphology.

6.3 Impact of Monsoon

The monsoon plays a crucial role in shaping the physical and ecological characteristics of India's coastline. The country experiences two principal monsoon systems—the southwest monsoon (June to September) and the northeast monsoon (October to January)—each exerting distinct influences on coastal processes.

6.3.1 Southwest Monsoon Impact

The southwest monsoon is the dominant climatic driver, bringing heavy rainfall and strong winds across much of India. It divides into two branches: (i) the Bay of Bengal branch, which moves northward and affects the eastern coast, and (ii) the Arabian Sea branch, which flows along the western coast and interacts with the Western Ghats, resulting in intense orographic rainfall.

This season is marked by high wave energy and strong littoral currents, leading to accelerated coastal erosion, particularly along open beaches and estuaries. The monsoon also enhances sediment transport from river discharge and marine reworking, contributing to deltaic expansion in areas such as the Sundarbans and Mahanadi delta. Temporary inundation and flooding during this period alter coastal wetlands and mangrove ecosystems. Cyclonic disturbances, especially in the Bay of Bengal, cause abrupt shoreline changes and sediment displacement

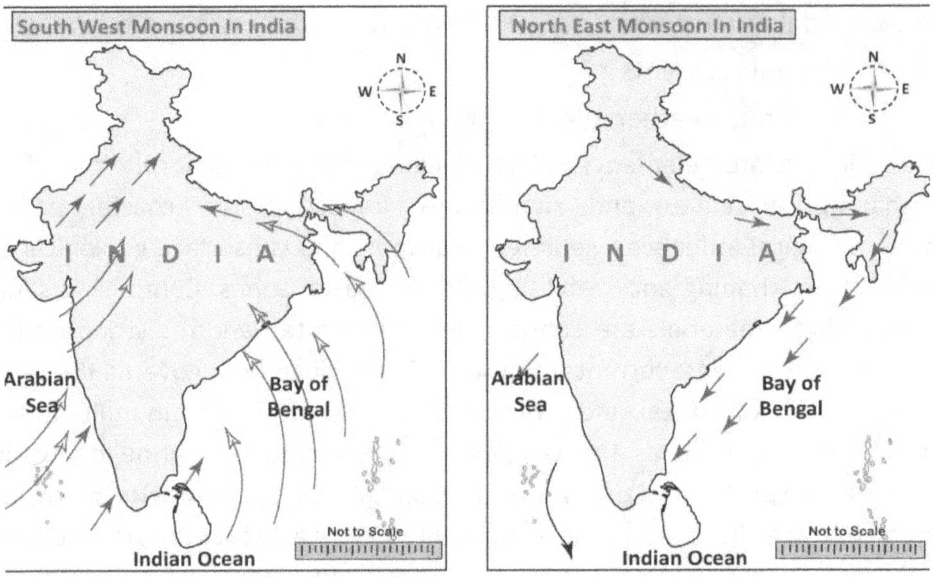

Figure 6.1: Wind Directions of the Southwest and Northeast Monsoons in India

6.3.2 Northeast Monsoon Impact

The northeast monsoon primarily influences the southeastern coast, including Tamil Nadu and parts of Andhra Pradesh. Although less intense than the southwest monsoon, it significantly contributes to localised sediment deposition, particularly in low-lying coastal plains. It also facilitates the formation and reshaping of sandbars, spits, and lagoons due to calmer wave conditions and sediment settling. Additionally, it aids in the recharge of coastal aquifers and estuarine systems, thereby supporting ecological productivity.

The northeast monsoon predominantly affects India's east coast, bringing consistent northeasterly winds and elevated atmospheric pressure over land. While the west coast is influenced solely by the southwest monsoon, the east coast experiences both southwest and northeast monsoons, resulting in a more variable and dynamic climate.

Monsoons are not the only forces influencing India's coasts—cyclones and storm surges also play a significant role. When storm surges coincide with spring tides, they can cause substantial damage, particularly in low-lying coastal regions and river mouths.

India's west coast usually has steady wave patterns throughout the year, except during the southwest monsoon when conditions become more intense. These wave and current patterns are influenced by the Inter-Tropical Convergence Zone (ITCZ), a calm area near the equator where winds from both hemispheres meet. As the ITCZ shifts north or south with the seasons, it changes wind and ocean conditions, which in turn affect coastal currents and how sediments move along the shore.

Please note, the ITCZ doesn't stay in one place—it **moves north or south depending on the season**. In summer, it shifts northward, and in winter, it moves southward. This movement affects **wind patterns, ocean currents, and weather conditions**, especially in tropical regions like India. For example, when the ITCZ moves north during the southwest monsoon, it helps bring heavy rains and stronger winds to the west coast of India.

Unlike the west coast, the east coast of India has a more variable climate. Between October and **January**, low-pressure systems frequently develop over the Bay of Bengal and move inland, often leading to cyclones. This region is particularly prone to such events, and cyclone activity in the northern Bay of Bengal has increased by approximately 26% over the past 120 years, making it a significant factor in shaping coastal conditions.

6.4 Impact of Wave

Waves play a major role in coastal erosion by constantly hitting the shoreline and moving sediments. When waves are strong—especially during monsoons—they carry more energy, which can break down rocks, wash away sand, and reshape beaches. On the west coast of India, high wave activity during the southwest monsoon leads to noticeable erosion along open beaches and estuaries. On the East Coast, strong waves during both monsoons cause even more frequent changes to the shoreline. The direction and strength of waves also affect how and where sediments are removed or deposited, which can lead to the loss of land in some areas and buildup in others. Over time, this process changes the shape and stability of the coast.

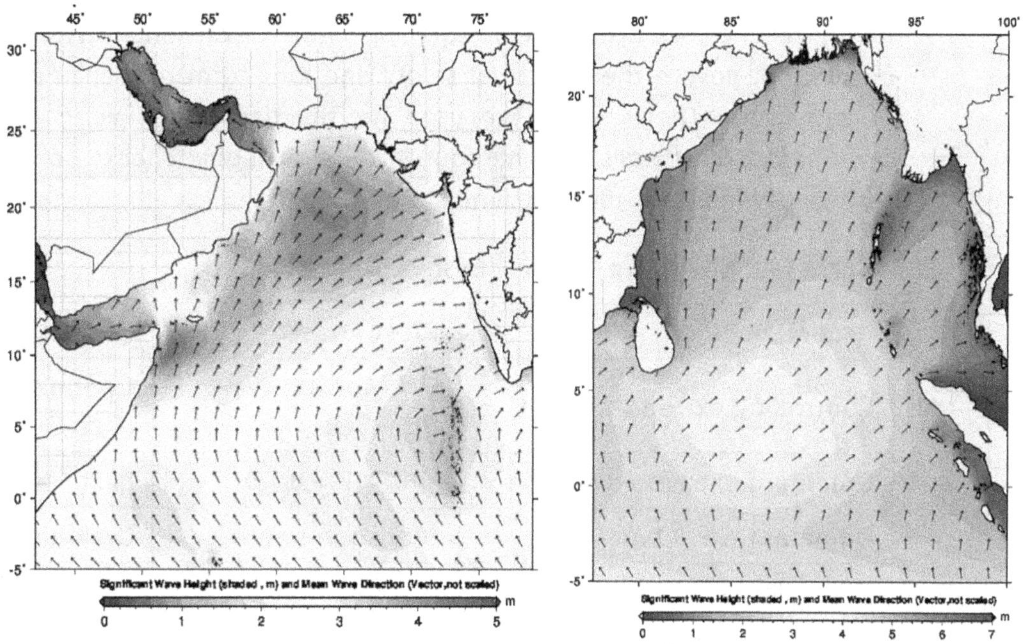

Figure 6.2: Indicative Wave Activity along the East and West Coastlines of India

Wave activity differs between India's east and west coasts. The west coast generally experiences steady wave patterns throughout the year, with stronger wave action mainly during the southwest monsoon. In contrast, the East Coast faces strong wave activity during both the southwest and northeast monsoons, making it more dynamic. The direction from which waves approach also varies by season and coast:

West Coast:

- During the southwest monsoon, waves come from the west and southwest.
- In the northeast monsoon, they approach from the west and west-northwest.

- In fair weather, waves typically come from the southwest.

East Coast:

- During the southwest monsoon and fair weather, waves approach from the southeast.
- In the northeast monsoon, they come from the northeast.seasonal shifts in wave

These directions and intensities influence coastal erosion, sediment movement, and shoreline changes along both coasts

6.4.1 Operational vs. Extreme Wave Conditions

For port and harbour design, the prevailing wave climate is categorised into operational and extreme wave conditions, each serving a distinct design purpose. Operational (working) waves represent the sea states under which routine port activities are carried out, including vessel berthing and unberthing, cargo handling, berth availability assessment, operational downtime estimation, and selection of feasible construction windows. These conditions define normal port operability and form the basis for harbour tranquillity evaluation and mooring performance analyses.

In contrast, extreme (design or survival) waves represent low-probability, high-energy events that govern the structural safety and long-term resilience of coastal and marine infrastructure. These waves are used for the design and verification of breakwaters, revetments, quay walls, and harbour layouts, including checks on overtopping, stability, resonance limits, and ultimate limit states.

Table 6-1: Key Wave Details (East vs West Coast)

Aspect	West Coast	East Coast
Controlling extreme season	SW Monsoon (Jun–Sep)	Cyclone season (Oct–Dec)
Dominant extreme direction	South-West	East / North-East
Dominant wave type	Monsoon swell	Cyclonic swell
Operational calm season	Dec–Mar	Feb–Apr
Design critical months	Jul–Aug	Oct–Nov

The controlling season, dominant wave direction, and governing wave type vary markedly between the West Coast of India (Arabian Sea) and the East Coast of India (Bay of Bengal). Accordingly, operational and extreme wave conditions for each coast must be defined and assessed independently to ensure both functional efficiency and structural robustness of port facilities.

Generally, West Coast ports are designed primarily for monsoon-dominated wave attack, with long-period swell directly impacting breakwaters and harbour entrances. East Coast ports require robust design against short-duration but severe cyclonic waves, often approaching from directions different than operational waves.

Separation of operational wave criteria and extreme wave criteria is essential for avoiding over-design of operational facilities and under-design of critical structures.

6.4.2 Waves Affecting Sediment Transport

Waves play a crucial role in sediment transport along coastlines by influencing both the energy available for the movement of sediments and the direction in which they are carried. When waves approach the shore, their energy—determined by factors like wave height and period—mobilises sediment particles from the seabed or beach. The direction of wave approach, especially when oblique to the shoreline, generates longshore currents that transport sediment parallel to the coast. The combined effect of wave energy and direction thus governs the dynamic redistribution of sediments in coastal environments. These two factors are described below;

6.4.3 Wave Energy

Waves carry energy that can break down, move, or drop sediment depending on how strong and frequent they are. High-energy waves, like those during monsoon storms, cyclones, or storm surges, create strong turbulence and powerful currents that can stir up and shift large amounts of sand, silt, and other particles. For example, waves from tropical cyclones in the Bay of Bengal or storm waves hitting the western coast of India during the southwest monsoon often cause serious coastal erosion and move sediment around. These strong waves can reshape beaches, wash away dunes, and push sediment either offshore or along the coast. On the other hand, low-energy waves, usually seen during calm weather or in protected bays, don't have enough force to keep sediment floating, so it settles down and builds up over time. Thus, the balance between erosion and deposition is closely tied to the energy levels of incoming waves.

Generally, the sediment deposition zones are spots where water slows down enough for particles to settle and pile up. We may find these in places like beaches, deltas, lagoons, estuaries, and offshore bars. For example, sheltered bays or the protected sides of structures like breakwaters often become deposition zones due to reduced wave action.

The kind of sediment that settles—whether it's fine silt or coarse sand—depends on how much energy is in the water and how much sediment is available. These zones are important for keeping coastlines stable, creating habitats, and helping land grow, but they can shift due to seasonal changes, storms, or human activities like dredging and construction.

6.4.4 Wave Direction

When waves approach the shore at an angle, they generate a current known as a longshore current, which flows parallel to the coastline. This current acts like a conveyor belt, moving sand and other sediments along the shore. For instance, longshore currents along the beaches of Goa, especially between Sinquerim and Baga, have been observed to shift sand under varying tidal conditions. Similarly, Krishnapatnam on India's east coast experiences strong longshore currents due to angled wave approaches, which influence sediment movement and affect the stability of the coastal inlet. Apart from their strength, the direction of longshore currents is mainly controlled by the angle at which waves hit the shore, playing a key role in reshaping beaches. To manage erosion and sediment deposition, coastal structures like groynes and jetties are often constructed to interrupt or redirect these currents.

6.4.5 Wave Impact Along the Indian Coast

East Coast (Bay of Bengal): Seasonal Sediment Changes Along the Indian Coast. Sediment movement along the Indian coastline changes with the seasons, mainly due to shifting wind patterns and wave directions during the monsoon cycles. During the southwest monsoon (June to September), strong winds and high-energy waves—especially on the east coast—push sediment northward, causing siltation in river mouths and erosion of open beaches. Ports like Chennai and Visakhapatnam often face heavy sediment buildup during this time, requiring regular dredging. In contrast, during the northeast monsoon (October to December), wave directions shift, and sediment may move southward, leading to deposition in some areas and erosion in others. These seasonal changes also affect local ecosystems—for example, sediment buildup can block mangrove roots or reduce water quality, while erosion can damage turtle nesting sites and fish breeding grounds. Understanding these seasonal patterns is important for managing coastal infrastructure, protecting habitats, and planning port operations.

West Coast (Arabian Sea): On the west coast of India, wave activity is mainly influenced by the southwest monsoon, which brings strong southwest waves. These waves push sediment southward, causing beach erosion and sediment buildup in harbours, which can affect navigation and port operations. Along the Kerala coast, seasonal wave action reshapes beaches significantly—sand is often washed away during the monsoon and slowly returns afterwards. To maintain beach stability, artificial nourishment (adding sand back to the shore) is sometimes needed. In the Gulf of Khambhat, strong tidal waves stir up large volumes of sediment, which can impact estuarine ecosystems by reducing water clarity, disturbing fish habitats, and affecting mangrove health. These wave-driven changes not only alter the coastline but also influence local livelihoods, fishing, and coastal biodiversity. Here's a comparison of wave impacts on sediment transport along the east and west coasts of India:

Table 6.2: Comparison of wave impacts on sediment transport along the east and west coasts of India

Aspect	East Coast (Bay of Bengal)	West Coast (Arabian Sea)
Dominant Wave Direction	Southeast waves during the southwest monsoon	Southwest waves during the southwest monsoon
Sediment Movement	Northward along the coast	Southward along the coast

Aspect	East Coast (Bay of Bengal)	West Coast (Arabian Sea)
Key Effects	- Siltation in river mouths - Erosion of open beaches	- Beach erosion - Sedimentation in harbours
Port Examples	Chennai, Visakhapatnam, Paradip	Kochi, Mormugao
Seasonal Changes	Strong monsoonal impact; sediment shifts north during the SW monsoon	Beaches reshape seasonally; nourishment is needed post-monsoon
Tidal Influence	Moderate tidal influence	Strong tidal impact (e.g., Gulf of Khambhat)
Ecosystem Impact	Affects mangroves, turtle nesting sites, and fish habitats	Affects estuarine ecology, water clarity, and mangrove health
Management Measures	Dredging, coastal protection structures	Artificial beach nourishment, groynes, jetties

6.5 Effect of Wind

Wind patterns and ocean currents along India's coasts change with the seasons, especially during the monsoon cycles. During the northeast monsoon (October to January), wind speeds are moderate, around 20 km/h, while the southwest monsoon (June to September) brings stronger winds, often reaching 35–55 km/h. These seasonal winds influence ocean currents, which in turn affect wave energy, sediment movement, and coastal erosion.

On the west coast, currents generally flow clockwise for most of the year, but reverse to counterclockwise for 2–4 months after the southwest monsoon. This shift causes beach erosion and harbour siltation at ports such as Mumbai, Kandla, Mormugao, and Kochi. For example, Kandla Port in Gujarat often experiences sediment buildup due to strong tidal currents, requiring regular dredging to maintain navigability.

On the east coast, currents also flow clockwise most of the year, but during the southwest monsoon, strong counterclockwise currents develop. This leads to northward sediment drift, causing siltation in river mouths and erosion of beaches, especially near ports like Chennai, Ennore, Tuticorin, Visakhapatnam, and Haldia

In the Gulf of Khambhat, strong tidal waves stir up large volumes of sediment, affecting estuarine ecosystems and reducing water clarity, which impacts fish habitats and mangrove health. These wind-driven changes are critical for coastal management, as they influence port operations, fishing activities, and the health of coastal ecosystems.

These changes in wind and current direction affect wave energy, sediment movement, and coastal erosion, making them important factors in coastal management.

6.6 Salinity and Water Movement in the Northern Indian Ocean

The northern Indian Ocean has a mix of climates—dry regions like the Arabian Peninsula in the west and rainy areas like Southeast Asia in the east. This causes more evaporation in the west and more rainfall in the east. As a result, the Arabian Sea has saltier water, while the Bay of Bengal (BoB) is much fresher. The BoB receives the highest rainfall in the world and has the lowest salinity in the eastern Indian Ocean, especially north of the equator, due to heavy river inflow.

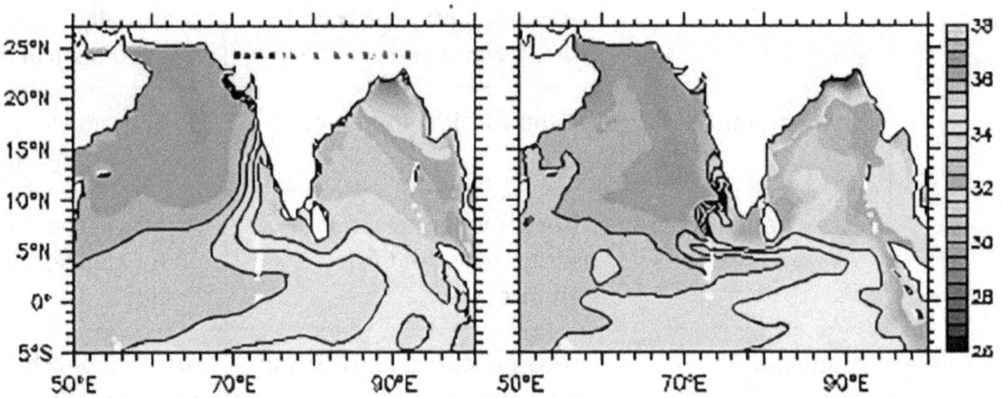

Figure 6.3: Indicative Salinity Profile along the east and west coasts of India

To balance this, freshwater from the BoB moves westward into the Arabian Sea and southward toward the equator, helped by ocean currents. In winter, the Northeast Monsoon Current (NMC) carries fresher water from BoB to the Arabian Sea. From June to August, the Southwest Monsoon Current (SMC) reverses the flow, pushing saltier water from the Arabian Sea into BoB. During transition seasons, strong eastward currents called Wyrtki Jets (WJs), driven by equatorial winds, move salty water across the equator in the upper 80–100 meters of the ocean.

The southern Indian Ocean stays more stable and is less affected by monsoons. On India's west coast, high evaporation and low rainfall make the water saltier, while on the east coast, heavy rains and river runoff during both monsoons keep the water less salty.

6.6.1 Wyrtki Jets in the Indian Ocean

Wyrtki Jets are strong, eastward-flowing equatorial ocean currents in the Indian Ocean that occur during the inter-monsoon periods—typically in April–May and October–November. Named after oceanographer Klaus Wyrtki, these jets are driven by equatorial westerly winds and are confined to the upper 80–100 meters of the ocean.

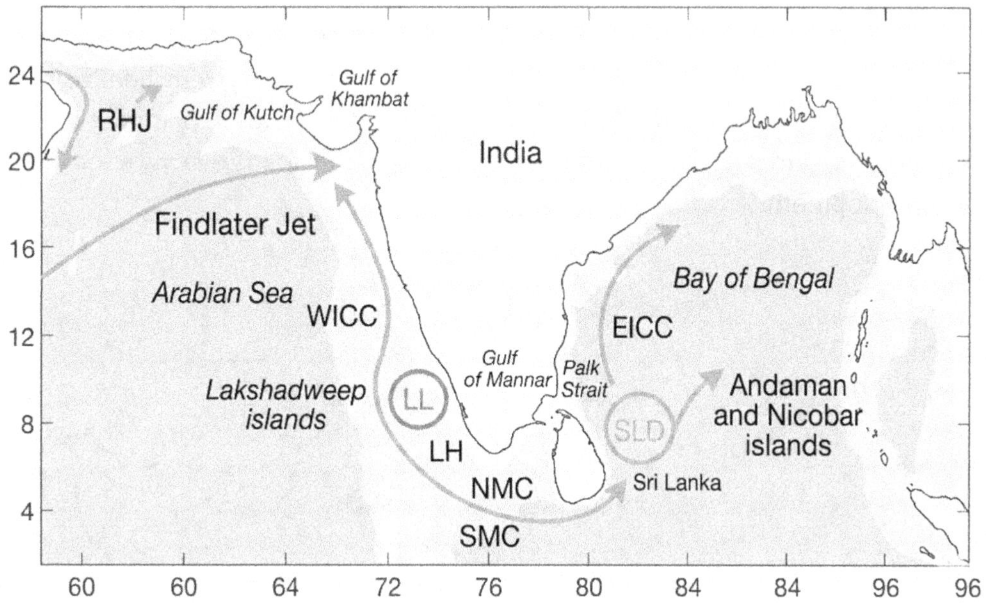

Figure 6.4: Direction of ocean currents (Wyrtki Jets in the Indian Ocean)

Key Characteristics:

- Direction - Eastward along the equator
- Speed - Up to 1–1.5 m/s
- Depth - Surface layer (top 80–100 m)
- Duration - A few weeks during transition seasons
- Seasonality - Twice annually (pre- and post-monsoon)

Significance:

i) Redistributes low-salinity water from the Bay of Bengal toward the Arabian Sea.

ii) Influences sea surface temperatures and regional climate.

iii) Affects ocean-atmosphere interactions linked to monsoon strength and timing.

iv) Alters nutrient flow, affecting marine productivity and fisheries.

6.7 Tidal Dynamics in Coastal Planning

Tides along the Indian coastline vary greatly from place to place, and this variation plays a key role in port planning and coastal dynamics. For example, in the Gulf of Khambhat near Bhavnagar, the tidal range can reach up to 8.5 meters, one of the highest in India. Such strong tides stir up large volumes of sediment, which can lead to harbour siltation, requiring frequent dredging to keep shipping channels clear. In contrast, places like Kanyakumari have a much smaller tidal range—around 0.5 meters—which means less sediment movement and simpler port maintenance. Further north, areas like the Hooghly River estuary and Haldia experience tidal ranges over 6 meters, which significantly affect estuarine flow, sediment transport, and navigation safety.

For port planners, understanding tidal dynamics is crucial for designing berths, docks, and access channels that remain functional during both high and low tides. Tidal patterns also influence coastal erosion, flooding risks, and ecosystem health, especially in regions with mangroves and mudflats. Accurate tidal data helps in planning coastal defences, flood barriers, and shoreline protection structures, making it a key factor in sustainable coastal development.

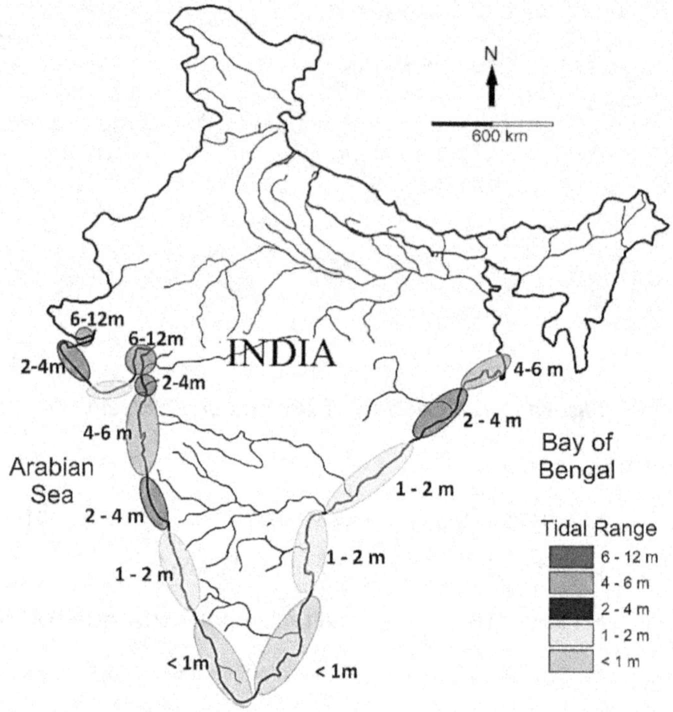

Figure 6.5: Indicative Tidal Map of India

6.8 Current Conditions Along the Coast of India

Sea currents along the Indian coast can be very different from place to place and change a lot with season and weather. They are mainly affected by monsoon winds, tides, coastal shape, river flow, and cyclones. Because of this, the West Coast, East Coast, and Gujarat Coast each behave differently, and ports along these coasts need to be planned with these differences in mind.

West Coast of India: Along the West Coast of India, strong currents mainly occur during the south-west monsoon. These currents are driven by steady monsoon winds and usually flow along the coast. They are not sudden or short-lived but continue for several weeks or even months. Even though the current speeds are moderate to high, their long duration makes them important for port planning. They affect ship movement, navigation channel alignment, breakwater stability, dredging works, and how long port operations may be restricted during the monsoon season.

East Coast of India: On the East Coast of India, extreme currents behave very differently. Here, the strongest currents usually happen during cyclones and storm surges, especially in the post-monsoon season. These currents are short-lived but very strong and can quickly change direction. Even though they may last only a few hours or days, they are critical for design because they can cause heavy loads on structures, scouring at harbour entrances, and sudden problems for moored vessels. For East Coast ports, these currents are treated more as emergency or survival conditions rather than normal seasonal behaviour.

Gujarat Coast: The Gujarat Coast, although part of the west coast, needs special attention. This region experiences very strong tidal currents due to its large tidal range. Currents here reverse direction between ebb and flood tides and can become extremely strong during spring tides, especially in channels, creeks, and near port entrances. In many Gujarat ports, tidal currents can be more important than monsoon-driven currents when it comes to design. Navigation safety, mooring forces, dredging stability, and scour protection are key concerns in this region.

Table 6.3: Design Perspective of Extreme Current Conditions

Aspect	**West Coast (Arabian Sea – General)**	**East Coast (Bay of Bengal)**	**Gujarat Coast (Northern Arabian Sea)**
Dominant extreme current driver	South-west monsoon winds combined with tides	Cyclones, storm surges, and river discharge	Strong astronomical tides combined with south-west monsoon winds
Typical extreme current speeds	~1.0 – 1.5 m/s	~1.5 – 2.0 m/s (event-based)	~1.5 – 3.2 m/s locally in channels and creeks
Duration of extreme currents	Long-lasting, persistent (weeks to months)	Short-duration but intense (hours to days)	Persistent tidal reversals, peak extremes during spring tides and monsoon
Directional stability	Relatively consistent, mainly alongshore	Highly variable and rapidly changing	Strong bidirectional (ebb–flood), with directional dominance controlled by tidal phase
Interaction with tides	Moderate	High during storm surge	Very high; tides dominate the overall current regime
Design control mechanism	Seasonal extreme regime (monsoon-controlled)	Event-based extreme regime (cyclone-controlled)	Tidally controlled extreme regime with superimposed monsoon effects
Key port planning sensitivity	Channel alignment, entrance orientation,	Structural safety, scour protection,	Navigation safety, berth orientation, mooring loads,

	and seasonal operability	and emergency operability	dredging stability, scour
Design-critical periods	June – September (peak in July–August)	October – December (cyclone season)	Spring tides throughout the year, intensified during June – September

Overall, when planning ports along the Indian coast, it is important to clearly separate normal operating currents from extreme currents that control design safety. Just as important, currents should always be considered together with waves and tides, because strong currents can make waves steeper and more dangerous, increasing loads on ships and structures. Treating each coastal region according to its dominant current behaviour helps ensure ports are safe, efficient, and resilient over the long term.

6.8.1 Tidal Influence on Coastal Currents

In open coastal waters, the speed of these currents is usually around 1.4 m/s. However, inside narrow and semi-enclosed regions like the Gulf of Khambhat, the tidal water gets squeezed into a smaller space. Because of this, the water flows much faster, and current speeds can go up to 3.2 m/s.

These strong tidal currents can have a big impact on the coastline. They affect how sand and silt are carried, where erosion happens, and where sediment gets deposited. This is especially important near ports, harbours, and river mouths, where strong currents can lead to navigation issues and the need for regular dredging.

6.8.1.1 Effects of Tidal Currents on Ports

Strong tidal currents can have a big impact on how ports operate and how they are designed. In areas like river mouths or gulfs, where tides move large amounts of water in and out every day, ports have to deal with fast-moving currents. One of the biggest effects is sedimentation and siltation. Tidal currents carry sand and silt along with them. When the water slows down near port entrances or inside harbour basins, this sediment settles on the seabed. Over time, this can reduce water depth, making it difficult for ships to enter or leave the port. As a result, regular dredging is often needed to keep navigation channels open.

Strong currents also make ship movement and berthing more difficult. When currents are fast, especially during peak tides, it becomes harder for ships to line up properly while entering the harbour or docking at the berth. This increases the risk of vessels drifting off course, touching structures, or colliding with other ships. Port operations often have to be planned carefully around tidal timings to ensure safe movement of vessels. In addition, tidal currents can cause erosion around port structures such as quay walls, piles, breakwaters, and training walls. Fast-flowing water can scour the seabed around foundations, weakening structural support if not properly designed. This means ports in high-current areas need stronger foundations and protective measures to prevent long-term damage.

Maintenance is another concern. Strong tides increase wear and tear on mooring systems, fenders, and navigation aids such as buoys and markers. These components may need more frequent inspection, repair, or replacement, which increases maintenance costs.

Because of all these factors, ports located in areas with strong tidal currents must be carefully planned and designed. Engineers consider tidal flow direction and speed when deciding port layout, berth orientation, entrance alignment, and dredging strategies. Operational rules, such as restricting ship movement during peak tidal flow, are also commonly used to improve safety.

6.9 Sediment transport along the Indian coast

Sediment transport along the Indian coast is notably dynamic and varies by region. In areas such as southern Kerala and Odisha, annual sediment movement is estimated at approximately 1 million cubic meters. The East Coast predominantly experiences northward sediment transport, whereas the West Coast generally sees sediment moving southward. A notable exception to this pattern occurs along the southern Gujarat coast, where localised variations alter the general transport direction.

The Ganga-Brahmaputra river system contributes significantly to sediment influx, discharging an estimated 1 billion tonnes of sediment into the Bay of Bengal annually. The total suspended sediment load entering the Bay is projected to be around 1.4 trillion tonnes, underscoring the scale of fluvial contributions to coastal sediment dynamics.

Certain coastal segments, including southern Odisha, northern Tamil Nadu, southern Kerala, northern Karnataka, and southern Gujarat, experience high sediment mobility, with annual transport volumes ranging from 1.5 to 2 million cubic meters. In contrast, regions such as southern Tamil Nadu and Maharashtra exhibit relatively lower sediment movement, averaging around 0.5 million cubic meters per year.

Overall, the sediment transport regime along India's coastline is governed by a consistent directional pattern — northward along the east coast and southward along the west coast — with localised deviations influenced by geomorphological and hydrodynamic factors.

Annual Sediment Movement Estimates

- Southern Kerala and Odisha continue to see ~1 million cubic meters of sediment movement annually.
- High-mobility zones such as south Odisha, north Tamil Nadu, south Kerala, north Karnataka, and south Gujarat report 1.5 to 2 million cubic meters of sediment transport per year.
- Lower-mobility regions like southern Tamil Nadu and Maharashtra experience ~0.5 million cubic meters annually.

Directional Patterns

- On the East Coast, sediment predominantly moves northward.
- On the west coast, the general trend is southward, with localised deviations such as those observed along the south Gujarat coast.

Table 6.4: Regional Sediment Transport

Region	Direction	Volume (Million m^3/year)	Notes
South Kerala	Southward	~1.0	Moderate movement due to wave action
Odisha	Northward	~2.0	High sediment influx from rivers
North Tamil Nadu	Northward	~2.0	Strong longshore currents
Maharashtra	Southward	~0.5	Relatively low mobility
Andhra Pradesh	Northward	~1.8	Medium-energy coast, increasing transport
South Gujarat	Southward	~2.0	Exception to the West Coast trend

6.10 Effect of Sea Level Changes

6.10.1 Sea Level Changes and Coastal Evolution

Over time, the coastal zone has gone through cycles of flooding and exposure due to changes in sea level, especially during the late Quaternary period. Sea levels dropped during ice ages (glacial periods) and rose during warmer times (interglacial periods). During the Last Glacial Maximum, sea levels along India's coast were over 120 meters lower than today. As the climate warmed in the early Holocene, sea levels rose quickly, flooding low-lying areas and pushing inland through rivers, forming estuaries.

Evidence of past higher sea levels—like raised marine terraces, beach rocks, oyster beds, corals, and tidal deposits—has been found along parts of the Saurashtra, Konkan, Malabar, Coromandel, and Andhra coasts. For example, marine shells and coral reefs have been spotted 1–2 meters above the current high tide line in places like Rameswaram and Kanyakumari. Beach rocks in Manori (Mumbai) and Mirya Bhatti (Ratnagiri) are even higher—about 3–6 meters above today's sea level.

Along the Indian coastline, long-term measurements show that sea level has been rising at an average rate of about 1.3 mm per year during the 20th century. Although this increase looks small each year, it adds up over time and leads to a noticeable rise in sea level. Climate studies suggest that if global warming continues, sea level along the Indian coast could rise by about 450–500 mm by the end of this century, with an average estimate of around 480 mm above the present level.

Rising sea levels can strongly affect India's coastline, especially when combined with coastal processes like waves, tides, storm surges, and the movement of sand along the shore (littoral drift). When the average sea level goes up, waves and tides start acting from a higher level, which can cause:

- More frequent flooding of coastal areas,
- Faster erosion of beaches and shorelines, forcing them to move inland,
- Higher risk to low-lying coastal regions, river deltas, and estuaries,
- Increased entry of salty seawater into freshwater sources and rivers near the coast.

Sea level rise also changes the shape and depth of the nearshore seabed. This affects how waves behave and how sand moves along the coast. Even if the main direction of sand movement along the shore does not change, deeper water and stronger waves reaching the coast can lead to more erosion, faster sand movement, and more frequent dredging near ports, harbours, and shipping channels.

For coastal and marine structures, these changes mean that future conditions must be considered during design. Water levels used for design—such as mean sea level, high tide levels, storm surge levels, and wave run-up heights—should include the expected sea level rise over the planned life of the structure. If this is not done, structures may suffer damage sooner than expected, lose their function, or face higher repair and maintenance costs during extreme weather events.

Therefore, sea level rise due to global warming should be treated as an important design factor when planning, building, and maintaining coastal infrastructure in India. Design measures such as higher structure levels, extra safety margins, stronger shoreline protection, and flexible maintenance plans are needed to ensure long-term safety, performance, and durability under changing climate conditions.

6.11 Coastal Landforms of India

India's coastline looks different depending on where you are — and it's split into three main parts: the west coast, the east coast, and the southern tip called Cape Comorin.

The west coast is mostly made up of land that's sunk a bit, forming narrow strips of land between the Arabian Sea and the Western Ghats. This makes it great for natural ports like Kandla, Mumbai's JLN Port, Marmagao, Mangalore, and Cochin. Rivers here are short and fast, so they don't form big deltas like on the East Coast. The sea floor (continental shelf) is wider here, especially near Mumbai, and the amount of sediment (sand and mud) is moderate, usually not more than 3 km thick. The west coast also has lots of estuaries and backwaters, shaped by rivers like the Narmada, Tapi, Mandovi, and Periyar. These areas form five major sediment zones: Kachchh, Saurashtra, Mumbai Offshore, Konkan, and Kerala.

From Gujarat down to Kerala, the west coast is divided into:

- Kachchh and Kathiawar coast (Gujarat)
- Konkan coast (Maharashtra and Goa)
- Malabar coast (Karnataka and Kerala)

The land is narrow in the middle and gets wider toward the north and south.

Now, the East Coast is quite different. It has long sandy beaches with dunes and big river deltas. Rivers like the Ganga, Mahanadi, Godavari, Krishna, and Kaveri flow here and create huge delta areas. These rivers also form four major sediment zones: Bengal, Mahanadi, Krishna-Godavari, and Kaveri, with sediment layers going over 5 km deep in some places.

The East Coast is divided into:

- Sundarbans Delta (West Bengal)
- Utkal Plains (West Bengal and Odisha)
- Northern Circars (Odisha and Andhra Pradesh)
- Coromandel Coast (Tamil Nadu)

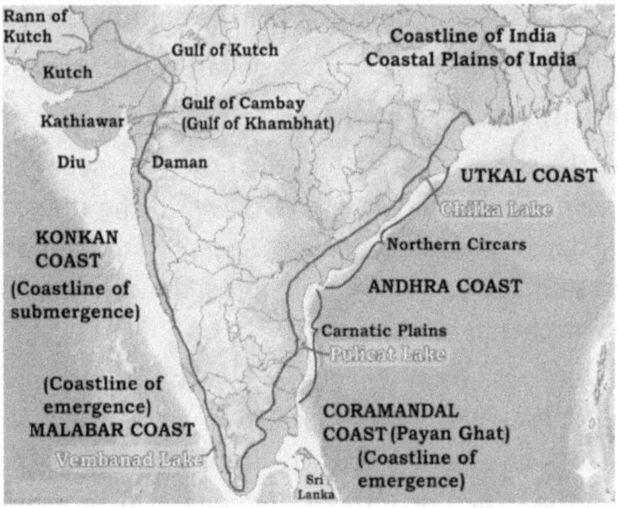

Figure 6.6: Coastal Divisions of India based on Geomorphological Features

And finally, Cape Comorin is at the very bottom of India. The various coastal divisions of India are shown in Figure 6.6.

6.12 The Gujarat Coast

Gujarat has the longest coastline in India—about 1600 km—and it's packed with all kinds of coastal features thanks to its diverse landscape and natural processes. The wetlands here vary a lot depending on the climate, soil, and terrain.

There are two major gulfs along the coast: the Gulf of Kachchh and the Gulf of Khambhat. The Gulf of Khambhat has seven estuaries that bring in loads of water and sediments, while the Gulf of Kachchh gets very little from rivers. On the far western edge near Kori Creek, the land is part of the lower Indus delta. It's full of tidal creeks and small channels, and it supports a big area of mangroves growing on sandy and silty ground.

6.12.1 Gulf of Kachchh

The Gulf of Kachchh is a shallow stretch of water surrounded by dry to semi-dry land. Most of the southern coast and the Rann of Kachchh are covered with loose sediments from the Quaternary period, with older rock layers mixed in. Rivers here barely flow into the Gulf, and most have dams, so the Gulf doesn't get much freshwater except during the short monsoon season from July to September—making it a region with more water loss than gain.

North Gujarat, especially Kachchh and Saurashtra, often gets hit by cyclones. These storms bring huge tidal waves that flood the coast temporarily and can damage structures near the shore. There's even geological proof of past extreme wave events. The Kachchh coastline, which faces the Arabian Sea to the west and the Gulf to the south, has landforms shaped by tides and waves. In the west, you'll find beaches, dunes, and swampy areas, while the east has wide mudflats and low-lying areas.

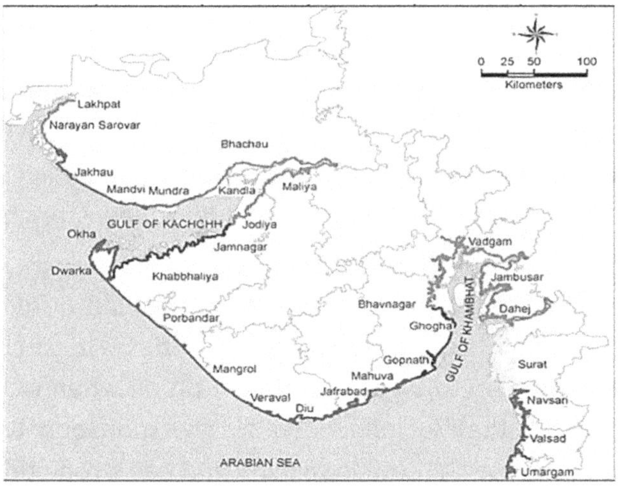

Figure 6.7: Coastline of Gujarat

These features sit about 2–3 meters above the high tide line and are bordered by dune ridges or ridge-runnel systems, depending on the area. These natural barriers help block incoming waves and trap sediments during extreme wave events, preserving them in the landscape.

6.12.2 Gulf of Khambhat

The Gulf of Khambhat stretches from south to north, cutting into the western coast of India between the Saurashtra peninsula and mainland Gujarat. The Saurashtra coast is fairly straight, but from Gopnath Point to Diu, you'll find cliffs, mudflats, beaches, estuaries, and creeks. Past Diu, the coast features sandy beaches, spits, bars, marshes, and small estuaries. The area around the Gulf is shaped by estuaries and has lots of mudflats, dunes, and sandy patches—all thanks to the Gulf's high tidal range. South Gujarat's coastline is more uniform, with only a few breaks and narrow sandy beaches near river mouths.

At the northern tip, between the mouths of the Sabarmati and Mahi rivers, the Gulf narrows to just 5 km between Ghogha and Dahej, then widens like a funnel as you move south, reaching its widest point near Gopnath. It's about 115 km long and covers around 3,120 square kilometers, mostly made up of mudflats with some rocky areas. Rocky beaches are common from Mahuva to Gopnath but become less frequent near Ghogha and Bhavnagar. There are also a few sandy spots here and there. The Gulf is fed by several rivers—including the Narmada, Tapi, Mahi, Sabarmati, and Shetrunji—which form estuaries and bring in a lot of sediment. Near Shetrunji, between Gopnath and Ghogha, there's even a small delta. The Gulf's ecosystem is rich, with mangroves, estuaries, creeks, and vast mudflats that support a wide variety of plants and animals.

6.12.3 Tidal and Current Activities

Tides: The tides along the Gujarat coast are a mix of semi-diurnal types, meaning there are two high and two low tides a day, but they're not all equal in size. Because Gujarat is close to the Tropic of Cancer, it gets some of the highest tides in India. The funnel-like shape of the coast, especially in places like the Gulf of Khambhat, causes the tide to rise sharply as it moves inland. In the Gulf of Kachchh, the tide enters from the west and gets stronger as it moves upstream due to the narrowing shape and natural resonance.

In the Gulf of Khambhat, spring tides average around 4.7 meters at Mahuva Bandar, go up to 6.5 meters at Gopnath Point, and reach a massive 10.2 meters at Bhavnagar. The highest recorded tide there is 12.5 meters, second only to the Bay of Fundy in Canada, which hits around 17 meters. These high tides create huge intertidal zones, stretching from 1.5 to 5 km wide, making them the largest along India's coast.

On the open Arabian Sea coast, longshore currents dominate, but inside the gulfs, strong tidal currents take over, especially during flood and ebb tides. These currents can reach speeds of 3 to 4 knots, and even up to 6 knots during mid-tide when wave energy is high. Though the currents are tidal, they're influenced by the monsoon winds and change direction with the seasons. Because the Gulf is shallow and has strong tides, water doesn't stay long—it moves in and out quickly.

In the Gulf of Kachchh, tide heights vary: from 3 meters at Kori Creek to 7.3 meters at Navlakhi, then drop to 3.5 meters at Okha. Other places like Kandla Harbour and Dahej Bandar also see high tides. Moderate tide zones include Mandvi, Sikka, and Valsad, while places like Porbandar and Veraval have lower tides. Miyani has the lowest, around 0.98 meters.

During high tide, areas like the Little Rann of Kachchh get flooded, but they dry out completely at low tide, leaving behind tidal pools in shallow spots. As you move upstream, the intertidal zone gets wider—from 0.5 to 2 km on the northern coast and from 1 to over 5 km on the southern coast. Storms and high tidal waves hit coral reefs on the northern and western sides, which helps reduce their energy before they reach the shores of Saurashtra.

Current

The water movement in the Gulf is mostly driven by tides and the shape of the sea floor (bathymetry), though wind plays a smaller role too. The strongest currents usually happen around mid-tide—about 2 to 3 hours before and after high or low tide. During spring tides, the currents are about 60–65% stronger than during neap tides. Surface currents are generally moderate, ranging from 0.7 to 1.2 meters per second, but they can speed up to 2.0–2.5 meters per second in the central part of the Gulf. Bottom currents also get strong at times and tend to follow the shape of the sea floor.

Tidal currents here are quite powerful and flow in two main directions—upstream during rising tide (flood) and downstream during falling tide (ebb)—creating a back-and-forth motion. This kind of movement doesn't fully flush out the Gulf, so some areas always have swirling water. In the middle part of the Gulf, especially between Vadinar and Bedi, the water moves in an elliptical pattern during each tidal cycle, with the size of the loop ranging from 5 to 20 km depending on the tide and location. Near the shore, the flow pattern changes a lot because of the presence of creeks and small islands.

6.12.4 Geomorphology of Gujarat Coast

The eastern part of Saurashtra's coastline stretches up to Bhavnagar, and beyond that, it becomes part of the Gulf of Khambhat. This stretch of coast is quite diverse, with estuaries, creeks, islands, mudflats, and salt marshes. Thanks to the high tidal range, it has the largest intertidal zone in India. Because of the Gulf's funnel-like shape, the northern mudflats can stretch up to 5 km inland, while the southern ones are mostly found on the eastern side. Altogether, the mudflats cover around 2,588 square kilometers.

Between Bhavnagar and Khambhat lies a salty lowland area called the 'Bhal' region, which used to be an ancient mudflat. From Gopnath to Ghogha, the coast is rocky with narrow, non-calcareous sandy beaches. But from Ghogha to the mouth of the Sabarmati River, the coastline becomes muddy, with widespread mudflats and mudbanks. The stretch from Khambhat to Bharuch is made up of river deposits and features steep, cliff-like river mouths

6.13 Konkan Coast Overview

The Konkan coast stretches across Maharashtra, Goa, Karnataka, and Kerala, each with its own unique coastal features.

6.13.1 Maharashtra Coast

This coastline is known for small, crescent-shaped beaches tucked between rocky cliffs made of Deccan basalt. In the southern part, you'll find estuaries and mangrove patches. Popular spots like Harihareshwar and Murud Janjira have wave-cut platforms, while Ratnagiri has sea caves. Mudflats are common near creeks and rivers like Thane, Panvel, Vasai, and Kundalika. Coastal dunes with Casuarina trees are found near Malvan and Devgarh, and the Malvan Marine Sanctuary is rich in corals, seaweeds, and mangroves.

6.13.2 Goa Coast

Goa's northern coast has long, wide beaches, while the southern part is more rocky. Sandy beaches and dunes are found near Cape Rama, Vasco Da Gama, and Calangute. Mudflats appear around the Mandovi estuary and Marmugoa Bay. The coastline is a mix of bays and headlands, with major estuaries from the Mandovi and Zuari rivers. Famous beaches like Baga, Calangute, and Colva are known for their white sands and palm trees.

6.13.3 Karnataka Coast

Karnataka's coast is about 300 km long and mostly narrow and straight, except near estuaries. The northern part (Uttara Kannada) is rocky, with high cliffs that stick out into the sea. Waves hit these cliffs and help form nearby beaches. The southern part (Dakshina Kannada) has long, straight beaches. Estuaries from rivers like Netravati, Kalinadi, and Sharavati are prominent, with Kundapur Estuary being the largest. Sandbars are common at estuary mouths. Notable features include wave-cut platforms at Dona Paula, cliffs at Someshwar, columnar basalt at St. Mary's Island, and crescent beaches near Gokarna. Beaches like Surathkal, Panambur, and Udupi are formed from sediments mostly coming from Deccan basalt.

6.13.4 Kerala Coast

Kerala's coastline is about 590 km long and mostly straight, though some promontories break the line. Sandy beaches are found at places like Chittari, Kappad, Ponnani, Calicut, Cochin, Alleppey, and Kovalam. These beaches have mixed sand and broken shells. Crescent-shaped beaches appear at Ezhimala, Dharmadom, Tellicherry, Kadur Point, and Ealthur. Rocky coasts made of laterite or ancient crystalline rocks like Khondalites and Charnockites are seen near Bekal, Ezhimala, Azhikode, and Kadalur Point in the north, and Vizhinjam, Varkala, and Tangasseri in the south.

6.14 East Coast of India

The east coast of India Lies between Sagar Island, West Bengal and the Eastern Ghats. It extends from the Sundarbans Delta to Kanyakumari and it is marked by deltas of rivers like the Ganga, the Mahanadi, the Godavari, the Krishna, and the Cauveri and has varying geomorphological features. Based on geomorphological features, the east coast may be divided into the following divisions;

6.14.1 Sundarbans, West Bengal

Deltas form where rivers meet the sea, shaped by both river and ocean forces. Their size and shape depend on things like geology, climate, and location. Most of India's big deltas are on the eastern coast along the Bay of Bengal.

The Sundarbans is a tidal delta created by the Ganga and Brahmaputra rivers. It's full of tidal creeks, inlets, and islands, and is known for its swampy terrain and dense mangrove forests. You'll find mudflats, swamps, dunes, estuaries, creeks, beaches, and islands here. The winding tidal channels between the islands are fed by the Bay of Bengal's backwaters.

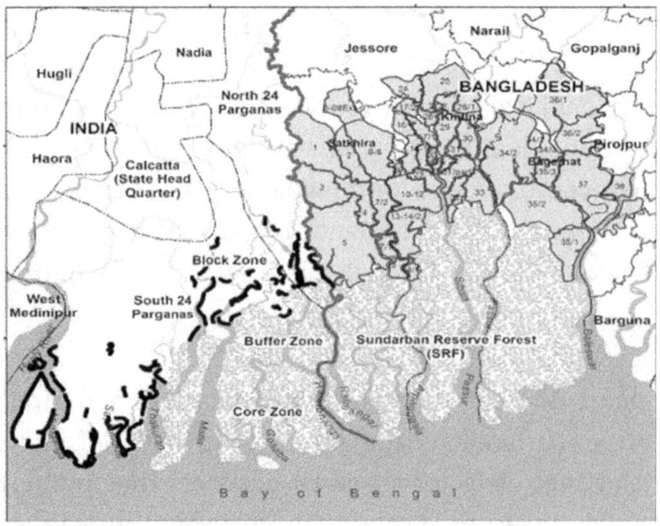

There are a few sandy beaches, but most of the area—especially around the Hooghly estuary and Sagar Island—is covered in muddy shores. Rivers like Bartala, Muri Ganga, Saptamukhi, Thakuran Matta, Ichhamati, Kalinadi, Harinbari, Bidya, and Gosaba bring freshwater from the Himalayas and help form islands at the river mouths.

Figure 6.8: Coastline of Sundarbans Delta

Further south near DIgha, the coast has a dune belt inland and sloping sandy areas near the sea. The sediments in the intertidal zone vary a lot in size. This region also faces cyclones every monsoon season, with wind speeds reaching 45–50 knots.

Tides: Near Haldia Dock, the main tidal range varies between 5.70 m (MHWS) and 1.41 (MLWS). The high tidal difference produces a large extent of tidal flats, and it has two distinct morphological features,

(a) a sandy sloping near shoreward zone with an average width of 30 m, and

(b) a wide, silty flat matted with ripples, having an average width of 1.5 km.

Towards the south near Digha, the main tidal range varies between 4.89 m and 1.87.

6.15 Coastline along Odisha

Odisha's coastline stretches about 480 km and features long sandy beaches and muddy shores. The coastal plain here is known as the "hexadeltaic region" because it's shaped by six rivers—Subarnarekha, Budhabalanga, Baitarani, Brahmani, Mahanadi, and Rushikulya. Along the Mahanadi Delta, several old shoreline marks (strandlines) have been identified.

The coast is mostly built up by deposits from the Mahanadi and Brahmani-Baitarani rivers. Fine soil and silt dominate areas like Balasore, the Devi estuary, and around Chilika Lake. Between the Dhamra River and Chandipur, there's a wide stretch of subtidal mudflats. Sandy beaches run from the mouth of the Dhamra River to the Mahanadi, and from the Devi River to the Bahuda estuary near Andhra Pradesh. The beach near Gopalpur is rich in rare minerals like monazite, ilmenite, zircon, rutile, and sillimanite, while Puri's beach is a popular tourist spot.

Odisha's muddy coast contains more sand and low-plasticity silt. Well-developed spits are found near major estuaries. Near the mouth of the Mahanadi River, mangroves grow along a network of creeks—Luna, Jambu, Kharnasi, Khola, and Batighar Jora—which mostly run parallel to the shoreline.

6.15.1 Mahanadi River and its Distributaries

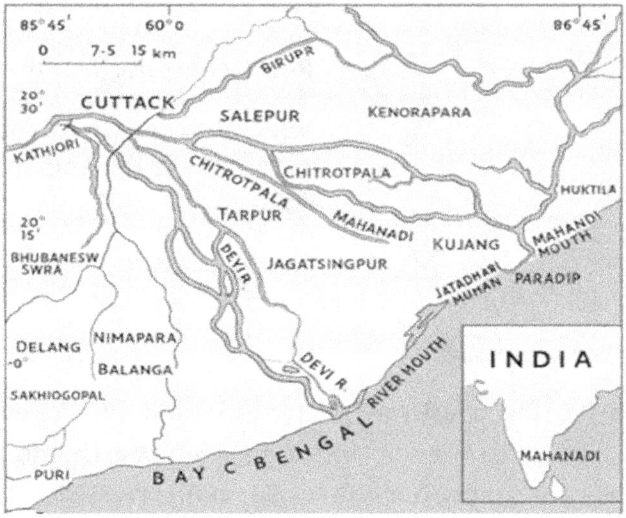

Figure 6.9: Mahanadi River and its Distributaries

The coastal basins here have thick layers of sediment—between 1500 and 2500 meters—based on deep seismic surveys, which show that the land has been sinking over time due to active faults. During the Holocene period, the shoreline moved outward as the delta kept building up.

The Mahanadi delta coast is microtidal, meaning it has a small tidal range—about 1.29 meters—and follows a semi-diurnal cycle (two high and two low tides daily).

During the southwest monsoon, this coast is mainly shaped by waves, while in the rest of the year, it's influenced by both waves and tides. The northeast monsoon (December to early January) is much calmer. Southwest monsoon winds create strong waves—sometimes over 3 meters high—that hit the shore at an angle, pushing sand north-eastward. This longshore drift moves around 1.5 million cubic meters of sand every year.

During monsoon floods, fine sediments carried by rivers settle in floodplains, grassy swamps, mangrove areas, tidal flats, and even deeper offshore zones. In the dry season, leftover estuarine currents help spread these fine sediments around the delta and estuary areas.

Over time, the delta has grown outward thanks to a steady supply of sediments from inland areas, supported by good rainfall, sea level changes, and land subsidence. Whether the delta continues to grow or starts to shrink depends mainly on how much sediment keeps reaching the shore.

6.16 Andhra Coast

The Andhra coast stretches about 300 km and is mostly shaped by the Godavari-Krishna delta system, along with long sandy beaches. In the northern part, especially between Revupolavaram and Visakhapatnam, you'll find rocky sea cliffs, shore platforms, and sea caves. The region's geology includes rocks like Khondalites and garnet-rich gneisses.

Figure 6.10: Coastline of Andhra

The coastline is mostly smooth, with some flooding in the far south near Pulicat Lake and between the Godavari and Krishna deltas. Based on terrain and vegetation, the coast is divided into three zones:

- **Rocky coast** north of the Godavari delta
- **Vegetated coast** around the Godavari and Krishna deltas and the area between them
- **Sandy coast** south of the Krishna delta down to Pulicat Lake

North of the Godavari delta, the coast has tidal creeks and small wetlands near river mouths. Hills and ridges lie close to the sea, and some even stretch into the water. Rocky outcrops and small beaches are common in the Visakhapatnam district. Though rivers like Bahuda, Vamsadhara, Gosthani, and Sarda flow here, they don't form major deltas. Between the Godavari and Krishna deltas, the coast features bays, tidal creeks, mudflats, spits, mangrove swamps, and marshes. East of Coringa, there are large intertidal mudflats dissected by tidal creeks, with dense mangrove forests on both sides.

South of the Krishna delta, sandy beaches, dunes, and dune sands dominate the landscape. These stretch from Suryalanka to Tutipalem, Kuruvatipalem, Durgarajupatanam, and Shriharikota Island. Wide sandy plains are found in areas like Chirala, Vetapalem, and Chinagunjam, with sand dunes at Vetapalem, Krishnapatanam, Kotapalem, and Gopallapalem.

Further south, the Penner Delta and Pulicat Lake—India's second-largest coastal lagoon after Chilika—are key features. This region also hosts several major marine ports.

6.16.1 Visakhapatnam Area

The coastline around Visakhapatnam is rocky, with several headlands sticking out into the sea, creating a mix of bays and cliffs. You'll find sea cliffs, wave-cut platforms, and sea stacks along the shore. Coastal red sands are also spread across many spots in this region.

6.16.2 Kakinada Area

This stretch of coast runs about 105 km, from just north of the Vainateyam estuary to north of Kakinada town. Kakinada is the biggest town here and has a major seaport. The coast is mostly covered with mudflats, which are used for salt production and aquaculture. Old beach ridges—remnants of former shorelines—cover a large part of the area.

6.16.3 Nizampatnam Area

This part of the coast features a mix of mudflats, beach ridges, backwaters, and mangroves. The Krishna delta spreads out here with its four distributaries. Several smaller rivers like Romperu, Musi, Gundlakamma, Palleru, and Manneru also drain into this area, especially south of the Krishna delta.

6.16.4 Krishnapatnam Area

The Krishnapatnam Port area sits on a gently sloping coastal plain with a mix of sandy and silty soils. The region is geologically part of the Eastern Ghats mobile belt, with underlying rocks like khondalites and charnockites in the inland areas, though the immediate coastal zone is mostly made up of recent alluvial and marine deposits. The shoreline is relatively straight and low-lying, with some tidal creeks and backwaters nearby. The port area has been shaped by both natural coastal processes and extensive human intervention, including dredging, land reclamation, and shoreline protection structures. Littoral drift (longshore sand movement) is active here, and the port has implemented breakwaters and shore protection measures to manage sediment transport and prevent erosion

6.17 Coast along Tamil Nadu

Tamil Nadu's coastline stretches about 950 km and is shaped by around 46 rivers that flow into the sea, bringing loads of sediment that influence coastal changes. The type of mud varies—it's sandy in places like Pichavaram and Nagapattinam, and more clay-like in areas like Pattikudi, Pamban, Vedaranyam, Point Calimere, and Muttupet. Sandy beaches are found in spots like Ennore, south of Chennai, Covelong to Mahabalipuram, Manakkanam to North Pondicherry, South Cuddalore, and near Kollidam and Rameshwaram. Some beaches like Vattakatti and Muttam have black sand mixed with red patches. A large spit is seen at Pulicat Lake's mouth, with smaller ones near Mahabalipuram, Kollidam, and Rajmatam.

Coral reefs are found near Rameshwaram in Palk Bay and in the Gulf of Mannar. The Gulf of Mannar is rich in marine life, with coral reefs, seagrass beds, and mangroves surrounded by swampy areas. Palk Bay also has vibrant coral reefs.

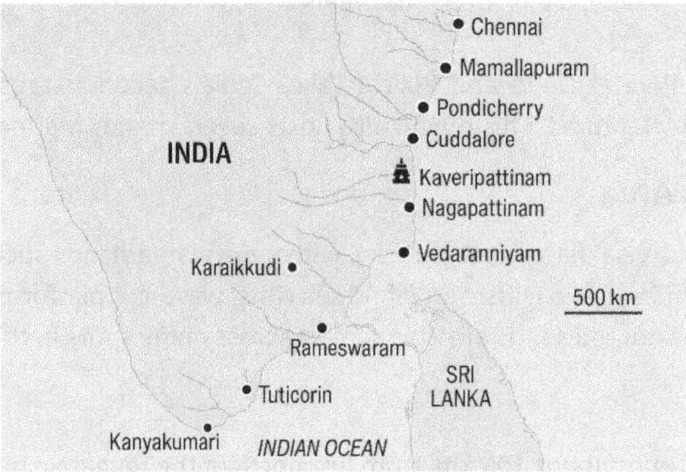

Figure 6.11: Coastline of Tamil Nadu

Notable features along the Tamil Nadu coast include Marina Beach in Chennai, coastal dunes near Mahabalipuram, the Kaveri Delta, and Pulicat Lagoon. Rameswaram and Mandapam have beach rocks, while Mandapam and Tiruchendur show sea caves, wave-cut platforms, and marine terraces.

6.18 Andaman and Nicobar Islands

This island group has over 570 islands and islets, made up of volcanic and sedimentary rocks like ophiolites, sandstones, tuffs, and limestones from the late Cretaceous to Pleistocene periods. Narcondam and Barren Island are two volcanic islands in the Andaman Sea. The islands feature sandy beaches, mangrove-lined creeks, tidal inlets, bays, fringing coral reefs, and rocky shores. Many of these coastal features—especially beaches, tidal flats, mangroves, and coral reefs—were heavily damaged during the 2004 tsunami.

6.19 Working Knowledge of Geology in Civil Engineering

Before starting any geotechnical investigation—whether at a green-field or brown-field site—it is important to understand the basic geological conditions. A working knowledge of geology helps engineers predict how soil and rock will behave during construction and under long-term loads, and how they will interact with foundations.

This knowledge does not require specialist expertise, but it does require the ability to identify common soil and rock types, understand that ground conditions vary with depth and location, and recognise features such as layering, joints, faults, and weathering. These factors can strongly affect strength, settlement, stability, and groundwater flow.

Geology also plays a key role in understanding groundwater conditions, which is essential for dewatering, controlling uplift and seepage, and ensuring the long-term performance of foundations. Overall, a working knowledge of geology supports effective site investigations, accurate interpretation of ground conditions, and safe, economical geotechnical design.

6.20 Tectonic Framework of the Indian Plate

Now that we've looked at the coastal landforms of India, let's talk about the tectonic setup. The Indian Plate—sometimes called the India Plate—is a smaller tectonic plate in the Eastern Hemisphere that crosses the Equator. It broke off from the ancient supercontinent Gondwana about 100 million years ago and started drifting north. Today, it includes most of South Asia, parts of the Indian Ocean floor, and regions like Ladakh, Kohistan, and Balochistan.

Around 40–50 million years ago, the Indian Plate collided with the Eurasian Plate, which led to the rapid rise of the Himalayas. This was a classic example of a convergent plate boundary. The collision began in the late Middle Eocene and triggered the uplift of the Himalayan mountain belt. Later, in the Late Oligocene, the northeastern edge of the Indian Plate collided at an angle with the Burmese Plate, forming the Assam-Arakan mountain belt. These collisions are still ongoing, with the Indian Plate continuing to slide under both the Eurasian and Burmese Plates. In the northeast, this oblique collision is gradually closing off the remaining ocean basins, like the Bengal Basin and parts of the Bay of Bengal.

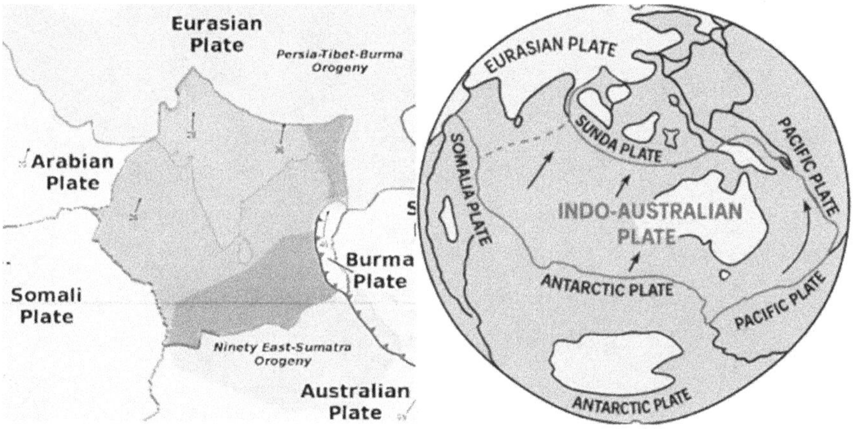

Figure 6.12: Tectonic setup of the Indian Plate

Currently, the Indian Plate is moving northeast at about 5 cm per year, while the Eurasian Plate moves north at around 2 cm per year. This difference causes the Indian Plate to compress against the Eurasian Plate at a rate of about 4 mm per year.

Because of this movement and the stress it creates, different parts of the Indian Ocean's edge have developed two main types of landforms: uplifted and sunken blocks. In the stable shield areas like peninsular India, normal faulting has created rift valleys and block-faulted zones. Outside these shield areas, the crust has sagged to form deep basins that were later filled with sediments.

As the Indian Plate drifted northward with a slight counterclockwise twist, older rift faults were reactivated. This caused sideways (trans-tensional) movements, which can be seen in the broken folds and uplifted areas like the Kutch region.

Modernising Maritime Infrastructure

Deterioration in Concrete Structures: in Maritime Environment

Chapter Summary

RCC structures rely on both concrete and reinforcement steel for strength. During design and construction, maintaining proper concrete quality and ensuring protection of reinforcement steel are crucial. Over time, deterioration is typically caused by physical and chemical factors such as impact, temperature changes, and chemical reactions. In marine environments, durability depends on chloride levels, exposure conditions, construction quality, and protective measures. These factors must be evaluated to ensure long-term structural performance. Under ideal conditions, the rate of deterioration is time-dependent, and once damage to concrete and steel reaches a critical level, the structure's service life is compromised. Routine structural health assessments, along with identifying and documenting damage causes and extent, are essential for enhancing durability and enabling early defect detection.

Learning Objectives

- Understand how concrete deteriorates
- Learn the corrosion process in steel
- Examine the impact of chloride concentration and exposure in marine environments
- Assess the effect of sea proximity on chloride attack
- Identify stages of RCC element degradation

7.1 Introduction

Structural engineering is the science and art of planning, **designing**, and constructing an economical structure that can safely resist all applicable forces. Structures are designed and detailed to sustain the intended loads and actions likely to occur within their life span, and the structure is expected to maintain stability, fulfilling the durability criteria. The principal goal for designing a general engineering structure is to remain stable and fit throughout its useful life.

When installed properly, concrete is assumed to be one of the most durable and long-lasting products, especially in marine environments. Concrete has long been known as a reliable construction material. However, it does not always behave as it is intended to. Some of the undesirable behaviour and signs of deterioration, such as spalling, cracking, delamination, wear and tear, or excessive deflection, can be seen during its design service period, and there are a variety of factors that influence it. Defective materials and inferior workmanship during construction can also lead to deterioration and possible failure of a concrete structure. It is also possible that the original plans and specifications of the structure did not place sufficient emphasis on the effect of unknown loading conditions, which may likely occur. Therefore, the factors responsible for structural behaviour are;

- Design criteria and presumed loads,
- materials properties,
- Workmanship adopted during construction,
- actual loads being applied during service,
- prevailing environment or exposures of structure.

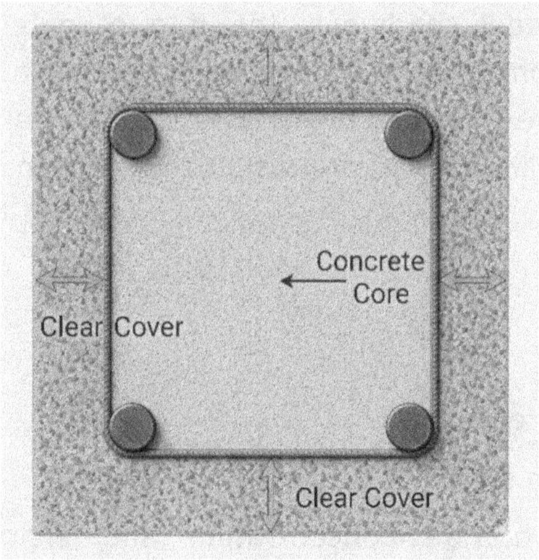

When designing an RCC structure for a specific purpose and load, it is assumed that the overall strength of the structure will be governed by the composite effect of Concrete and steel. This only works well if the concrete is made properly—things like the water-cement mix, how tightly it's packed, how it's cured, how easily water can seep through it, and how much protective cover is given to the steel inside all need to follow standard construction practices.

Figure 7.1: Key Elements of an RCC Structure

In this setup, the **concrete core** gives the structure its strength, while the **concrete cover** around the steel bars protects them from environmental damage like moisture, which can cause rust.

So, the quality and condition of that concrete cover are super important for keeping the structure healthy over time. Any deficiency—whether in the design, materials, construction work, curing process, or maintenance and operating loads can affect how long the structure lasts and how safe it is, which can be explained very clearly by the service life model of RCC structure.

7.1.1 Service Life Model of RCC Structure

The service-life model shown below helps explain how a reinforced concrete structure behaves and gradually deteriorates over time. It shows how the structure slowly loses strength and functional performance due to environmental exposure, loading, and natural ageing. This deterioration happens in clear stages—from the structure being fully healthy at the start, to reaching its serviceability limit, and finally approaching a point where structural failure becomes possible.

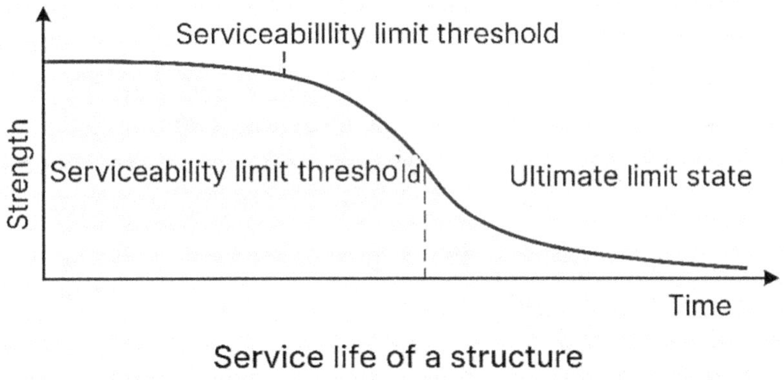

Figure 7.2: Service Life Model of RCC Structure

The main stages include:

- **Initial Performance Phase (Healthy State)**

  At the beginning, the structure works perfectly and shows no signs of distress, even though slow chemical or physical changes may already be starting inside.

- **Initiation Phase of Deterioration**

  This is the early stage where deterioration has begun internally, but nothing is visible yet. The structure still performs normally and looks fine from the outside, but hidden damage is starting to develop.

- **Propagation Phase of Deterioration**

  Once corrosion starts in the reinforcement, the deterioration speeds up. Steel begins to lose its cross-sectional area, and corrosion-driven cracks in the concrete widen and spread. This leads to a faster drop in overall strength.

- **Serviceability Limit State (SLS)**

  At this point, deterioration becomes visible. Cracks, spalling, deflection, and other surface issues start affecting the usefulness and appearance of the structure, although it may still be safe.

- **Ultimate Limit State (ULS)**

  If the earlier warning signs are ignored, the structure enters a rapid decline. Strength reduces significantly, and the structure approaches a point where it may no longer safely carry loads. This is the stage where the risk of failure or collapse becomes real.

Understanding these stages helps in planning inspections, maintenance, and timely repair actions. Intervening before deterioration crosses critical limits can extend the structure's life, reduce repair costs, and maintain safety and performance.

7.1.2 Deterioration–Repair–Extended Service Life Model for Structure

A reinforced concrete structure is designed to handle specific loads and environmental conditions so that it stays strong, stable, and durable throughout its intended service life. However, over time, it may start showing serviceability problems such as cracking, surface damage, or other signs of distress. Even though structures are designed to last, several real-world factors can affect their performance much earlier than expected. These include poor workmanship during construction, overloading or misuse, neglect of maintenance, mechanical damage, and exposure to aggressive environmental conditions. Any of these can trigger early deterioration long before the structure reaches its expected life.

To keep the structure performing as intended, it is important to identify the reason behind this early deterioration and take steps to mitigate it. Repairing, strengthening, or rehabilitating the affected areas becomes necessary to restore the structure's health. A proper maintenance approach starts with a condition survey, identifying the root cause of deterioration, and assessing how severe and widespread the damage is. Based on this assessment, suitable repair or strengthening measures can be planned to restore the structure's strength and durability.

The conceptual model shown below illustrates how early deterioration occurs and how appropriate strengthening measures can help recover the structure's performance.

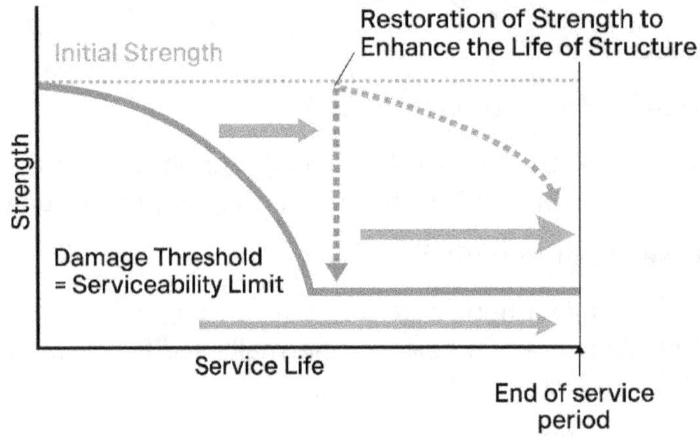

Figure-7.3: Deterioration–Repair–Extended Service Life Model for Structure

Based on the sign of Distress and outcome of the condition survey of the structure, a suitable repair, strengthening, rehabilitation and methodology for upgradation are formulated according to the demand of the structure. For a general overview, these terms are defined below, which are further discussed in detail in chapter -11.

7.2 Assessment of Defects and Obsolescence

Marine and port infrastructure—including quay walls, jetties, wharves, berths, dolphins, and waterfront retaining structures—operates in highly aggressive environments and is subjected to complex structural and operational demands. Over time, these structures are affected by defects, deterioration, and obsolescence, which can compromise safety, serviceability, operational efficiency, and regulatory compliance. Systematic assessment and appropriate strategies for repair, rehabilitation, renovation, and structural or functional upgradation are therefore essential to ensure continued performance and extended service life of port assets.

7.2.1 Defects and Deterioration in Marine Structures

Defects and Deterioration are distinct yet interrelated conditions impacting the structural integrity and durability of marine and port structures.

- **Defects** are generally *visible* and *construction- or workmanship-related*, such as shrinkage and temperature cracks in concrete decks, settlement cracks in quay walls, misalignment of piles or coping beams, honeycombing in marine concrete, and surface abrasion due to berthing and cargo operations.
- **Deterioration** is *time-dependent* and results from *chemical and physical mechanisms* typical of marine environments, including chloride-induced corrosion of reinforcement, sulphate and carbonation attack, abrasion from vessel contact or sediments, and overstressing due to increased operational loads.

Deterioration represents a reduction in structural adequacy relative to original design assumptions and differs from obsolescence. It is primarily addressed through maintenance and repair measures such as removal of deteriorated concrete, protection and reinstatement of corroded reinforcement, improvement of cover systems, and application of durable repair materials suitable for marine exposure.

7.2.2 Obsolescence in Port Infrastructure

Obsolescence refers to reduced functional or structural adequacy of port structures relative to current standards, vessel characteristics, and operational requirements.

Common causes include:

- Increased operational loads and berthing forces exceeding original design capacity,
- Reduced safety margins under present-day vessel sizes and cargo handling practices,
- Inability of older infrastructure to comply with current safety, environmental, and operational standards.

i) Functional Obsolescence

Functional obsolescence occurs when a marine structure can no longer perform its intended operational role, despite remaining structurally standing. Examples include quays unable to support modern quay cranes or decks lacking sufficient width, clearance, or load capacity for contemporary cargo handling equipment. Such structures require repair, upgrading, or modification to remain safely usable.

ii) Technological Obsolescence

Technological obsolescence results from advancements such as larger vessel sizes, increased berthing energies, deeper drafts, higher crane capacities, and modern automation systems. Existing infrastructure may become inadequate unless upgraded. This form of obsolescence is influenced not only by physical constraints but also by economic competitiveness, operational efficiency, and evolving port user expectations.

iii) Alternatives for Reversing Deterioration

For marine and port structures, alternatives for addressing deterioration include:

- Protective coatings, cathodic protection, and surface treatments to limit chloride ingress;
- Localised patch repairs to decks, piles, and quay wall elements;
- Rehabilitation through structural strengthening or partial replacement;
- Renovation involving broader functional and operational improvements to port facilities.

The selection depends on exposure severity, remaining service life, operational disruption, and future port development plans.

7.2.3 Repair, Renovation, and Rehabilitation

i) Repair (Patching)

Repairing is a localised intervention addressing small deteriorated zones such as spalls in quay decks or limited pile damage. Repairs are typically short-term, may need repeated intervention, and are governed by safety and serviceability requirements.

Repair strategy principles emphasise accurate diagnosis of damage causes, concurrent treatment of ongoing deterioration mechanisms, and a systematic, step-by-step approach to ensure structural safety, durability, and extension of service life in aggressive marine environments.

ii) Renovation

Renovation focuses on updating port structures to meet current codes, operational practices, and berth requirements. It balances preservation of existing assets with adaptation to modern needs and extends service life without necessarily increasing original structural capacity.

iii) Rehabilitation

Rehabilitation is more extensive than repair and involves systematic removal of deteriorated or chloride-contaminated concrete, cleaning and protection of corroded reinforcement, and restoration of structural integrity. In port structures, rehabilitation often includes strengthening to accommodate higher berthing forces or heavier equipment and to extend operational life.

iv) Rehabilitation Strategies and Methodology for Marine Structures

Rehabilitation strategies typically combine:

- **Preventive measures**, such as marine-grade coatings, cathodic protection systems, and improved detailing to limit aggressive exposure;
- **Corrective measures**, including structural repairs, strengthening, or retrofitting due to corrosion, increased berthing loads, or revised design criteria.

A successful rehabilitation programme requires a structured methodology comprising:

- Condition assessment via visual inspection and non-destructive testing (e.g., cover meter, half-cell potential, ultrasonic testing);
- Identification of root causes such as chloride ingress, abrasion, or overloading;
- Evaluation of deterioration extent and severity;
- Determination of appropriate intervention level;
- Selection of repair materials compatible with marine exposure;
- Proper preparation, execution, and curing of repair works to ensure durability.

v) Upgradation Perspective in Ports

Upgradation strategies for port infrastructure range from minor repairs to full reconstruction, often classified as *very soft*, *soft*, and *hard* interventions. The chosen approach depends on structural condition, foundation capacity, operational downtime constraints, remaining service life, and future port development plans.

7.2.4 Structural and Functional Upgradation of Marine Structures

i) Structural Upgradation

Structural upgradation aims to restore or enhance the load-carrying capacity, stability, stiffness, and durability of marine structures. It addresses deficiencies due to material degradation, increased vessel sizes, higher berthing energies, seismic requirements, and updated design codes.

Typical measures include strengthening of quay walls and decks, addition of anchors or piles, foundation improvement, replacement of severely deteriorated members, and enhancement of durability systems.

ii) Functional Upgradation

Functional upgradation improves operational performance and relevance of port structures. It is driven by changes in vessel characteristics, cargo types, crane capacities, and port logistics requirements.

Measures may include modification of deck layouts and clearances, strengthening to support heavier cranes, upgrading fenders, bollards and utilities, improving access and safety facilities, and adapting structures for future port operations.

iii) Relationship Between Structural and Functional Upgradation

In port projects, structural upgradation often enables functional improvements by safely accommodating higher operational loads. Conversely, functional limitations—such as inadequate berth length or insufficient deck capacity—may highlight the need for structural modification even where visible distress is limited.

iv) Assessment and Decision Framework

A comprehensive framework includes:

- **Structural assessment** through inspection, testing, and verification against current port loading criteria;

- **Functional assessment** considering vessel mix, operational scenarios, equipment requirements, and compliance with port safety and environmental regulations.

Marine and port infrastructure is continuously exposed to harsh environmental and operational conditions. Defects, deterioration, and obsolescence are inevitable over time, but systematic assessment combined with appropriate repair, rehabilitation, renovation, and upgradation strategies can restore safety, extend service life, and maintain operational relevance. A structured, condition assessment-driven approach provides a technically sound, economical, and sustainable alternative to premature replacement of critical port structures.

7.3 Physical Signs of Defects and Deterioration in Concrete Structure

Physical defects in an RCC structure usually show up through clear visual cues such as cracking (hairline, structural, or shear-related), spalling where concrete flakes off due to rusting steel, and exposed or corroding reinforcement indicating loss of cover. You may also notice damp patches, water stains, or white powdery efflorescence from moisture ingress, along with honeycombing or surface erosion caused by poor compaction. Other warning signs include sagging or noticeable deflection of slabs or beams, rust stains on the surface indicating internal corrosion, gaps or failures at joints, and hollow-sounding delaminated areas. All these symptoms signal that the structure is experiencing stress, environmental attack, or material degradation and needs timely assessment and repair. That's why it's important to **spot any defects early**.

Normally, one or a number of physical defects can occur in structures; therefore, it is necessary to identify them properly. Before proceeding with a condition survey of the structure, one needs to understand and identify the different defects properly in order to get a more realistic evaluation. The most common structural defects are;

7.3.1 Scaling

Scaling refers to the **local flaking, peeling, or** loss of the finished surface portion of **hardened concrete**, where the top layer of concrete starts to come off in small pieces, making the surface look rough and patchy. We might see bits of stone (called aggregates) showing through.

This usually happens because of weather changes or from using certain chemicals like de-icers. It can also be caused by mistakes during construction. Over time, these things weaken the surface, causing it to peel or flake away. Thus, it is a physical or environmental action that usually leaves the aggregates clearly exposed.

Figure 7.4: Scaling and Peeling on RCC Surface

7.3.2 Disintegration

Disintegration is the physical deterioration (such as scaling) or breaking down of the concrete into small fragments or particles. This can happen in small ways, like the surface flaking off (called scaling), or in more severe cases where the concrete crumbles into tiny pieces or particles. It usually happens because of weather, poor materials, or long-term wear and tear, and it weakens the structure over time.

7.3.3 Abrasion

Abrasion of concrete is a progressive loss of concrete mass due to mechanical degradation such as friction, grinding action, impact, overloading, and local crushing. It happens when the surface of concrete slowly wears away because of repeated rubbing, scraping, or pressure. This kind of damage is caused by things like people walking on it, vehicles driving over it, or heavy objects grinding against it. Over time, these actions remove small layers of concrete, making the surface thinner and weaker. The worst kind of abrasion usually comes from vehicles that use studded or chain tyres, which can tear into the concrete more aggressively.

7.3.4 Delamination

Figure 7.5: Delamination on RCC Surface

Concrete delamination is the separation of the thin layers of concrete from the concrete surface, creating an unbonded concrete layer similar to blisters. This creates a weak, loose layer that isn't properly attached to the rest of the concrete. Once this layer breaks away, the exposed surface is left open to damage from things like weather, foot traffic, and chemicals, which can make the structure even weaker.

This problem can happen for several reasons:

- **Corrosion of steel bars** inside the concrete
- **Bleeding**, where water rises to the surface during setting
- **Air trapped** under the surface when the concrete is poured too early

Delamination not only reduces the strength and durability of the concrete but also makes it look uneven and damaged. This phenomenon tends to destroy a coating's strength and durability, as well as the material's aesthetic appearance.

7.4 Corrosion of Reinforcement

In reinforced concrete structures, steel reinforcement is covered by a thin passive film of oxide that protects it from oxygen and water, which may cause corrosion and produce rust. This passive layer can only be maintained at high pH values (i.e. where the pH is greater than 11). The passive layer can be destroyed due to 'depassivation', when chloride ions penetrate the concrete and then accumulate to a critical level on the surface of steel reinforcement or the pH value of concrete is reduced due to carbonation. As the corrosion on the steel surface proceeds, the corrosion products (rust) cause considerable volume expansion. This generates internal stress and causes cracking and spalling or delamination as shown in the figure below.

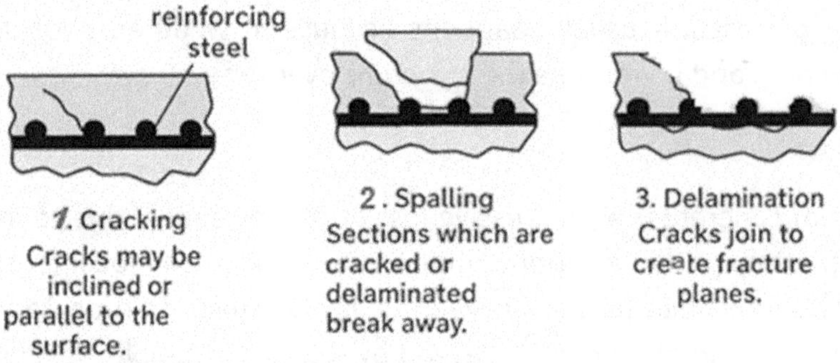

Figure 7.6: Process of Delamination in RCC Surface

7.4.1 Evaluate the Corrosion of Reinforcement

Steel in concrete develops a protective oxide film that provides a barrier to the transformation of the iron to rust. The passive oxide is similar to the "mill scale" that forms on hot-rolled steel when it cools to room temperature. It is well known that the mill scale provides a protective barrier to steel. The oxide, however, is brittle and is easily removed, as evidenced by the rapid corrosion of the bent portions of reinforcing bars. In reinforced concrete, a passive oxide coating forms because of the alkaline conditions (high pH, normally about 12.5) in the pores of the cement paste. There are two major causes for the breakdown of the passive oxide coating: carbonation and the presence of chloride ions.

Carbonation refers to reactions between carbon dioxide and alkalis in the pore solution of the cement paste. Typical reactions are as follows:

$$Ca(OH)_2 + CO_2 \rightarrow CaCO_3 + H_2O$$

$$2NaOH + CO_2 \rightarrow Na_2CO_3 + H_2O$$

As a result of these reactions, the pH of the pore solution decreases (even below a value of pH 9), and conditions are no longer favourable for maintaining the passive coating on the reinforcing steel. The steel becomes susceptible to corrosion. The penetration of the "carbonation front" depends on the quality of the cover concrete (water-cement ratio and degree of hydration) and the degree of saturation of the pores in the cement paste. The rate of penetration is low for dry concrete and saturated concrete, and it is a maximum when the concrete is in equilibrium with ambient relative humidity between about 40% and 80%.

The presence of chloride ions also breaks down the passive coating. The exact mechanism for this breakdown is not known with certainty, but it is known that it happens when the chloride ion content reaches a "threshold value." As pointed out by Rosenberg et al., there are many factors that affect the threshold chloride ion concentration,

Electrochemical methods are used to evaluate the corrosion activity of steel reinforcement. As is the case with other non-destructive test (NDT) methods, an understanding of their underlying principles and inherent limitations is needed to obtain meaningful results. In addition, an understanding of the factors involved in the corrosion of steel in concrete is essential for reliable interpretation of data from this type of testing. This chapter provides basic information about three commonly used methods:

i) Half-cell potential

ii) Concrete resistivity

iii) Polarisation resistance

Corrosion is a complicated topic, but here's a simple explanation for understanding purposes.

7.4.2 Principles of Corrosion

7.4.2.1 Electrolytic Cell

Corrosion is an electrochemical process, which means that it involves chemical reactions (gain and loss of electrons) and electrical current (motion of charges through a conductor). To maintain electrical neutrality and sustain the chemical reactions, electrical charge transfer occurs through external and internal pathways connecting the two electrodes where these reactions take place. Electrons flow through the external connection, and ions flow between the two solutions surrounding the electrodes.

7.4.2.2 Corrosion of Steel in Concrete

If the passive coating on the steel is lost, due to the carbonation effect or the presence of chloride ions above the critical concentration, conditions are set up for corrosion. Heterogeneities in the surface of the steel, such as differences in grain structure and composition, and local differences in the electrolyte, due to the heterogeneous nature of concrete, cause a region of the bar to act as an anode and another region to act as a cathode.

In summary, when steel in concrete loses its passive layer, corrosion occurs due to naturally occurring anodic and cathodic sites. However, the rate of corrosion depends on the electrical resistance of the concrete surrounding the anodic and cathodic sites and on the availability of oxygen. If the paste has low porosity, the resistance will be high, the oxygen diffusion coefficient will be low, and the corrosion rate will be low.

A conceptual model is provided by ***Tutti*** to represent the service life of a reinforced concrete structure that is susceptible to steel corrosion. Figure 7.7 represents the state of corrosion, which is a function of time.

During the initiation period, there is no corrosion, but as chloride ions or the carbonation front penetrate from the surface to the reinforcement, the corrosion process is set to begin. When the steel loses its passive layer, the propagation phase begins, and the degree of corrosion increases rapidly with time.

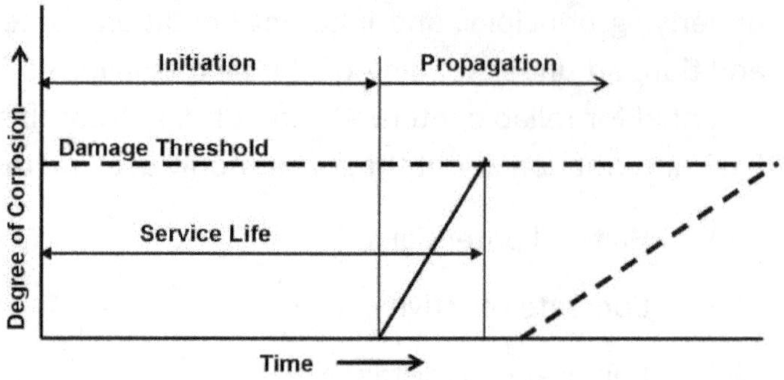

Figure 7.7 Model proposed by Tutti2 to represent the corrosion process of steel in concrete.

The rate of corrosion during the propagation phase is affected by the factors listed above. The service life is denoted by the time when the degree of corrosion has reached a level where the strength or serviceability of the structure is reduced below a critical level. The dashed line in Figure 7.7 represents a structure with a longer service life because of a longer initiation period and reduced rate of corrosion during the propagation phase. With this simple model, it is easy to explain why increased concrete cover and low water–cement ratio in concrete are two effective means for increasing service life.

7.5 Causes for Deterioration in Concrete Structures

According to ACI, deterioration in concrete structure can be caused by physical, mechanical, and chemical factors and can be induced by factors external and internal to the concrete structure. Physical factors have to do with forces acting on the structure, including those caused by temperature variations. The effect of impact or vibrations on structures, equipment operation, earthquakes, and ship or heavy equipment impact could cause damage to a concrete structure.

Chemical factors are commonly associated with the intrusion of aggressive waters containing mainly chlorides and sulphates. Alkali-aggregate reaction, the reaction between the alkalis in Portland cement and certain carbonate and siliceous rocks and minerals, in advanced stages, can cause physical damage to the concrete. Evidence of alkali-aggregate reaction includes cracking in random or regular patterns, expansion, gelatinous discharge, spalling, and chalky surfaces or rings around aggregates. However, it is often difficult to ascertain if the concrete has been or is being subjected to a chemical attack. Evidence of a chemical attack is generally a progressive disintegration of concrete. Chemical and Physical processes can impact the durability and structural performance of the concrete. If the cause of the attack can be identified, it should be recorded on the sketches on which defects have been recorded. The extent of the attack area should also be described. Coring can be used to check the depth and lateral extent of damage.

There are different types of causes under these four main divisions that cause deterioration; each of them is explained below.

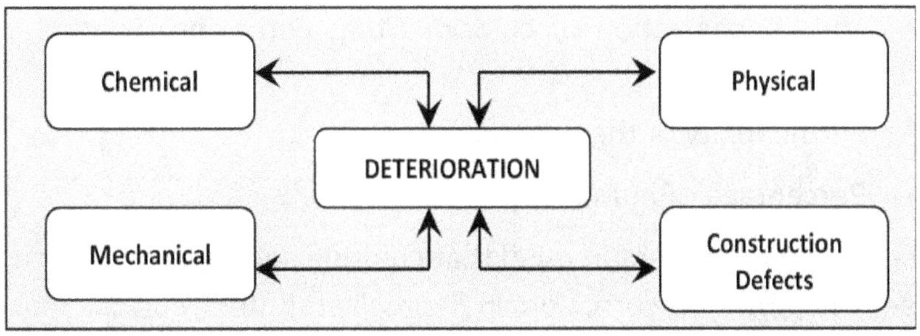

Figure 7.8 Agents Causing Deterioration in Concrete Structures.

In this chapter, we will discuss some of the most common types of deterioration, their mechanisms, and their cause in concrete.

7.5.1 Aggression from Chemicals

Chemical attack is one of the more common causes of deterioration of concrete in marine structures. A chemical attack involves the dissolution of substances or chemical reactions between substances and components of the concrete. Reaction products might cause problems due to dissolution or expansion. The deterioration of RCC is basically related to the loss of water tightness of cover concrete and migration of aggressive chemicals through interconnected porosity, which in turn chemically attacks its constituents, i.e., hydrated cement gel, aggregate, and the reinforcement as under:

i) Aggression by chlorides

ii) Aggression by carbon dioxide

iii) Aggression by sulphates

iv) Alkali-aggregates reaction

7.5.2 Aggression by chlorides

Chemical deterioration is mostly influenced by the environment and exposure conditions. The performance of concrete structures is affected especially in the long term, and chloride-induced corrosion is the major threat to the durability of maritime and coastal concrete structures. In reinforced concrete structures, steel reinforcement is covered by a thin passive film of oxide that protects it from oxygen and water. This passive layer can only be maintained at high pH values (i.e. where the pH is greater than 11). The passive layer can be destroyed by a process known as 'depassivation', when chloride ions penetrate the concrete and accumulate to a critical level on the surface of steel reinforcement

Aggression of chlorides on concrete occurs if it is in contact with environments with a high chloride content, such as seawater. Once the chloride penetrates into the concrete and reaches the reinforcement bars, it eliminates the passivating ferrous oxide film on the reinforcement steel so that they are exposed to the corrosion process. Penetration starts on the surface and continues deeper into the concrete. The penetration time of chloride ions depends on:

- The concentration of chlorides that come into contact with the surface of the concrete;
- Permeability of the concrete;
- Percentage of humidity present.

Further, it should be kept in mind that chloride aggression in concrete causes depassivation of steel and promotes corrosion. In depassivated steel, corrosion occurs with a combination of oxygen and water. The part of marine structures, completely immersed in seawater, will have higher chloride content, but oxygen will not be able to penetrate. The part of the marine structure, exposed to the splash zone, is aggravated by chloride in the presence of moisture and oxygen. Once the process starts, wherever the structure is located, corrosion proceeds faster because access routes may be found more easily. The mechanism of chloride aggression is described below.

The electrochemical action of chloride in the presence of oxygen and moisture (optimum relative humidity is 70 - 80%) is the most favourable condition for corrosion.

Chloride-induced corrosion at the steel surface involves the following reaction (Neville, 2008)

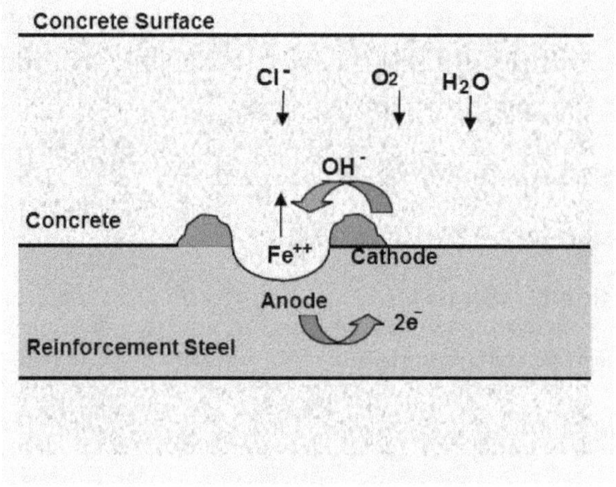

$$Fe^{++} + 2Cl^{-} \rightarrow FeCl_2$$

$$FeCl_2 + 2H_2O \rightarrow Fe(OH)_2 + 2HCl$$

Figure 7.9 Mechanism of Chloride Aggression on Concrete Surface

The de-passivated area of steel becomes an anode, while the passivated surface becomes a cathode. At the anode of the cell, the reaction can be presented as:

$$Fe \rightarrow Fe^{++} + 2e^{-}$$

$$Fe^{++} + 2(OH)^{-} \rightarrow Fe(OH)_2$$

$$4Fe(OH)_2 + 2H_2O + O_2 \rightarrow 4Fe(OH)_3$$

and the reaction at the cathode is given by:

$$4e^{-} + O_2 + 2H_2O \rightarrow 4(OH)^{-}$$

During the corrosion process, two major products – $Fe(OH)_2$ (ferrous hydroxide) and $Fe(OH)_3$ (ferric hydroxide) – are produced as rust/corrosion, as shown in Figure 7-9. Table 7-1 (by Thoft-Christensen,

2001) indicates these corrosion products cause considerable volume expansion. This generates internal stress and, ultimately, causes cracking, delamination, and spalling. The failure process may be further accelerated by external loading.

Table 7-1: Corrosion Products (Thoft-Christensen, 2001)

Corrosion product	Colour	Volume (1 cm³ Fe)
Fe3O4	Black	2.1
$Fe(OH)_2$	White	3.8
$Fe(OH)_3$	Brown	4.2
$Fe(OH)_3$, $3H_2$	Yellow	6.4

Thus, chloride-induced corrosion mainly depends on three factors, i.e., diffusion coefficient, surface chloride concentration, and critical chloride level beyond which corrosion of reinforcement will be initiated.

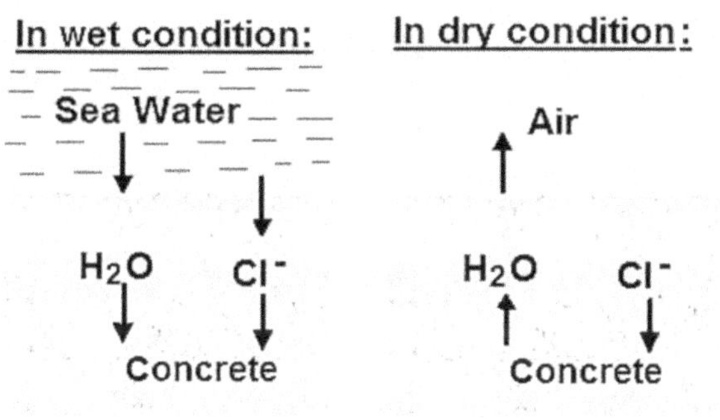

A previous study shows that the service life of concrete structures under chloride corrosion attack is more sensitive to cover depth than the diffusion coefficient, and more sensitive to chloride concentration than the critical chloride level. (Khatri et al, 2004).

Figure 7.10: Mechanism of Chloride Ingress via Diffusion in RCC Structures

The corrosion process will slow and may eventually stop when water or oxygen is limited. Very high humidity in concrete may reduce the diffusion of oxygen to the corrosion area and slow the corrosion of the reinforcement, while a shortage of water in dry concrete also reduces corrosion activity.

For determining the rate of chloride penetration into concrete and the resistance of concrete to the penetration of chloride ions, refer to ASTM C1543-02 and ASTM C1202, respectively.

7.5.3 Aggression by carbon dioxide or Carbonation of Concrete

Carbonation in concrete is essentially caused by the penetration of atmospheric CO_2. It significantly reduces alkalinity, which, in turn, increases the vulnerability of steel reinforcement to corrosion due to depassivation. The direct consequence of carbonation is shrinkage, as well as a decline in pH, which is left unchecked, will eventually cause corrosion and cracking, followed by spalling.

Cement in concrete normally includes tricalcium silicate (C3S), dicalcium silicate (C2S), tricalcium aluminate (C3A), and calcium aluminoferrite (C4AF). When cement is mixed with water, the hydration process that takes place produces crystalline calcium hydroxide, $Ca(OH)_2$, hydrated gel (CSH), and other products with heat evolution.

In actual practice, CO_2 present in the atmosphere permeates into concrete and carbonates it naturally, reducing the alkalinity of the concrete. The pH value of pore water in the hardened cement paste, which was around pH 13, will be reduced to around pH 9.0. When all the $Ca(OH)_2$ becomes carbonated as shown in the reaction below,

$$\mathbf{Ca(OH)_2 + CO_2 \rightarrow CaCO_3 + H_2O}$$

The pH value of concrete reduces to about 8.3, at which level the passive film on the steel is not stable. The loss of passivity usually occurs around pH 11.

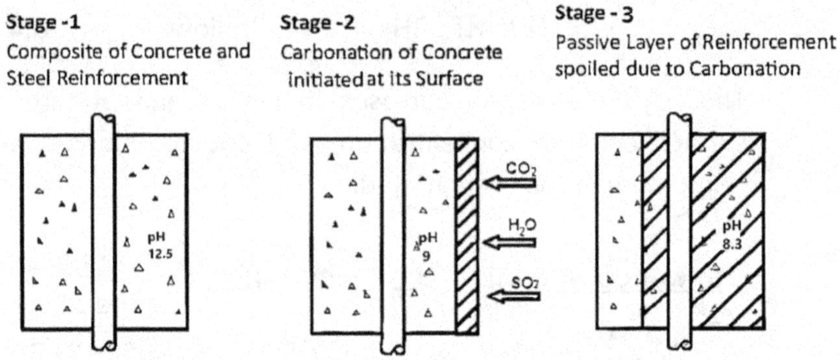

$$Ca(OH)_2 + CO_2 = CaCo_3 + H_2O$$

Calcium Hydroxide + Carbon Dioxide = Calcium Carbonate + Water

Figure 7.11: Mechanism of the carbonation of concrete

Further, the carbonation process is dependent on the relative humidity of the concrete. The highest rates of carbonation occur when the relative humidity is maintained between 50 percent and 75 percent. Below 25 percent relative humidity, the degree of carbonation that takes place is considered insignificant. Above 75 percent relative humidity, moisture in the pores restricts CO_2 penetration. Carbonation-induced corrosion often occurs in areas of the structure exposed to rainfall, moist or humid air shaded from sunlight, and having low concrete cover over the reinforcing steel.

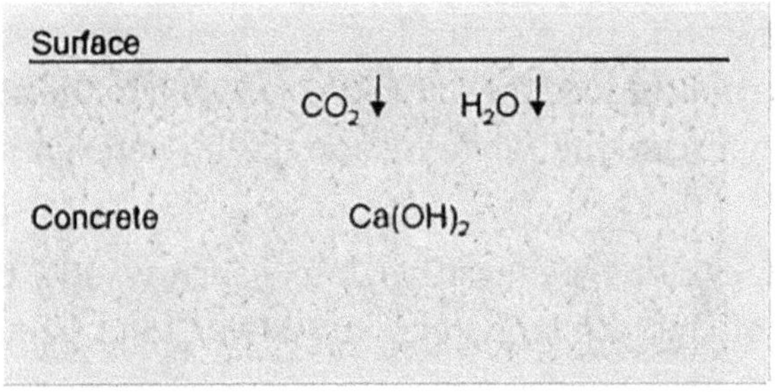

Figure 7.12: Carbonation caused by the penetration of CO2 and water

Carbonation is considered relevant to the concrete having a water/cement ratio, more porosity, and it proceeds in proportion to the square root of time (Ho et al. 1987). Carbon dioxide penetrates below the exposed surface of concrete extremely slowly. The time required for carbonation can be estimated by knowing the concrete grade and using the following equation:

$$t = (d/k)^2$$

where,

t is the time for carbonation,

d is the concrete cover,

k is the permeability

Typical permeability values are shown in Table -7.2.

Table -7.2: Permeability Values versus Concrete Grade

Grade of Concrete	Permeability *x* 10^{-9} cm/sec
15	17
20	10
25	6
30	5
35	4
40	3.5

However, the rate of carbonation can vary and depends on the following factors of the concrete:

i) relative humidity around concrete;

ii) grade of concrete;

iii) permeability of concrete;

iv) whether the concrete is protected or not;

v) depth of cover; and

vi) Time (period of exposure).

7.5.3.1 Repairing for Carbonation in Concrete

Repairing for carbonation damage should be carefully planned to prepare a specification for repair. A simplified summary of the recommended steps is:

1. **Hammer Test**: Tap the concrete surface to identify any loose or hollow areas. These areas are structurally weak and must be removed to ensure the repair bonds well to solid concrete. Leaving them in place can lead to future failure.
2. **Remove Loose Concrete**: Break away all loose concrete around exposed steel reinforcement. Corrosion often spreads beyond visibly damaged areas. Removing all loose concrete ensures that repairs are made on a sound base and that the steel is fully accessible for treatment.

3. **Grit Blasting**: Clean the steel bars thoroughly using grit blasting to remove rust and other deposits, including hidden surfaces and intersections.
4. **Apply Steel Primer**: Coat the cleaned steel with a protective primer to protect it from further corrosion, as it acts as a barrier against moisture and oxygen. This may be:
    - Alkali-based: Helps restore the passive layer.
    - Encapsulating (e.g., epoxy resin): Seals the steel completely.
    - Sacrificial (e.g., zinc-rich paint): Corrodes in place of the steel, offering protection.
5. **Patch Repair**: Rebuild the damaged concrete using a mix of cement, sand, and polymer dispersion. Cementitious materials restore structural integrity, while polymers improve adhesion, flexibility, and resistance to shrinkage or cracking. This ensures a durable and long-lasting repair.

7.5.4 Aggression by Sulphate

Soluble sulphates of sodium, potassium, magnesium, and calcium, which often exist in groundwater, may react with hydrates such as calcium hydroxide, $Ca(OH)_2$, from cement hydration, producing expansive products inside the cement paste matrix and near the aggregate interface, resulting in the exposure of aggregates. Sulphate aggregation generally causes a pattern of micro-cracks in the concrete. As a result, surface scaling and disintegration set in, followed by mass deterioration.

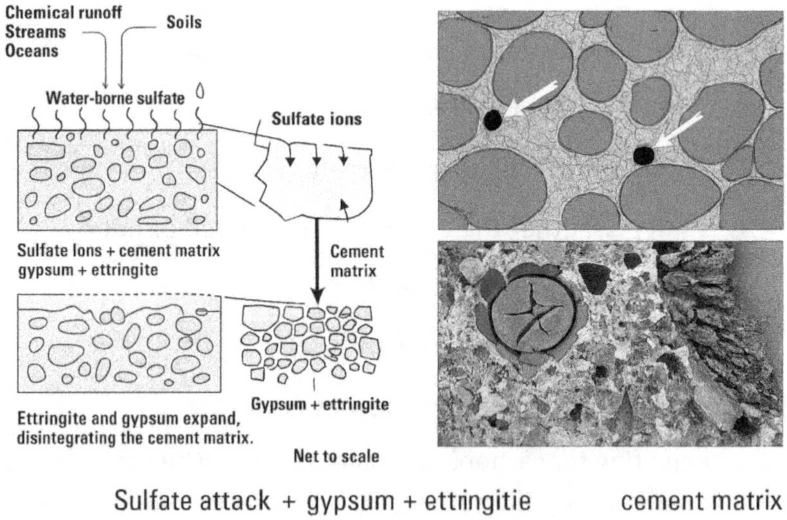

Figure 7.13: Mechanism of Sulphate Aggression in Concrete

When sulphate enters concrete, it combines with hydrated calcium aluminate and/or the calcium hydroxide component of hardened cement paste and begins destroying the paste that holds the concrete together. As sulphate dries, new expansive compounds are formed, often called ettringite. These new crystals occupy empty space, and as they continue to form, they cause the paste to crack, further damaging the concrete. The volume increase and the needle-like shape of ettringite crystals are responsible for creating internal stresses that cause cracking. This is known as a chemical process. A sulphate attack can also be the result of a physical process when water-soluble salts travel through the concrete and get deposited at the surface, causing it to scale and eventually spall. External sources of sulphate are more common and usually are a result of high-sulphate soils and ground waters, or can be the result of atmospheric or industrial water pollution. Seawater is considered to have a "moderate" level of exposure to water-soluble sulphate ions.

Sulphate attack can be visually identified by spotting salt buildup on concrete surfaces, surface scaling, spalling with traces of salts, de-bonding between the aggregates and paste, and the change of texture of concrete becoming like toothpaste. This happens at a developed stage of the Sulphate Attack. This phenomenon is mainly observed in buried or close-to-the-surface concrete members, such as piers, columns, slabs, foundations, and basements, where sulphate contamination is found in the surrounding soil or water.

7.5.5 Alkali-aggregates reaction

An alkali-aggregate reaction may cause considerable expansion and serious deterioration of concrete structures. Aggregate those that contain reactive silicon; they react with alkalis contained in the cement, and due to the porous concrete cover, it reacts with seawater containing sodium chloride (NaCl). This reaction forms a gel that is highly expansive if exposed to humidity, and the gel creates forces that break the concrete around the aggregates. The alkali-aggregates reaction shows up when it starts in the cortex of the concrete, with micro and macro cracks on the surface, or even causing small portions of the concrete above the reactive silicon aggregate to lift (pop-out phenomenon) as shown in **Figure 7.14**.

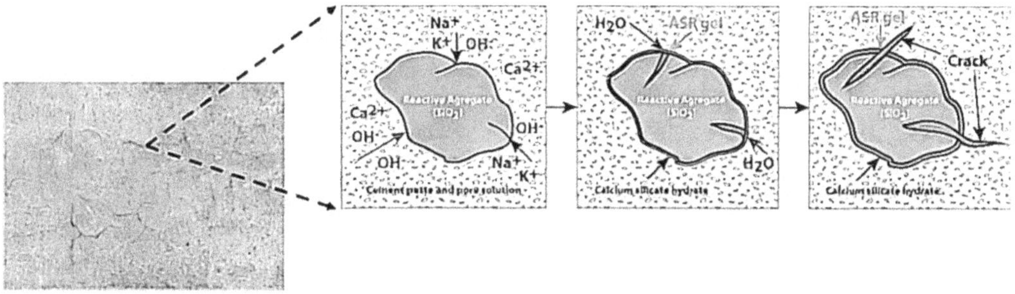

Figure 7.14: A sample picture showing a crack due to the Alkali-Aggregate reaction

7.6 Abrasion of concrete

Abrasion of concrete is a progressive loss of concrete mass due to mechanical degradation such as friction, grinding action, impact, overloading, and local crushing. Vehicular movement and pedestrian traffic cause abrasion. The worst effect of abrasion is caused by vehicles with studded/chain tyres. Similarly, in industrial structures, the concrete floors are subjected to impact loads and scratching effects.

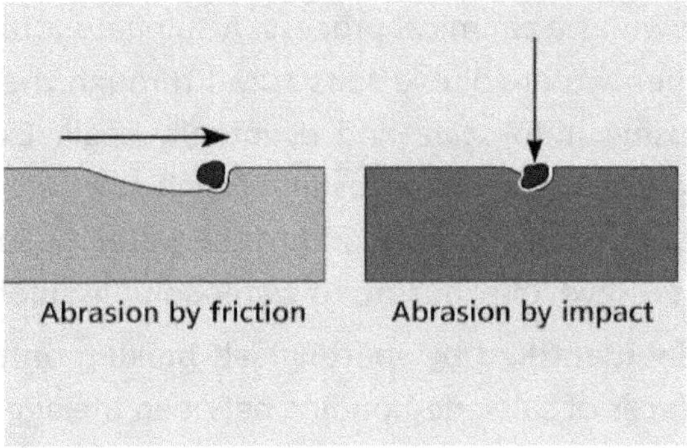

The abrasion resistance of concrete depends upon the surface hardness, aggregate hardness, and the bonding between cement and aggregate. In general, the higher the compression strength of concrete, the higher its abrasion resistance. But, strength alone cannot be taken as the parameter to estimate the abrasive behaviour.

Figure 7.15: Mechanism of Abrasion on Concrete

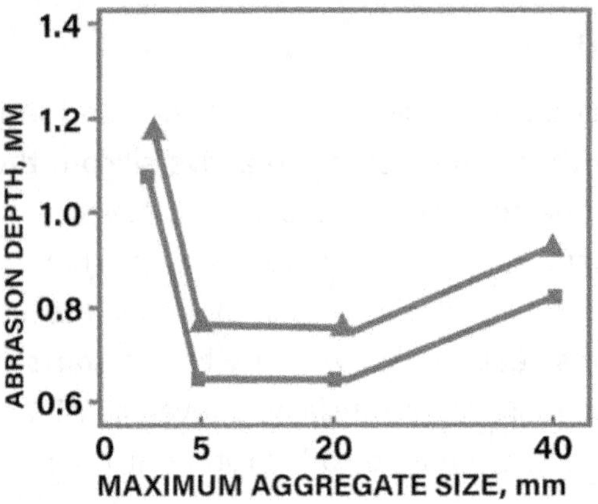

In an extensive study made by Dhir et al, the size of aggregate and curing time were observed to be dominant variables besides the compression strength (or w/c ratio). It was observed that the size of the aggregate plays an important role in the rate of abrasion (mainly in the case of impact loading). Due to the higher stiffness of smaller aggregates (10 - 20mm size), they are less prone to damage in impact loading.

Figure 7.16: Variation in Abrasion Depth with Aggregate Size

The general concrete works with aggregate sizes less than 10mm or more than 20mm could be damaged easily, as indicated in Figure 7.16

7.6.1 Variation of abrasion depth with aggregate size

The porosity of the top layer is another main factor controlling the abrasion of concrete (Sadegzahed, 1987). A dense, well-cured surface has higher hardness. It is also common practice to coat the abrasion-prone surface with a resistant material. For example, in bridges, a flexible material (bitumen) is used, while in industries a harder surface is made with the addition of finer particles (e.g. silica, ironite, etc.). Other methods to achieve abrasion-resistant surfaces are as follows:

a) Surface finishing
b) Extensive curing
c) Aggregate proportioning
d) Cement proportioning
e) The addition of fly ash up to 30% has been found to increase abrasion resistance

f) Addition of fibres (steel, polypropylene)

7.7 Cracking of Concrete Structures

Cracks are the most commonly observed defects in RCC structures and may range from minor surface imperfections to indicators of serious structural or durability-related problems. They can arise from design assumptions, material behaviour, construction practices, environmental exposure, or service and loading conditions. In some cases, cracking may also indicate more critical issues such as overstressing of members, impact or abuse, restraint-induced stresses, differential foundation settlement, or structural instability.

Concrete is inherently prone to cracking due to its low tensile strength and its response to shrinkage, temperature changes, and sustained loading. However, not all cracks are structurally benign. The characteristics of cracks—such as type, orientation, width, length, location, depth, and rate of progression—provide important insight into the internal stress state and structural behaviour.

Minute observation of cracks—considering their geometry, distribution, and interaction with structural elements—is therefore essential. Such observations help in distinguishing between cracks caused by material-related phenomena (such as plastic shrinkage, drying shrinkage, or thermal effects), structural actions (flexure, shear, torsion, axial force), or environmental effects (carbonation, chloride ingress, moisture variation, freeze–thaw cycles, or chemical attack). Different environmental exposure conditions can significantly influence crack development, propagation, and the rate of associated deterioration, particularly reinforcement corrosion. From a structural mechanics perspective, crack patterns often reflect stress flow, load paths, restraint conditions, and redistribution of internal forces within the structure.

In line with IS 15988 and relevant EN guidelines, cracks should not be treated as isolated surface defects. Effective repair, rehabilitation, or strengthening measures can only be designed and implemented when the underlying mechanisms of cracking are clearly understood, rather than merely addressing the visible symptoms on the concrete surface.

7.7.1 Causes and Mechanisms of Concrete Cracking

Cracks in concrete are common and can occur due to various reasons at different stages of a structure's life. These cracks vary in appearance, severity, and underlying causes.

Hairline cracks are very fine and usually appear on the surface due to early shrinkage or minor thermal movements. They often form during the initial setting of concrete and may not affect structural integrity, but can allow moisture ingress over time.

Shrinkage cracks occur when concrete loses moisture and shrinks during curing. If the concrete is restrained by reinforcement or adjacent elements, tensile stresses develop, leading to cracking. These are typically surface-level and can be minimised with proper curing practices.

Settlement cracks happen when the concrete settles unevenly, especially around reinforcements or changes in section thickness. Poor compaction or excessive bleeding can cause the concrete to separate and crack vertically. Lastly,

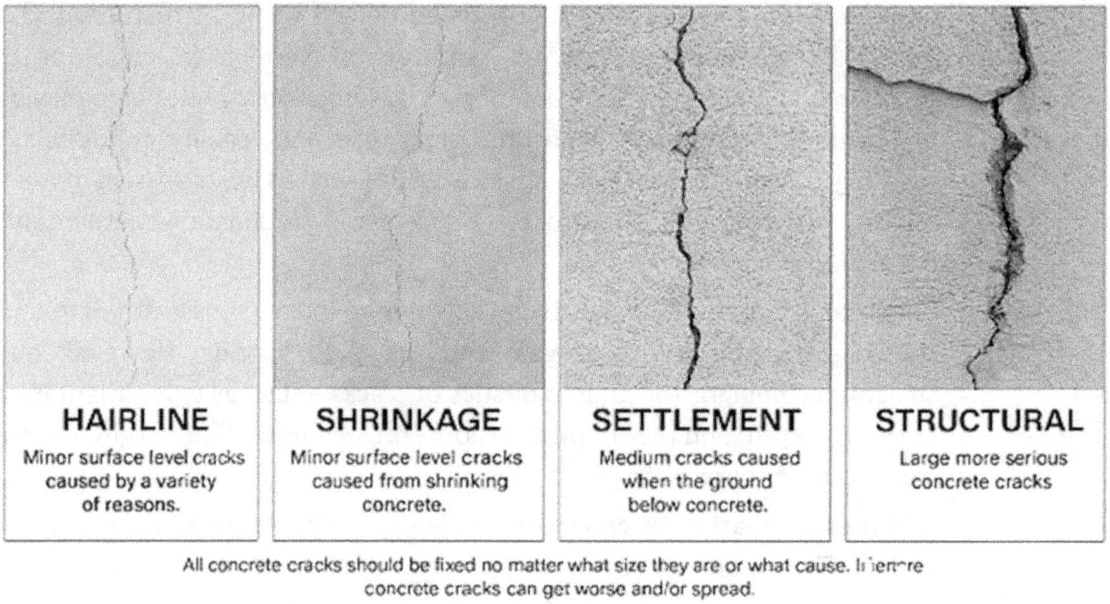

Figure 7.17: Various types of cracks in concrete

Structural cracks are more serious and result from excessive loads, foundation movement, or design flaws. These cracks often extend deeper into the concrete and can compromise the safety and durability of the structure if not addressed promptly. Understanding the type and cause of each crack is essential for accurate diagnosis and effective repair strategies.

Hairline cracks are very fine, narrow surface-level fractures that typically appear in concrete during its early life. These cracks are usually less than 0.1 mm wide and may not be immediately visible without close inspection. They often form due to **plastic shrinkage**, which occurs when moisture evaporates too quickly from the surface of freshly placed concrete, especially under hot, windy, or dry conditions. As the surface dries and shrinks faster than the underlying concrete, tensile stresses develop, leading to cracking. Hairline cracks can also result from **thermal movements**, where temperature changes cause expansion and contraction in the concrete.

Though these cracks are generally not structurally significant, they can become pathways for water, chlorides, and other aggressive agents, potentially leading to long-term durability issues such as corrosion of embedded reinforcement. Proper curing, controlled placement conditions, and use of shrinkage-reducing admixtures can help minimise the occurrence of hairline cracks.

7.7.1.1 Hairline Cracks in Concrete

Hairline cracks are very fine, narrow surface-level fractures that typically appear in concrete during its early life. These cracks are usually less than 0.1 mm wide and may not be immediately visible without close inspection. They often form due to **plastic shrinkage**, which occurs when moisture evaporates too quickly from the surface of freshly placed concrete, especially under hot, windy, or dry conditions. As the surface dries and shrinks faster than the underlying concrete, tensile stresses develop, leading to cracking. Hairline cracks can also result from **thermal movements**, where temperature changes cause expansion and contraction in the concrete. Though these cracks are generally not structurally significant, they can become pathways for water, chlorides, and other aggressive agents, potentially leading to long-term durability issues such as corrosion of embedded reinforcement. Proper curing, controlled placement conditions, and use of shrinkage-reducing admixtures can help minimise the occurrence of hairline cracks.

7.7.1.2 Plastic Shrinkage Cracks in Concrete

Shrinkage of plastic concrete during hardening is the most common reason for cracking. Cement concrete mix does not require much water for hydration; however, to increase the workability of concrete, if too much water is added, it causes shrinkage in concrete. When concrete is still in its plastic state (before hardening), it has excess water. When that water eventually leaves the slab, it leaves behind large voids between the solid particles. These empty spaces make the concrete weaker and more prone to cracking. This type of cracking happens frequently and is referred to as "plastic shrinkage cracking".

While plastic shrinkage cracks can happen anywhere in a slab or wall, they almost always happen at restraint corners or with circular objects in the middle of a slab (pipes, plumbing fixtures, drains, and manholes). Since concrete cannot shrink around a corner, stress will cause the concrete to crack from the point of that corner. Plastic shrinkage cracks are typically very narrow in width and barely visible. They don't just exist on the surface but extend throughout the entire thickness of the slab. An excessively wet mix is a contributing factor to shrinkage in concrete. When the mix contains too much water, the slab will shrink more than if the correct amount of Water was used.

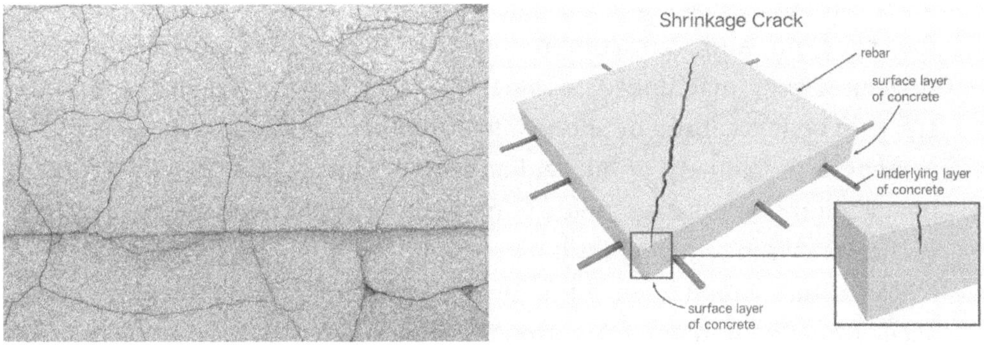

Figure 7.18: Pattern of shrinkage crack in concrete

- Hot weather is another big reason for plastic shrinkage cracks.
- Control joints can be incorporated into the slab to prevent shrinkage cracking. The joints will open up as the concrete slab contracts (shrinks).

7.7.1.3 Crack due to Thermal Expansion of Concrete

Concrete expands when exposed to heat, and this thermal expansion can lead to cracking if not properly managed. As the temperature rises, the concrete tries to increase in volume. If this expansion is restrained—either by adjacent structural elements, fixed supports, or the absence of properly designed expansion joints—internal stresses build up within the concrete. When these stresses exceed the tensile strength of the material, thermal cracks can develop.

Figure 7.19: Crack due to thermal expansion in concrete

These cracks are often seen in long, continuous concrete elements such as pavements, walls, or slabs, especially when exposed to direct sunlight or high ambient temperatures. The cracks typically appear as straight or slightly curved lines and may run across the width or length of the element. The risk of thermal cracking is higher in mass concrete structures where temperature gradients between the core and surface are significant. To prevent such cracking, it is essential to incorporate expansion joints, use temperature control measures during curing, and consider thermal movement in the design phase.

7.7.1.4 Structural Cracks in Concrete

Structural cracks are serious defects that occur when concrete is subjected to stresses beyond its design capacity. These cracks typically result from factors such as excessive loading, foundation movement, poor structural design, inadequate reinforcement, or construction errors. Unlike superficial cracks, structural cracks often penetrate deep into the concrete and can compromise the load-bearing ability and overall safety of the structure. They may appear as vertical cracks in columns due to axial overload, diagonal cracks in beams caused by shear failure, or horizontal cracks in walls due to lateral pressure or poor reinforcement detailing. For example, in multi-storey buildings, diagonal cracks at beam-column joints may indicate inadequate anchorage or seismic stress. Similarly, wide vertical cracks in retaining walls can result from soil pressure exceeding design limits. These cracks often widen over time and allow moisture and aggressive agents to enter, accelerating reinforcement corrosion and further deterioration. Timely identification, structural assessment, and appropriate repair methods—such as epoxy injection, strengthening, or retrofitting—are essential to restore integrity and prevent long-term damage

7.7.1.5 Cracks caused by premature drying of Concrete

Premature drying of concrete refers to the rapid loss of moisture from the surface before the concrete has properly set and gained sufficient strength. This typically occurs under hot, dry, or windy conditions, especially when curing practices are inadequate or skipped altogether. When moisture evaporates too quickly, the surface of the concrete shrinks faster than the inner layers, creating tensile stresses that lead to **plastic shrinkage cracks**. These cracks are usually shallow, irregular, and appear within the first few hours after placement. They often form in slabs, pavements, and large exposed surfaces. Premature drying can also reduce the overall strength and durability of the concrete, making it more vulnerable to long-term issues like scaling, dusting, and reinforcement corrosion. To prevent such cracking, it is essential to implement proper curing methods—such as covering with wet burlap, using curing compounds, or maintaining surface moisture—and to avoid placing concrete during extreme weather conditions without protective measures.

7.7.1.6 Cracks due to Improper Workmanship during Construction

During construction, environmental effects should be studied in relation to the following:

i) Material Issues

If the concrete or the materials used to make it—like cement, sand, aggregates, or water—aren't stored, mixed, or handled properly either at the source or on the construction site, it can lead to weak spots in the structure. These weak areas are more likely to crack or fail over time.

ii) Bad Curing

Concrete needs to stay moist and protected while it hardens. If it dries out too quickly or isn't cured properly, it can shrink unevenly and develop cracks. Proper curing helps the concrete gain strength and durability.

iii) Improper Handling of Precast Elements

Precast concrete parts (those made off-site and brought to the construction site) must be lifted, moved, and supported carefully. If they're mishandled or not supported correctly during installation, they can crack or get damaged even before being used.

iv) Overloading Too Soon

Concrete takes time to reach its full strength. If heavy loads—like machinery, building materials, or structural pressure—are applied before it's ready, the concrete can get damaged or cracked because it's still too weak to handle the stress.

v) Removing Formwork Early

Formwork (the moulds that hold concrete in shape while it sets) should only be removed once the concrete is strong enough. If it's taken off too early, or if heavy materials or equipment are placed on the structure too soon, it can overload certain parts and cause cracking.

vi) Uncontrolled Movements or Impacts

Sudden movements, rough handling, or accidental impacts—like dropping heavy tools or bumping into the structure—can cause unexpected stress. This can lead to cracks, especially if the concrete hasn't fully hardened or is already under load.

7.7.1.7 Patch Accelerated Corrosion in concrete

Commonly referred to as "ring anode corrosion" or "halo effect", patch accelerated corrosion is a phenomenon specific to concrete restoration projects. When repairs are completed on corrosion-damaged structures, abrupt changes in the concrete surrounding the reinforcing steel are created.

Typical concrete repair procedures call for the removal of the concrete around the full circumference of the reinforcing steel within the repair area, cleaning of corrosion by-products from the steel, and refilling the cavity with new, chloride-free, high pH concrete.

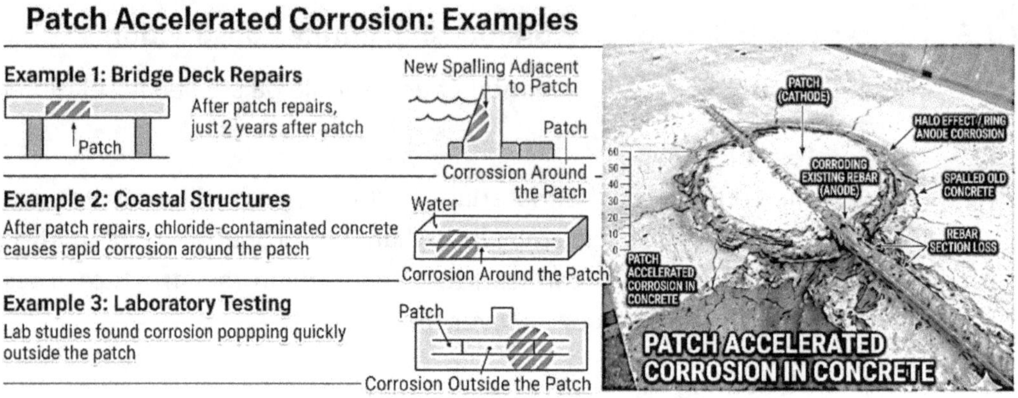

Figure 7.20: Patch Accelerated Corrosion

This procedure leaves the reinforcing steel embedded in adjacent environments with abruptly different corrosion potentials. This difference in corrosion potential (voltage) is the driving force for new corrosion sites to form in the surrounding contaminated concrete. The evidence of this activity is the presence of new concrete spalling adjacent to previously completed patch repairs.

7.8 Deterioration of Concrete Structures in Marine Environments

7.8.1 Marine environments

Marine environments normally encompass several different exposure zones, and each one requires individual consideration, mostly due to the variation of water level and/or dissolved oxygen concentration. The environments where port and harbour steel structures are applied are roughly classified into five zones, namely, atmospheric zone, splash zone, tidal zone, submerged zone, and mud zone. In cases where long steel products, such as steel sheet pile and steel pipe pile extend through multiple environments (tidal zone, submerged zone, and mud zone), the rate of corrosion is different according to the prevailing environment. The area where serious problems arise regarding the corrosion of steel is the splash zone and just above the mean high water level (M.H.W.L.). Corrosion tendencies by corrosive environment are indicated in **Figure 7.21**.

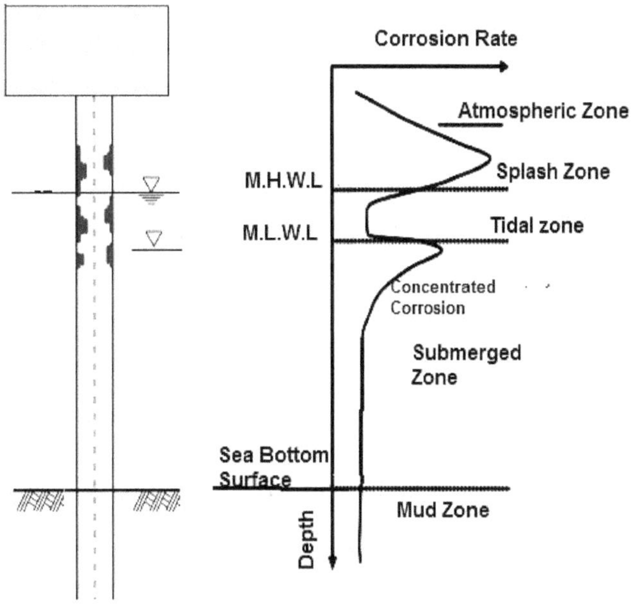

Figure 7.21: Representative figure for the rate of deterioration in the vertical direction of structures in the marine environment

In marine environments, the durability of concrete structures is influenced by several key factors that must be carefully assessed and analysed. These include:

i) The chloride ion concentration present in saline water, which can accelerate corrosion of reinforcement;

ii) The exposure conditions of the structure, such as tidal zones, splash zones, or submerged areas;

iii) The quality of concrete materials and the workmanship during construction, which directly affect the structure's resistance to aggressive elements; and

The additional protective measures and maintenance practices applied, such as coatings, sealants, or cathodic protection systems, which help extend the service life of the structure.

7.8.2 Chloride Ion Concentration

In marine environments, chloride ions found in saline water are a major factor affecting the durability of reinforced concrete structures. These ions act as aggressive agents that penetrate the concrete cover through its pores and microcracks, eventually reaching the embedded steel reinforcement. When the chloride concentration exceeds a critical threshold, it initiates the corrosion process. This deterioration is further accelerated by environmental exposure—such as humidity, submersion, and the presence of oxygen—which facilitates the electrochemical reactions responsible for corrosion, as explained in the following section.

7.8.3 The Stages of Deterioration Due to Chloride Ion

The durability of concrete structures in marine environments is significantly influenced by the concentration of chloride ions in seawater, even under similar exposure conditions. The deterioration process typically progresses through the following stages:

Stage 1: Initiation Phase

When the chloride ion concentration remains below the critical threshold (generally less than 0.4 % by weight of cement), no visible damage occurs. The passive protective layer around the steel reinforcement remains intact, and corrosion does not initiate. At this stage, the structure maintains its durability.

Stage 2: Propagation Phase

As chloride concentration increases, diffusion into the concrete accelerates. Once the chloride threshold (typically 0.4–1.0% by weight of cement) is exceeded at the reinforcement level, the passive protective layer on steel starts breaking down. This leads to the initiation of corrosion, which marks the beginning of durability loss. At this stage, the durability of the structure has a moderate to high risk; microcracks may form, leading to reduced structural integrity.

Stage 3: Acceleration Phase

Due to the continuous diffusion process, with increasing chloride levels (typically more than 1.0% by weight of cement), the Corrosion rate of steel reinforcement increases, producing expansive rust. In other words, in higher chloride concentration, the propagation phase (from corrosion initiation to visible damage) becomes shorter, leading to the early onset of severe deterioration. This causes cracking, spalling, and loss of bond between steel and concrete. In this process, the structural integrity is compromised faster, reducing the service life of the structure.

The effect of concentration is summarised below;

- **Initiation Stage (< 0.4%)**

  Risk: Low

  Effect: No corrosion; passive layer on steel remains intact.

- **Propagation Stage (0.4 – 1.0%)**

  Risk: Moderate to High

  Effect: Corrosion begins; microcracks may form.

- **Acceleration Stage (> 1.0%)**

  Risk: Severe

  Effect: Cracking, spalling, and structural degradation.

At the aforementioned chloride concentration, typical signs of deterioration are listed in the table below.

Table 7-3: Deterioration of Reinforced Concrete w.r.to Chloride Ion Concentration

Chloride Level (% by cement weight)	Deterioration Symptoms	Structural Impact
< 0.4	No corrosion	Safe
0.4 – 1.0	Initiation of corrosion	Reduced durability
> 1.0	Cracking, spalling, delamination	Severe structural degradation

After observing the concentration level of chloride ion in concrete, the approximate service life of the structure can be stipulated as;

Table 7-4: Service life Reinforced Concrete w.r.to Chloride Ion Concentration

Chloride Concentration	Time to Corrosion Initiation	Durability Impact
< 0.4%	> 30 years	Long service life
0.4–1.0%	10–30 years	Moderate life
> 1.0%	< 10 years	Short service life

7.8.4 Marine Exposure Conditions

The presence of chloride ions in the environment or the exposure condition of the structure plays a major role in determining the durability of a structure and the likelihood of corrosion in its reinforcement.

As defined in BS EN 206, marine exposure conditions are classified into several zones, each with distinct mechanisms through which chloride penetrates concrete.

7.8.4.1 Atmospheric Zone (XS1)

This zone is located above the splash zone, so it is not directly affected by seawater or tidal action. It includes parts of a structure that are exposed to open air but are still close to the sea. Structures in this zone are exposed to marine aerosols—tiny droplets of seawater carried by the wind. These droplets contain chloride ions that settle on the concrete surface over time. Through diffusion, as explained earlier, the chloride ions slowly move into the concrete through its pores due to concentration differences.

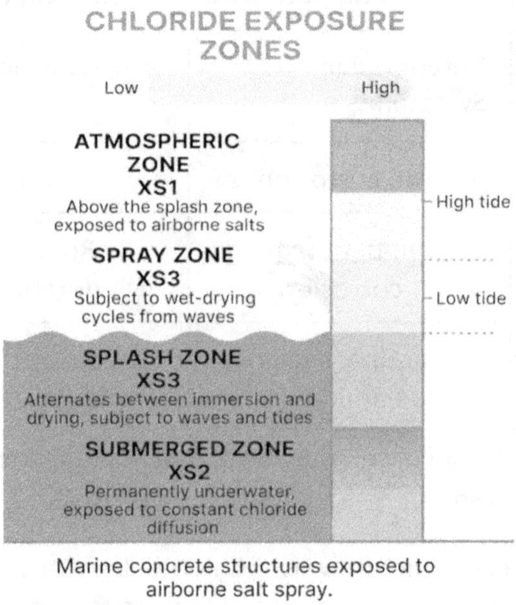

Although chloride ingress is slower here than in zones with direct seawater contact, long-term buildup can still cause reinforcement corrosion, especially if the concrete cover is thin or damaged. This presents a medium level of risk to the structure. In industrial areas, atmospheric CO_2 can also react with calcium hydroxide in concrete, leading to carbonation. This reduces the concrete's alkalinity and increases its vulnerability to chloride-induced corrosion.

Figure 7.22: Exposure Zone in Marine Environment

7.8.4.2 Splash Zone (XS3)

Splash zone located just above the high tide level, this zone is frequently exposed to wave action and splashes, causing repeated wetting and drying cycles. These cycles lead to significant chloride deposition. Chloride ingress occurs through capillary suction during wetting and diffusion over time. As concrete dries, its pores draw in chloride-rich water, allowing ions to migrate deeper. The continuous wetting and drying accelerate chloride buildup and enhance oxygen availability, making this zone highly susceptible to reinforcement corrosion and posing a high structural risk.

7.8.4.3 Tidal Zone (XS3)

Tidal zones are situated between high and low tide marks; this zone experiences regular cycles of immersion and exposure. Structures are alternately submerged and exposed to air, creating a dynamic chloride environment. Chloride ingress occurs mainly through diffusion during immersion. The alternating wet and dry conditions enhance chloride penetration and sustain oxygen levels, accelerating reinforcement corrosion and posing a high structural risk.

7.8.4.4 Submerged Zone (XS2)

The submerged zone, located permanently below the low tide level, this zone remains in constant contact with seawater, resulting in continuous chloride exposure. Chloride ingress occurs primarily through diffusion, driven by saturation. Although chloride penetration is ongoing, the absence of oxygen in submerged conditions greatly limits corrosion risk. However, if oxygen enters through cracks or marine organisms, corrosion may occur. Overall, the structural risk in this zone is moderate to low.

7.8.4.5 Spray Zone (XS3)

Spray Zone is located next to the splash zone; this area is affected by wind-driven sea spray. Structures here receive intermittent chloride-rich moisture from airborne droplets. Chloride ingress occurs through surface deposition followed by absorption into the concrete. Exposure levels depend on wind intensity and frequency. Over time, chloride buildup can lead to corrosion, particularly in porous or poorly protected concrete, resulting in a moderate to high structural risk.

7.8.5 Atmospheric Zone – Beyond the Sea Coast

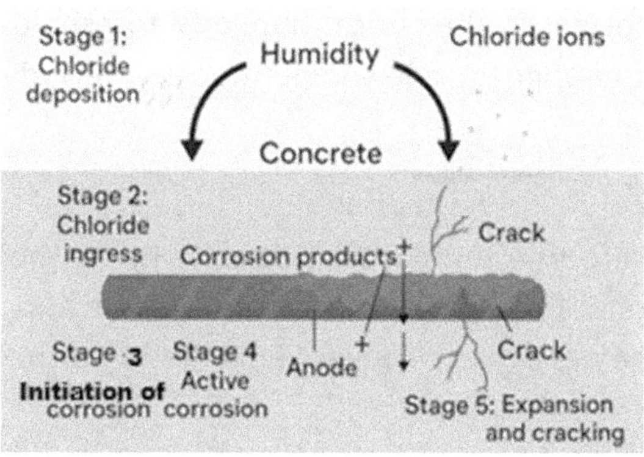

While several factors can influence the corrosion process in the marine environment, the most critical among them are moisture, oxygen, and chlorides—mainly from seawater. Moisture is the primary requirement, as it serves as an electrolyte that enables the flow of electrons between the anode and cathode. Oxygen is the next key factor, as it drives the cathodic reaction in the presence of moisture.

Figure 7.23: Stages of Corrosion in Marine Environment

In marine atmospheric conditions, humidity plays a critical role in the spreading of chloride ions through tiny air droplets and initiates the corrosion of reinforced concrete (RC) structures in coastal environments. Here's how:

7.8.6 Humidity Enhances Chloride Transport

High relative humidity (RH) increases the moisture content in concrete, which facilitates the diffusion and convection of chloride ions. In saturated conditions, chloride ions move primarily by diffusion, but in unsaturated environments, convection and chemical interactions also contribute significantly.

Table 7.5: Effect of the level of Humidity on Corrosion

Humidity Level	Effect on Chloride Spread	Impact on Structures
Low (< 60%)	Limited chloride mobility	Slower corrosion
Moderate (60–75%)	Increased diffusion	Moderate risk
High (> 75%)	Rapid chloride ingress	High corrosion probability

In coastal zones, especially wave splash and marine atmospheric zones, high humidity increases the probability of corrosion initiation. Various studies (using Monte Carlo simulations) show that variability in humidity can lead to significant differences in corrosion rates and structural degradation. Even within the same coastal region, local humidity variations due to wind patterns, rainfall, and structure orientation can cause non-uniform chloride distribution, leading to localised corrosion, and the effect of humidity level to propagate the corrosion is shown in the table below.

Humidity significantly influences the corrosion process of metals, especially in coastal environments where chloride ions are present. Here's a breakdown of how it works:

7.8.6.1 Role of Humidity in Corrosion Initiation

i) Level of Humidity

In areas where Relative Humidity (RH) > 60%, corrosion risk increases as moisture condenses on surfaces, forming a thin electrolyte layer.

Further, if RH > 75%, this is considered the critical humidity threshold. Above this, the electrolyte layer becomes thick enough to allow ionic mobility, including chloride ions, which are highly corrosive to steel.

ii) Chloride Ion Activation

Chloride ions from sea spray or airborne salt particles dissolve in the moisture present on the surface. These ions penetrate through the concrete surface and reach the steel reinforcement, breaking down the passive oxide layer and initiating corrosion.

iii) Accelerated Corrosion Mechanism

In high humidity, RH > 75%, the electrochemical reactions become more active. The Oxygen solubility in water increases to fuel the cathodic reaction. After the initiation of the corrosion level, at last the corrosion products (rust) expand, causing cracking and spalling in concrete to further accelerate the corrosion and degrade the strength and reduce the life of the structure.

iv) Seasonal and Diurnal Effects

Morning dew and seasonal rains in humid coastal areas can repeatedly wet surfaces of concrete and unprotected steel to sustain the corrosion cycles. The dry-wet cycles also.

7.8.6.2 Sea Proximity and Chloride Attack on Concrete

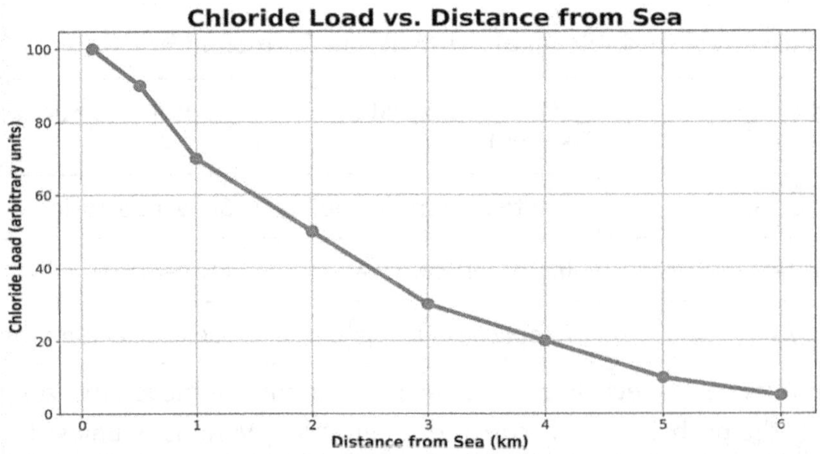

The proximity to the sea coast plays a significant role in determining the concentration of airborne chloride ions, which are a major contributor to the corrosion of reinforced concrete (RC) structures.

Figure 7.24: Risk of Chloride Attack in the proximity of the sea

The line chart of Chloride Load vs. Distance from Sea, as shown above, demonstrates how chloride concentration in the air decreases as the distance from the sea increases, influencing the corrosion potential.

The breakdown of **corrosion risk for each environmental zone** as you move progressively inland from the sea coast:

Zone 1: Splash Zone (0–500 m)

In this zone, structures may be located within the tidal area or in regions with extremely high relative humidity (RH), often approaching saturation. As a result, the concentration of chlorides in the moist air is exceptionally high, leading to a very severe corrosion risk. Therefore, structures in this zone require high-grade concrete and protective coatings to ensure durability.

Zone 2: Marine Atmospheric Zone (500 m–2 km)

In this zone, airborne salt particles are transported by wind. Humidity levels remain high, particularly due to coastal winds and fog, although the chloride concentration is lower than in Zone 1. Nevertheless, the corrosion risk to structures remains severe.

Zone 3: Transitional Zone (2–5 km)

In this zone, structures are occasionally exposed to salt-laden winds, depending on wind direction and local topography. Relative humidity ranges from moderate to high, and the chloride concentration in moist air is moderate. The corrosion risk is also moderate and tends to develop over extended periods. Therefore, the structural design life can be prolonged with appropriate moderate-level protective measures.

Zone 4: Inland Zone (>5 km)

Beyond 5 km from the sea coast, wind exposure to salt is minimal unless influenced by other sources such as industrial emissions. Humidity levels may vary depending on the climate, and the chloride concentration in moist air is low to negligible. Consequently, the corrosion risk is low, and standard concrete cover and materials are generally sufficient for protection.

Table 7-6: Corrosion Zones Based on Distance from Sea Coast

Zone	Distance from Sea	Environmental Conditions	Chloride Load	Corrosion Risk
Zone 1: Splash Zone	0–500 m	Very high RH, often near saturation and high salt concentration	Extremely high	**Very Severe**
Zone 2: Marine Atmospheric Zone	500 m –2 km	High humidity due to coastal winds and fog; airborne salt particles carried by wind	Lower than Zone 1	**Severe**

Zone	Distance from Sea	Environmental Conditions	Chloride Load	Corrosion Risk
Zone 3: Transitional Zone	2–5 km	Occasional salt-laden winds influenced by wind direction and topography; moderate to high humidity	Moderate	**Moderate** (may initiate over longer periods)
Zone 4: Inland Zone	>5 km	Minimal salt exposure unless affected by other sources (e.g., industrial emissions); climate-dependent humidity	Low to negligible	

Here's a summary of code compliance details for corrosion protection in coastal structures based on IS 456, ACI 318, and Eurocode 2:

Table 7-7: Provision of Concrete Covers by IS 456, ACI 318, and Eurocode 2 in a marine environment

Code	Exposure Class	Cover Range	Notes
IS 456	Severe to Extreme	45–75 mm	Based on general severity, not chloride-specific
ACI 318	C1, C2	50–75 mm	Specific to chloride exposure, bar size matters
Eurocode 2	XS1–XS3	30–60 mm	Based on exposure + structural class + tolerance

Notes: *Adjustments to the above standard values may be made based on structural classification, construction tolerances, and service life modelling. Design flexibility should be incorporated in accordance with performance requirements.*

7.9 Effect of Grade of Concrete on Durability of Structures

The grade of concrete plays a crucial role in determining the durability and service life of such structures. Concrete grades are typically denoted by their compressive strength (e.g., M30, M40, M50), which directly influences:

- **Permeability**: Higher-grade concrete has lower permeability, reducing ingress of chlorides and moisture.
- **Density and Compaction**: Improved compaction and reduced porosity in higher grades enhance resistance to aggressive agents.

- **Durability**: Higher grades offer better resistance to chemical attack and corrosion.

Additionally, use of suitable admixtures like pozzolanic materials (e.g. silica fume etc) enhances durability.

Higher grades (M40 and above) are recommended for zones with severe chloride exposure, while moderate grades may suffice in inland or transitional zones with proper protective measures.

7.9.1.1 Chloride Penetration Depth Over Time by Concrete Grade

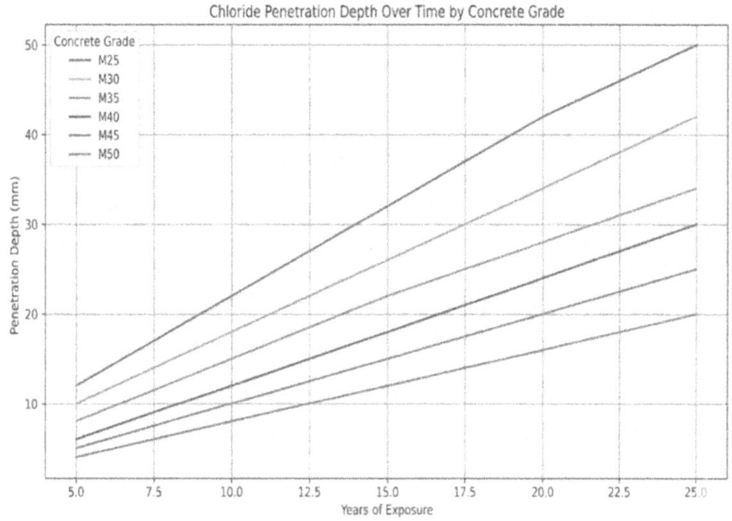

Figure 7-25: Chloride Penetration Depth Over Time by Concrete Grade

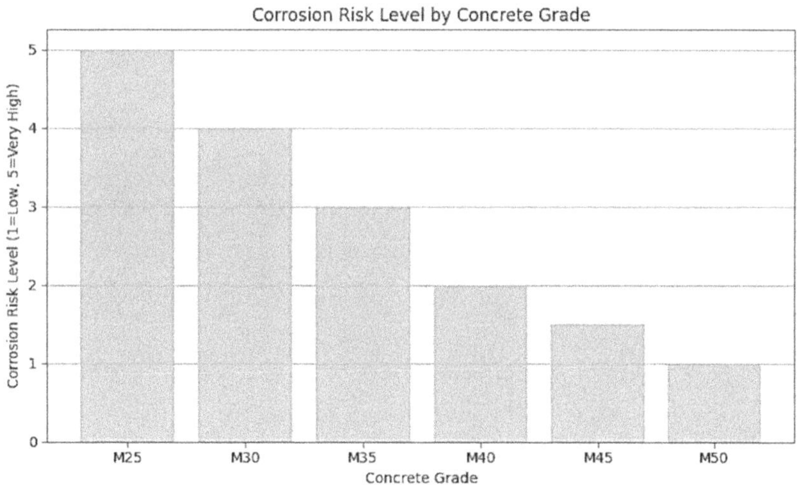

Figure 7-26: Corrosion Risk Level in various Grades of Concrete

This graph shows how chloride ions penetrate concrete over a 25-year period. Higher-grade concrete (e.g., M50) exhibits significantly lower penetration, indicating better durability.

7.9.2 Corrosion Risk Level by Concrete Grade

This bar chart represents the relative corrosion risk for each concrete grade. Lower grades (M25–M30) have a higher risk, while M45–M50 offer much better protection.

7.9.3 Corrosion Risk Levels Across Marine Zones

This bar chart shows how corrosion risk varies by proximity to the sea. It also includes recommended concrete grades for each zone.

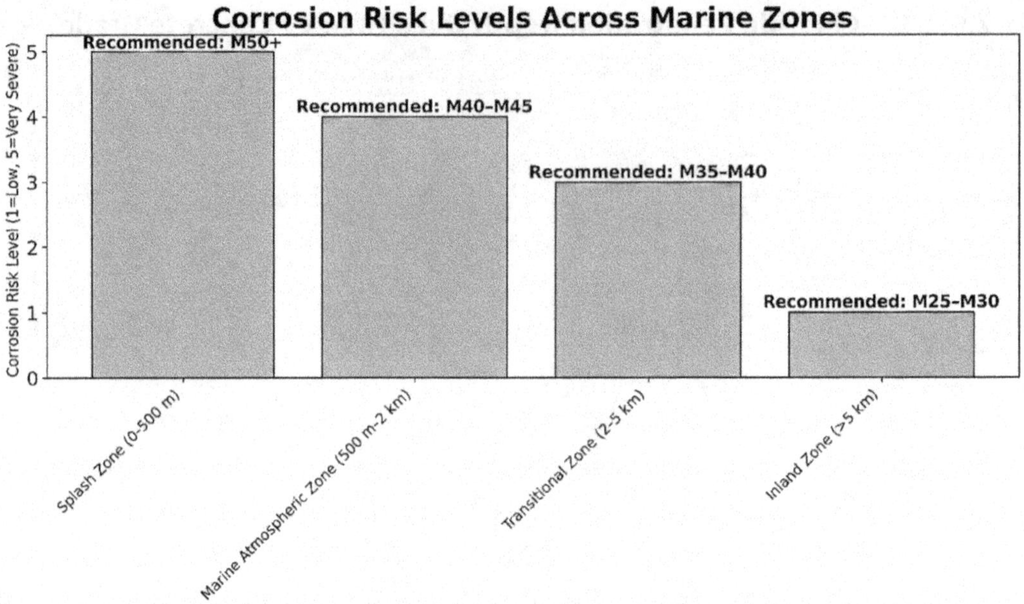

Figure 7-27: Corrosion Risk Level in various Marine Zones

Structural Condition Assessment Technique

Chapter Summary

The chapter presents a structured and comprehensive methodology for conducting a structural condition survey, which serves as a systematic process to evaluate the health and remaining service life of concrete and marine structures. It details a two-phase approach wherein Phase I focuses on preliminary visual inspection and identification of apparent deficiencies, while Phase II encompasses detailed investigations, including measurements, material testing, and structural stability analysis. The overall survey framework includes planning, data collection, field and laboratory evaluations, and the formulation of repair recommendations. These activities facilitate the assessment of structural strength, identification of internal defects, determination of corrosion activity, recognition of chemical deterioration, and understanding of material degradation mechanisms. Collectively, the chapter underscores the adoption of a rigorous, scientific, and safety-oriented methodology that enables accurate diagnosis of deterioration, reliable evaluation of material condition, and the development of appropriate rehabilitation strategies to ensure long-term structural performance and durability.

Learning Objectives

- Understand the purpose and scope of a structural condition survey
- Various Phases for Condition Survey and investigations
- Assess common forms of defects and deterioration
- Plan and execute a structured survey methodology
- Develop appropriate repair and rehabilitation recommendations

8.1 Introduction

The Structural Condition Survey is a process of assessing the present health of a structure, identifying the effects of actual working loads and forces, and finally assessing the stability & and safety of the structure to withstand its remaining life. A structural condition assessment is a process of collecting observations and data and systematically using them to evaluate and assess the condition of an existing structure. Typically, the structural condition assessment process entails the identification and basic inspection of critical members and areas, followed by a detailed inspection of larger portions of the structure, if deemed necessary by the basic inspection. The basic inspection consists of a document review to attempt to determine the original design information and criteria, followed by a visual condition assessment.

A condition survey of a civil works structure includes a comprehensive review of the design of the structure, construction techniques and materials, and operational and maintenance history. Information is obtained from available engineering data on the structure and on-site investigation. A weakness in the structural element may be some substantial inadequacy or failure of the structure. The investigators make a complete review of all available design data, conduct a detailed on-site investigation, and analyse all the available data, including the laboratory examinations or test results. A condition survey work should normally be executed in two phases, namely Phase I and Phase II. However, the key considerations to be followed during condition assessment are outlined below.

8.2 Objectives of Condition Assessment

The objectives of the condition assessment are to evaluate the existing condition, deterioration state, and performance capability of the structure by:

i) Assessing the condition of reinforced concrete members by identifying the concrete cover and concrete core as distinct functional components influencing strength and durability.

ii) Verifying the as-built workmanship and construction quality with respect to specified workmanship standards and design assumptions.

iii) Comparing design actions and loading assumptions with actual service loads, usage patterns, and any structural modifications made during the service life.

iv) Evaluating the adequacy of durability design criteria with respect to actual and evolving environmental exposure conditions.

The condition assessment, incorporating visual inspection, non-destructive testing (NDT), and diagnostic investigations, shall be carried out as a systematic and integrated process, rather than as isolated inspection stages. The assessment strategy shall establish a clear relationship between observed distress, material properties, exposure conditions, and in-situ structural performance.

i) Role of Concrete Cover in RCC Structures

Reinforced Cement Concrete (RCC) members consist of two functionally distinct components:

- Structural concrete core, responsible for load-bearing capacity and structural safety; and
- Concrete cover to reinforcement, which provides protection to embedded steel against environmental actions such as carbonation, chloride ingress, moisture penetration, and chemical attack that may initiate reinforcement corrosion.

In accordance with IS 15988 and EN 1504, the quality, thickness, continuity, and integrity of the concrete cover are key durability parameters governing service life, serviceability, and long-term performance. Therefore, the condition assessment should systematically check both the concrete cover and the concrete core at key and representative locations to identify the causes of deterioration and estimate the remaining service life of the structure.

ii) Limitations of Standard Quality Control Tests

During the structural design stage, it is assumed that specified material properties, prescribed workmanship standards, and quality control measures will be strictly complied with during construction. However, any deviations arising during construction or subsequent operation—such as deficiencies in material quality, concrete production, placement and compaction, formwork quality, reinforcement detailing and fixing, curing practices, maintenance regime, or applied service loading—may adversely affect structural reliability and reduce service life.

Concrete cube testing carried out under controlled laboratory conditions forms part of routine quality control; however, the results may not reliably represent the actual in-situ quality and performance of concrete placed in the structure. Inadequate site practices related to placement, compaction, curing, or workmanship can lead to localised defects, reduced durability, or compromised structural performance, even when laboratory test results are satisfactory.

Therefore, as part of the condition assessment, available construction records, quality control documents, and workmanship-related notes shall be reviewed and verified as a primary step. This shall be supplemented by detailed visual inspection and appropriately selected NDT and diagnostic tests to evaluate the actual physical condition of the structure and to realistically assess its durability, safety, and remaining service life.

iii) Effects of Overloading and Extreme Actions

Overloading, change in use, accidents, earthquakes, and harsh environmental conditions can speed up deterioration and cause damage, even in structures originally designed as per standards. These effects may result in cracks, excessive deflection, reduced durability, or loss of load-carrying capacity, making detailed assessment and timely repair necessary.

iv) Identification of Stages of Deterioration

Concrete deterioration happens gradually over time. Structures usually start their service life in good condition, with proper performance and no visible damage. As time passes, environmental exposure, applied loads, and material-related issues can trigger deterioration processes such as fine cracking, carbonation, chloride penetration, and corrosion of reinforcement.

Once deterioration starts, it gradually spreads and becomes visible as cracks, spalling, delamination, corrosion stains, or loss of concrete quality. As the damage progresses further, the structure may reach a stage where its usability, durability, or user comfort is affected, even though overall structural stability may still be maintained.

v) Outcome of Condition Assessment and Need for Timely Intervention

Identifying the present stage of deterioration helps in making clear decisions on how often inspections are needed and how maintenance and repairs should be planned. Detecting problems early and taking timely action can greatly increase the service life of the structure, lower overall repair costs, and ensure continued structural safety.

Based on the observed distress, test results, and overall findings of the condition assessment, suitable intervention measures—such as repair, rehabilitation, strengthening, protection, or upgrading—shall be selected in line with the principles of IS 15988 and EN 1504 to restore or enhance the required functional performance, durability, and safety of the structure.

8.3 Condition Survey of Structure

8.3.1 Phase I - Survey

The **Phase I objective** is to ascertain the general condition of the structure and determine the need for any additional engineering studies and analysis.

Problem and suspected deficiency identification is the main purpose of the Phase I work.

A preliminary field investigation, which is identified as Phase I work, is conducted to verify the existence of a known or suspected deficiency or an inadequate performance. The defect or inadequate performance is examined to ascertain its nature and to estimate its extent and effect on performance, service life, and safety of the structure.

It may include the following points:

a) Assessment of deflection or settlement, or both.

b) Visual examination of the exterior and interior of the concrete.

c) Underwater inspection of the concrete and foundation.

d) Examination of secondary chemical deposits in or on the concrete, or both.

8.3.2 Phase II Survey

The Phase II work may include additional visual examinations, measurements, structure and foundation explorations, material testing, and stress and structural stability analysis to evaluate the structural integrity. The challenges are not in simply defining the general nature of the original materials used in the structure, but the present characteristics of the material and the transformation of that material over time to the present. This transformation may involve a chemical or physical transformation of the material from years of environmental exposure, fatigue, one or more incidents or events of damage over the life of the structure, and any of these transformations to change the material properties or functionality of the existing structure. Non-destructive Test (NDT) is the most commonly used technique to evaluate the parameters of hardened concrete.

Non-destructive evaluation (NDT) is a broad category of evaluation techniques that allow engineers to predict the existing engineering properties of historic structures, materials, or assemblies. NDT is the examination of an object with technology that does not substantially affect the object's future usefulness. The methods can be as direct as visual inspection or as indirect as ultrasound or penetrating radar signal response. NDE includes methods that can detect internal and external imperfections; determine structure, composition, and material properties; and measure geometric characteristics of elements and deteriorated assemblies.

There are five major uses for NDT for the evaluation of existing structures:

- → Assessing and documenting the existing condition of the structure;
- → in-situ evaluation of materials, structural elements, and sub-assemblies;
- → determination of the quality of existing material, of the extent of the damage;
- → quality of workmanship adopted during the reconstruction; and
- → Determination of Physical Properties of existing materials for use in design, restoration, or rehabilitation.

While these five are not the only reasons to use NDT, they are the most prevalent reasons and key reasons to undertake NDT programs.

8.3.3 Planning the Survey

The condition survey of a concrete structure typically follows a structured sequence. Phase I generally includes all steps except laboratory investigation (step g), which is part of Phase II involving detailed field and lab analysis.

a. Meet with the client and define the problem

   Clarify the issue to be addressed, either at the office or on-site, to align expectations and understand project context.

b. Determine the objective and scope

   Set clear goals for the survey and define the extent of investigation required.

c. Select the evaluation team

   Choose qualified personnel with relevant expertise for conducting the survey.

d. Establish investigative procedures and techniques

   Decide on appropriate methods (e.g., visual inspection, NDT, sampling) based on the structure and objectives.

e. Collect and review engineering and geological data

   Gather existing design documents, construction records, and geological information to inform the investigation.

f. Conduct the field investigation

   Perform on-site assessments such as visual inspections, measurements, and non-destructive testing.

g. Conduct the laboratory investigation

    Analyse samples (e.g., core tests, chemical analysis) in a lab to validate field findings—typically part of Phase II.

h. Analyse findings

    Interpret data from field and lab work to assess the condition and identify causes of deterioration.

i. Make repair recommendations

    Propose suitable repair or rehabilitation measures based on the analysis.

j. Report findings

    Document all observations, analyses, and recommendations in a formal report for the client.

8.3.4 Condition Survey Team

Conducting a condition survey of quay walls and marine structures requires a multidisciplinary team with specialised knowledge in marine infrastructure diagnostics. The team must be capable of systematically identifying structural, material, and environmental issues while strictly adhering to safety protocols due to the inherently hazardous marine environment.

The team should consist of practical and dedicated diagnosticians who thoroughly investigate all observable indicators of structural behaviour. Members must possess a strong understanding of the failure mechanisms and deterioration processes specific to marine structures, including those caused by tidal forces, corrosion, and foundation instability.

The team should be composed of professionals with relevant technical expertise, ideally with hands-on experience in the design, construction, and operational monitoring of complex marine infrastructure such as quay walls, jetties, piers, and breakwaters.

Depending on the scope and complexity of the survey, the team may include 2 to 4 experts, ensuring a focused yet comprehensive approach.

8.3.4.1 Team Composition and Roles

1. **Civil Engineer**
    - Evaluates structural integrity, signs of deterioration (e.g., cracking, spalling), and load-bearing capacity.
    - Reviews design documentation and compares actual conditions with expected performance.
2. **Marine/Coastal Engineer**
    - Assesses tidal effects, wave loading, scour, and sedimentation.
    - Evaluates the interaction between marine forces and structural elements.
3. **Geotechnical Engineer**

- Investigates foundation conditions, soil-structure interaction, and subsurface stability.
- Conducts geotechnical testing and evaluates potential settlement or liquefaction risks.

4. **Materials Specialist**
   - Analyses concrete samples for chloride ingress, sulphate attack, corrosion, and other degradation mechanisms.
   - Recommends material-specific repair and protection strategies.
5. **Preferred Experience**
   - Team members should have prior experience in the design, construction, inspection, and maintenance of marine structures such as quay walls, jetties, piers, and breakwaters.
   - Familiarity with non-destructive testing (NDT), underwater inspection techniques, and marine construction practices is essential.
6. **Safety Considerations**
   - **Personal Protective Equipment (PPE):** All team members must wear appropriate PPE, including life jackets, helmets, safety harnesses, and non-slip footwear.
   - **Marine Access Safety:** Use certified vessels or platforms for overwater access. Ensure proper boarding procedures and emergency equipment availability.
   - **Weather Monitoring:** Surveys should be scheduled considering tide charts, wind conditions, and weather forecasts to avoid hazardous conditions.
   - **Confined Space and Diving Operations:** If underwater or confined space inspections are required, certified divers and confined space protocols must be followed.
   - **Emergency Preparedness:** A safety plan including first aid, communication protocols, and evacuation procedures must be in place.
7. **Team Size**

   Depending on the scope and complexity of the survey, the team may consist of 2 to 4 professionals, ensuring coverage of all critical technical domains while maintaining operational efficiency and safety.

8.3.5 Visual Inspection

Visual inspection is an extremely powerful tool and one of the most common and oldest on-hand non-destructive testing practices available. Visual inspection provides a lot of information with reference to the structure and its condition, but has certain limitations and requirements. During the field investigation, there should be a systematic approach to the visual examination. A walk-through of the entire project should be made; the environs, appurtenant structures, and the main structure are examined. The visual inspection of the atmosphere can be done with sketches, still photographic techniques, or video recordings.

Inaccessible areas can be inspected with zoom lenses or powerful low-light high-definition binoculars backed up by detailed photographs taken using flash guns (Browne, Doyle, and Papworth 1981). It is essential that any deficiency noted be located precisely and rapidly; this includes all categories of concrete as well as other affected elements in the project. This can be achieved by zoning and sub-zoning the particular element into inspection areas and locating the defects in the areas on a scale drawing or photograph. The methods for recording and presenting the mass of survey data will change from structure to structure depending on the complexity of the problem being examined.

Visual inspection can only be administered by a technically sound professional who has an understanding of structure, construction methodologies, and material science. Visual inspection just gives an impression of the visible problem, and the hidden problem stays unnoticed. It also doesn't provide us with any quantitative information in relation to the properties of the material. Because of these limitations, Visual inspection is not enough and has to be complemented by other non-destructive and partially destructive testing methods. With reference to BS1881: Part 201:1986.

During visual inspection, the survey team should record and collect independent photographic and documentary evidence of the site, before commencement of detailed or phase-II investigation. All prominent defects in the form of cracks, spalling concrete, settlement, movement, or other structural or cosmetic defects shall be recorded with digital photographs and compiled in the preliminary report.

Video evidence, although available, is not widely used due to a lack of ability to focus on specific areas. Video footage is typically utilised only to give wide-angle generalised views of large areas, but it cannot be documented with the same detail as high-resolution digital photographic evidence.

8.4 Non-Destructive Tests (NDT)

Non-destructive evaluation by NDT method is a broad category of evaluation techniques that allow engineers to predict the existing engineering properties of historic structures, materials, or assemblies. NDT is the examination of an object with technology that does not substantially affect the object's future usefulness. The methods can be as direct as visual inspection or as indirect as ultrasound or penetrating radar signal response. NDE includes methods that can detect internal and external imperfections; determine structure, composition, and material properties; and measure geometric characteristics of elements and deteriorated assemblies. This introductory paper presents the basic technologies by which the workshop launches its more detailed discussions on advancing technologies.

There are five major uses for NDT in the consideration of historic structures:

- assessing and documenting the existing condition of a building or structure;
- in-situ evaluation of materials, structural elements, and sub-assemblies;
- determination of the quality of existing material and the extent of the damage;
- quality assurance during the reconstruction of historic buildings; and

- determination of Physical Properties of existing materials for use in design, restoration, or rehabilitation.

While these five are not the only reasons to use NDE, they are the most prevalent reasons and key reasons to undertake NDE programs in any historic restoration.

A condition survey of a civil works structure includes a comprehensive review of the design of the structure, construction techniques and materials, and operational and maintenance history. Information is obtained from available engineering data on the structure as well as on-site investigation.

A weakness in the structural element may be some substantial inadequacy or failure of the structure. The investigators make a complete review of all available design data, conduct a detailed on-site investigation, and analyse all the available data, including the laboratory examinations or test results.

8.4.1 Rebound Method

8.4.1.1 Rebound Hammer by Schmidt

In 1948, a Swiss engineer, Ernst Schmidt, developed a test hammer for measuring the hardness of concrete by the rebound principle. A Rebound Hammer is one of the simplest or most handy instruments used to compute the elastic property or compressive strength of concrete, generally surface hardness and penetration resistance. The Schmidt rebound hammer works on the principle that the rebound of an elastic mass depends on the hardness of the surface against which the mass strikes. Once the plunger of the rebound hammer is pushed against the surface of the concrete, it will strike the concrete with a defined energy, and the spring-controlled mass rebounds.

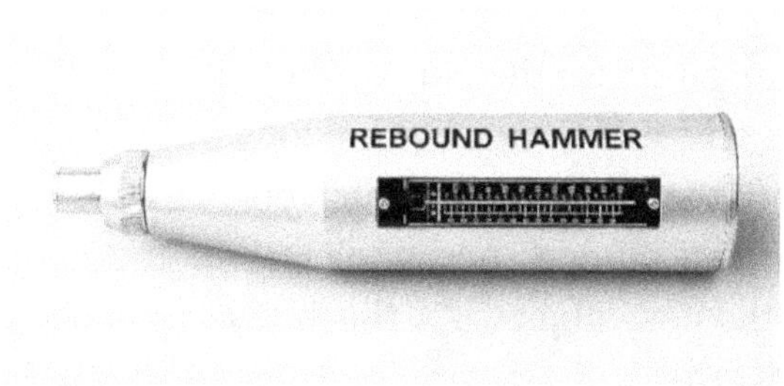

Figure-8.1: Picture of Rebound Hammer

The extent of such a rebound depends on the surface hardness of the concrete. The rebound value is read from the graduated scale and is designated as the rebound number or rebound index. With reference to the conversion chart, the rebound number can be used to find out the compressive strength. The procedure is as per IS 13311:1992 (Part-II).

8.4.1.2 Principle of the test

The test is based on the principle that the rebound of an elastic mass depends on the hardness of the surface upon which it impinges. When the plunger of the rebound hammer is pressed against the surface of the concrete, the spring-controlled mass rebounds and the extent of such rebound depends upon the surface hardness of the concrete. The surface hardness and, therefore, the rebound are taken to be related to the compressive strength of concrete. The rebound is read off along a graduated scale and is designated as the rebound number or rebound index.

8.4.1.3 Description

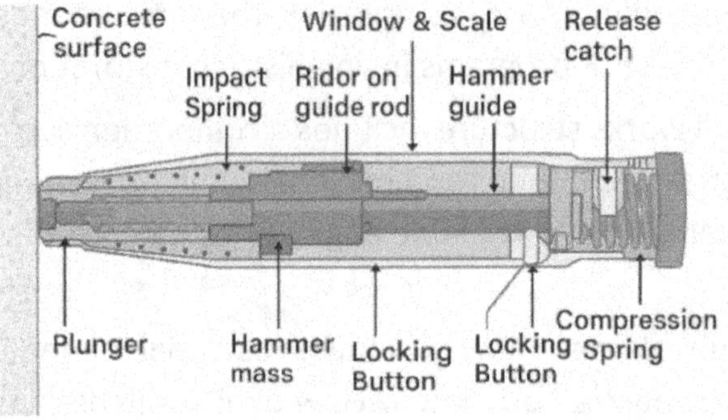

Figure-8.2: Basic Features of Rebound Hammer

The Schmidt rebound hammer is shown in Figure 8.1. The hammer weighs about 1.8 kg and is suitable for use both in a laboratory and in the field. A schematic cutaway view of the rebound hammer is shown in Figure 8.2.

The main components include the outer body, the plunger, the hammer mass, and the main spring. Other features include a latching mechanism that locks the hammer mass to the plunger rod and a sliding rider to measure the rebound of the hammer mass. The rebound distance is measured on an arbitrary scale marked from 10 to 100. The rebound distance is recorded as a "rebound number" corresponding to the position of the rider on the scale.

The rebound hammer contains a spring-loaded steel hammer, which, when released, strikes a steel plunger in contact with the concrete (ASTM C805-79/CRD-C 22-80). The amount of rebound of the plunger is measured on a linear scale attached to the instrument; in effect, the plunger rebounds depending upon the resiliency of the material struck.

8.4.1.4 Method of Testing

To prepare the instrument for a test, release the plunger from its locked position by pushing the plunger against the concrete and slowly moving the body away from the concrete. This causes the plunger to extend from the body, and the latch engages the hammer mass to the plunger rod (Figure 8.3A). Hold the plunger perpendicular to the concrete surface and slowly push the body toward the test object. As the body is pushed, the main spring connecting the hammer mass to the body is stretched (Figure 8.3B). When the body is pushed to the limit, the latch is automatically released, and the energy stored in the spring propels the hammer mass toward the plunger tip (Figure 8.3C). The mass impacts the shoulder of the plunger rod and rebounds. During rebound, the slide indicator travels with the hammer mass and records the rebound distance (Figure 8.3D). A button on the side of the body is pushed to lock the plunger in the retracted position, and the rebound number is read from the scale.

The test can be conducted horizontally, vertically upward or downward, or at any intermediate angle. Due to different effects of gravity on the rebound as the test angle is changed, the rebound number will be different for the same concrete and will require separate calibration or correction charts. Some advantages of the rebound hammer are that it is easily portable, easy to use, low in cost per test, and can be used quickly to cover a large surface area. The hammer is valuable as a purely qualitative tool.

8.4.1.5 Estimation of the strength

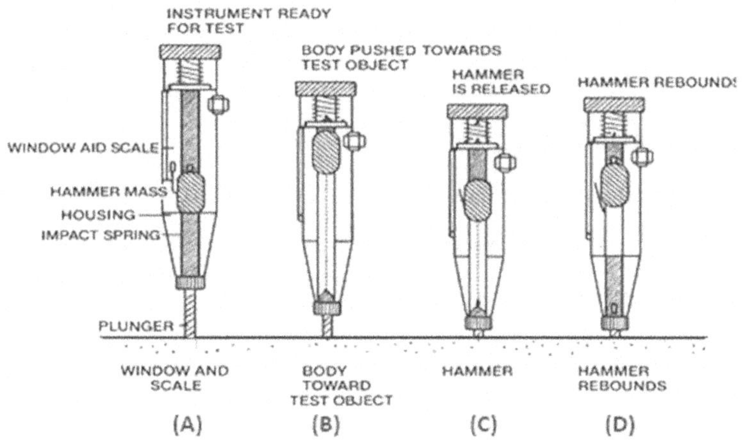

Estimation of the strength of concrete by the Schmidt hammer within an accuracy of ±15 to ±20 percent may be possible only for specimens cast, cured, and tested under identical conditions as those from which the calibration curves are established.

Figure-8.3: Working Methods of Rebound Hammer

The prediction of the strength of structural concrete by using calibration charts based on laboratory tests is not recommended. The user of the Schmidt hammer should make up their own calibration curves for the particular concrete under study.

8.4.1.6 Working of the rebound hammer

When the plunger of the rebound hammer is pressed against the surface of the concrete, a spring-controlled mass rebounds, and the extent of such rebound depends upon the surface hardness of the concrete. The rebound distance is measured on a graduated scale and is designated as a rebound number. Basically, the rebound distance depends on the value of kinetic energy in the hammer, prior to impact with the shoulder of the plunger, and how much of that energy is absorbed during impact. The energy absorbed by the concrete depends on the stress-strain relationship of the concrete. Thus, low-strength, stiff concrete will absorb more energy than high-strength concrete and will give a lower rebound number.

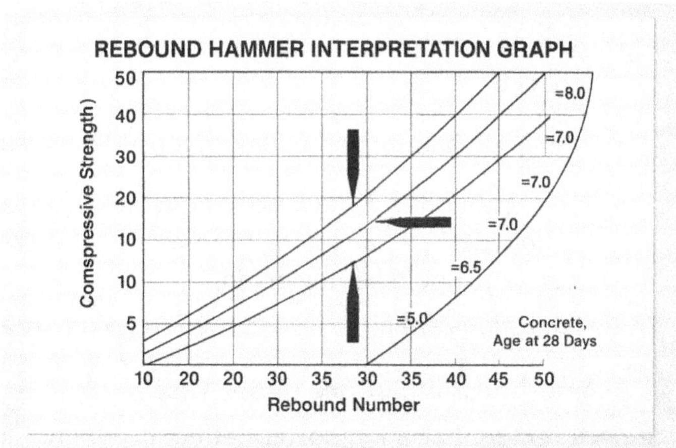

Figure-8.4: Sample Graph for Interpretation of Rebound Hammer Test Results

8.4.1.7 Limitations

Although the rebound hammer provides a quick, inexpensive means of checking the uniformity of concrete, it has serious limitations, and these must be understood clearly for the interpretation of test results.

8.4.1.8 Factors affecting rebound number

The results of the Schmidt rebound hammer are significantly influenced by the following factors

i) Smoothness of Test Surface

ii) Size, Shape, and Rigidity of the Specimen

iii) Age of Test Specimen

iv) Moisture Condition

v) Type of Coarse Aggregate

vi) Type of Cement

vii) Type of Mould

viii) Surface Carbonation

The influence of these factors has different magnitudes. Hammer orientation will also influence the measured values, although correction factors can be used to allow for this effect.

8.4.1.9 Precautions to be taken while using the rebound hammer

The following precautionary measures are taken while using the rebound hammer, which may help minimise error.

- The surface on which the hammer strikes should be smooth and uniform. Moulded faces in such cases may be preferred over Trowelled faces.
- The test hammer should not be used within about 20 mm of the edge of the specimen.
- Rebound hammer should not be used over the same points more than once.
- The rebound test must be conducted at closely placed test points; on at least 10 to 12 locations, while taking the average of extremely high and low values of the index number should be neglected.

These limitations are discussed in the foregoing order.

a. Smoothness of Surface under Test

Surface texture has an important effect on the accuracy of the test results. When a test is performed on a rough-textured surface, the plunger tip causes excessive crushing, and a reduced rebound number is measured.

b. Surface and Internal Moisture Condition of the Concrete

The degree of saturation of the concrete and the presence of surface moisture have a decisive effect on the evaluation of test hammer results. It has been demonstrated that well-cured, air-dried specimens, when soaked in water and tested in the saturated surface-dried condition, show rebound readings 5 points lower than when tested dry. It is suggested that, whenever the actual condition of the field concrete or specimens is not known, it would be desirable to saturate the surface several hours prior to testing and use the correlation for tests on saturated surface-dried specimens.

c. Carbonation of Concrete Surface

Surface carbonation of concrete significantly affects the Schmidt rebound hammer test results. The carbonation effects are more severe in older concretes when the carbonated layer can be several millimetres thick and, in extreme cases, up to 20 mm thick. In such cases, the rebound numbers can be up to 50% higher than those obtained on an uncarbonated concrete surface. Suitable correction factors should be established in such cases; otherwise, overestimation of concrete strength will result.

8.4.1.10 Rebound Number and Estimation of Compressive Strength

There is a general correlation between the compressive strength of concrete and the hammer rebound number. Under standard conditions, the result of a properly calibrated hammer generally varies between ±15 and ±20%. However, the probable accuracy of the estimation of concrete strength in a structure is ±25%.

8.4.2 Ultrasonic Pulse Velocity Test

The ultrasonic pulse velocity is probably the most widely used method for the non-destructive evaluation of in-situ concrete. It is a truly non-destructive method, as the technique uses mechanical waves, resulting in no damage to the concrete element being tested. It is done to evaluate the compressive strength and quality of concrete by introducing an ultrasonic wave pulse into the concrete. This method can be used for examining the quality of concrete and detecting internal cracking and other defects, as well as changes in concrete, such as deterioration due to an aggressive chemical environment. By using the pulse velocity method, it is also possible to estimate the strength of concrete empirically.

The test instrument consists of a means of producing and introducing a wave pulse into the concrete (pulse generator and transmitter) and a means of sensing the arrival of the pulse (receiver) and accurately measuring the time taken by the pulse to travel through the concrete. This method computes the time of travel of an ultrasonic pulse passing through the concrete. In test specimens, a relatively higher velocity is attained when concrete quality is good in terms of uniformity, density homogeneity, etc.

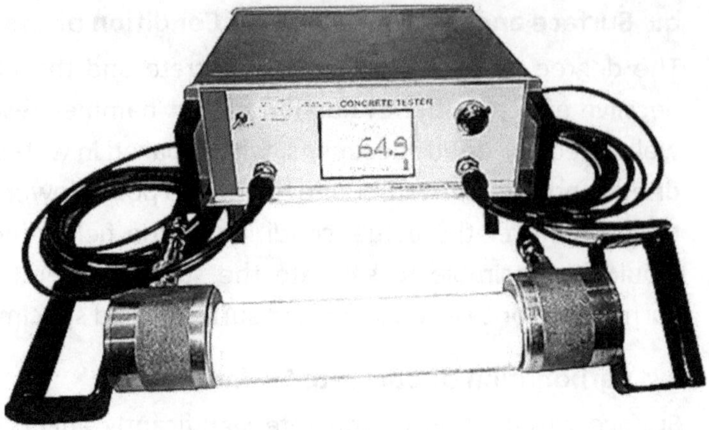

Figure-8.5: Schematic Diagram of UPV Test Instrument

A schematic diagram of the UPV test instrument is shown in Figure 8.5. The instrument comes with a set of two transducers, one for transmitting and receiving the ultrasonic pulse. Transducers with frequencies of 25 to 100 kHz are usually used for testing concrete. Transducer sets having different resonant frequencies are available for special applications: high-frequency transducers (above 100 kHz) are used for thin concrete sections, relatively short path lengths, or high-strength concrete, whereas low-frequency transducers (below 25 kHz) are used for larger sections and relatively longer path lengths, or concrete with larger size aggregates. These transducers primarily generate compressional waves at predominantly one frequency, with most of the wave energy directed along the axis normal to the transducer face. A commonly used instrument is shown in Figure 8.5.

To transmit or receive the pulse, the transducers should be in full contact with the concrete surface; otherwise, an air pocket between the test specimen and transducer may introduce an error in the indicated transit time. This error is introduced because only a negligible amount of wave energy can be transmitted through air. For proper surface contact, couplants such as grease and petroleum jelly shall be used. The coupling layer should be as thin. After applying the couplant between the test specimen and transducer, constant pressure is applied on the transducer, and repeated readings at that location are taken until a minimum value of transit time is obtained. If the concrete surface is very rough, thick grease may be required as a couplant.

When the pulse generated is passed into the concrete from the transducer by means of a liquid coupling material, it goes through multiple reflections at the boundaries of the different material phases inside the concrete. A complex structure of stress waves develops, which consists of both longitudinal and shear waves, and transmits through the concrete. The first waves to arrive at the receiving transducer are the longitudinal waves, which are transformed into an electrical signal by a second transducer. Electronic timing circuits enable the transit time T of the pulse to be measured. From known time and distance, the pulse velocity through the concrete can be calculated.

IS: 13311 (Part 1 (Indian Standard Code of Practice) gives criteria for concrete quality grading, part of which is given below:

Table 8-1: Pulse velocity vs Concrete Quality

- Pulse velocity - Concrete Quality
    a) Above 4.5 km/s – Excellent
    b) 3.5 km/s to 4.5 km/s - Good
    c) 3.0 km/s to 3.5 km/s - Medium
    d) Below 3.0 km/s - Doubtful

The pulse velocity for ordinary concrete is typically 3700 to 4200 m/s. Therefore, for a 300-mm path length, the travel time is approximately 70 to 85μs. It is obvious that the instrument must be very accurate to measure such a short transit time. The path length should also be carefully measured. Because the pulse velocity method is a wave propagation

n technique, any sources creating even the slightest wave motion in the element under test (e.g., jackhammers or vibration of an operating machine) should be eliminated during the time of the test. Many other factors also affect the pulse velocity. However, the equipment has sufficient power to penetrate through electro-acoustic signals 15 to 20 m of good continuous concrete and has a high data acquisition-to-cost ratio.

8.4.2.1 Method of Measurement

While testing a structure, there are three possible configurations in which the transducers may be arranged, as shown in Figure 8.6A to C. These are (1) direct transmission, (2) semi-direct transmission, and (3) indirect or surface transmission.

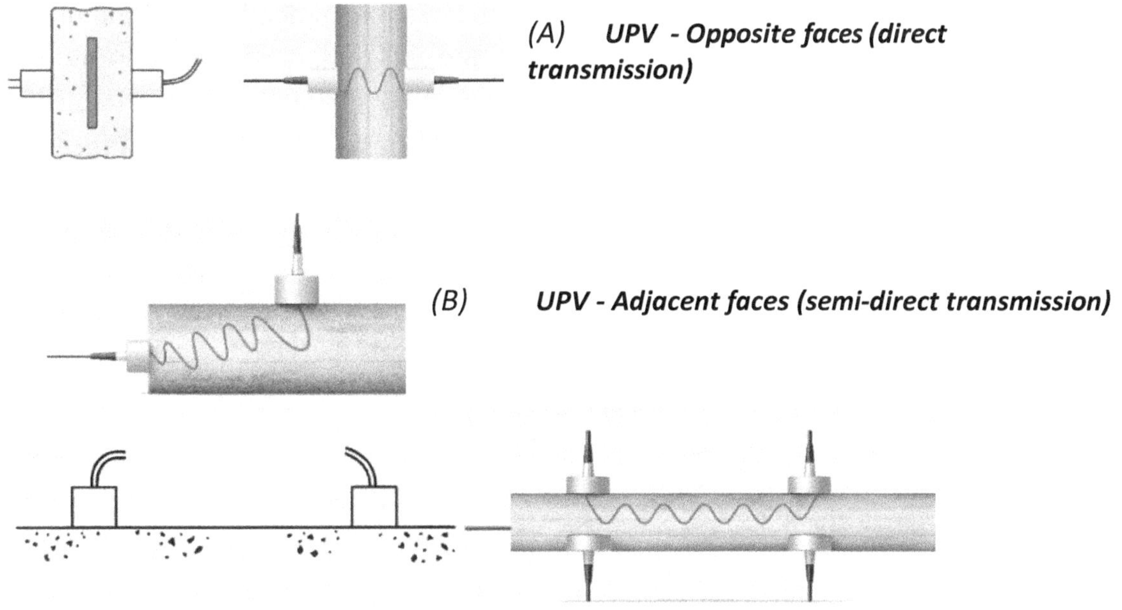

FIGURE 8.6 Pulse velocity measurement configurations. (A) Direct method. (B) Semi-direct method. (C) Indirect surface method.

The direct transmission method, Figure 8.6A, is the most desirable and the most satisfactory arrangement because the maximum energy of the pulse is transmitted and received with this arrangement. The semidirect transmission method, Figure 8.6B, can also be used quite satisfactorily. However, care should be exercised that the transducers are not too far apart; otherwise, the transmitted pulse might attenuate, and a pulse signal might not be detected. This method is useful in avoiding concentrations of reinforcements.

The indirect or surface transmission method, Figure 8.6C, is least satisfactory because the amplitude of the received signal is significantly lower than that received by the direct transmission method. This method is also more prone to errors, and a special procedure may be necessary for determining the pulse velocity.

8.4.2.2 Advantages and Limitations

The pulse velocity method is an excellent means for investigating the uniformity of concrete. The test procedure is simple, and the equipment is easy to use in the laboratory as well as in the field. The testing procedures have been standardised by ASTM and other organisations, and test equipment is available in the market. With the availability of small portable digital instruments, which are relatively inexpensive and easy to operate, ultrasonic testing adds a new dimension to quality control of concrete in the field.

Ultrasonic pulse velocity tests can be carried out on both laboratory-sized test specimens and concrete structures.

8.4.3 Impact Echo Test

In addition to the above electro-acoustic instrument, a second type of ultrasound technique is based on the response of a small impact hammer. Impact Echo ultrasound is a method that combines rebound methods and ultrasound. Impact Echo ultrasound uses a small hammer to impart a vibration in a structural member.

The vibration is then picked up with an ultrasound transducer or receiver.

The evaluation of the waveform can be conducted to determine the frequency of the echoed response and the time for the return wave, as shown in Figure 8.7.

Ultrasound may be the simplest of the advanced techniques of NDE. It is based on the theory that energy waves are transmitted through solids as a function of both elastic properties and density.

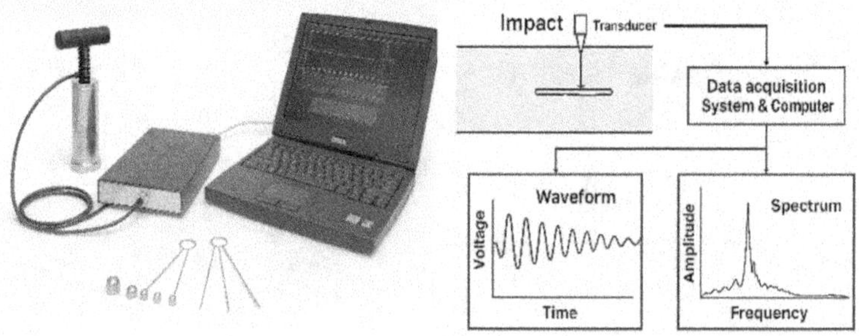

Figure 8.7: Impact echo method. (Impact Echo Instruments)

It is often used to identify internal defects, density changes, crack depths, and delaminations. The most common type of ultrasound uses a transmitter and a receiver array. The transmitter frequency is determined by the nature of the material that is being evaluated. A second type of ultrasound technique is based on the response of a small impact hammer. Impact Echo " ultrasound is a method that combines rebound methods and ultrasound. Impact Echo" ultrasound uses a small hammer to impart a vibration in a structural member. The vibration is then picked up with an ultrasound transducer or receiver. The evaluation of the waveform can be conducted to determine the frequency of the echoed response and the time for the return wave, as shown in Figure 8.7.

8.4.4 Half–Cell Potential Test

8.4.4.1 Introduction

Half-Cell Potential (HCP) is a non-destructive test method used to investigate the corrosion potential (tendency) of embedded reinforcing steel. In the marine environment, the state of steel plays a major role in the durability and longevity of the structure. Corrosion of steel reinforcement in concrete is a crucial problem in marine structures, as it poses a serious risk to the structural integrity of RCC structures. Therefore, a non-destructive inspection technique is needed to assess the corrosion of the reinforcement in order to maintain, protect, and modify the RCC structure so that it remains safe. However, the progress of corrosion cannot be visually assessed until a crack or a delamination appears. Therefore, during NDT, much attention is given to identifying the corrosion potential of reinforcement, and the corrosion process can be tracked using electrochemical techniques.

It is well known that the probability of corrosion in concrete structures depends on the ionic conductivity of the concrete electrolyte, the humidity, the temperature, and the quality of the concrete cover. Corrosion occurs simultaneously at the anode and cathode areas of steel when iron ions are released (at the anode areas) into the concrete to react and form deposits (at the cathode areas) as an electrolytic cell. Thus, potential readings of reinforcement measure the electrical activity of the corrosion process. The most popular method of *in situ* corrosion testing is the half-cell potential measurement. The use of this method and the interpretation of its results are described in ASTM C876.

It should be noted that half-cell potential values merely provide information about the probability of corrosion and not about the rate of corrosion, and they also have some inherent limitations. Therefore, corrosion potential measurements should be supplemented and corroborated with other non-destructive testing methods.

8.4.4.2 Principle & Procedure

The HCP method involves making one electrical connection between reinforcing steel and the positive terminal of a high-impedance voltmeter, and another connection between the negative terminal of the voltmeter and a reference electrode (RE). The RE consists of a copper electrode, enclosed in a plastic housing and surrounded by a saturated copper sulphate solution. The end of the cell is a porous disk, through which the copper sulphate solution can make electrical contact with the concrete. The most widely used half-cell RE is the copper-copper sulphate (CSE) reference cell, as shown in Figure 8.8.

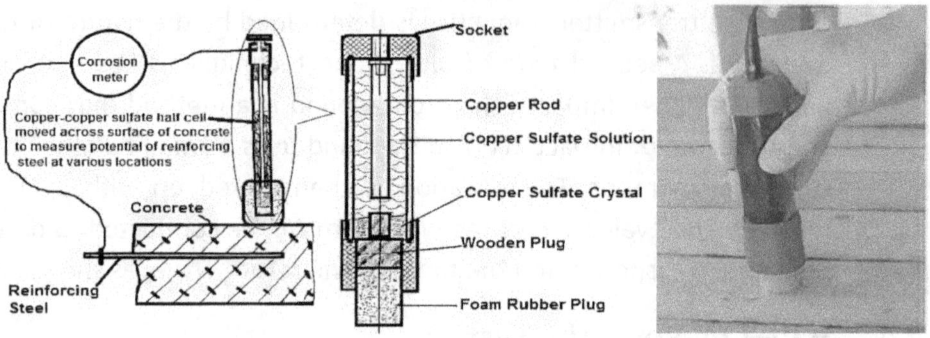

Figure 8.8: Schematic Details of HCP Measuring Device

The HCP instrument measures the potential and the electrical resistance between the reinforcement and the surface to evaluate the corrosion activity as well as the actual condition of the cover layer during testing. The electrical activity of the steel reinforcement and the concrete leads them to be considered as one-half of a weak battery cell, with the steel acting as one electrode and the concrete as the electrolyte. The name half-cell surveying derives from the fact that an electrode forms one half of the cell, and the reinforcing steel in the concrete forms the other. The electrical potential of a point on the surface of a steel reinforcing bar can be measured by comparing its potential with that of a copper sulphate reference electrode on the surface. The greater the potential difference between the electrodes, the higher the risk of corrosion in reinforcement. Practically, this is achieved by connecting a wire from one terminal of a voltmeter to the reinforcement and another wire to the copper sulphate reference electrode. Then generally readings taken are at the grid of 1 x 1 m for slabs, and walls, and at 0.5 m c/c for columns, beams.

The survey procedure is first to locate the steel and determine the bar spacing using a cover meter. The cover concrete is removed locally over a suitable bar, and an electrical connection is made to the steel. It is necessary to check that the steel is electrically continuous by measuring the resistance between two widely separated points. The reinforcing bar is connected to the half-cell via a digital voltmeter, see figure 8.8. Readings of half-cell potential are taken over a regular grid of points (say 1m apart) to give a potential map of the area.

Contour lines may be plotted between points of equal potential to indicate those areas that have the greatest risk of corrosion. Locally exposing and inspecting the reinforcement in areas where both high and low risks of corrosion are indicated can be used to calibrate the potential readings for the structure with respect to its present corrosion and the need for further investigation. In this process, measurements are quick as potential values only take a few seconds to stabilise before the next measurement can be taken.

The risk of corrosion is evaluated by means of the potential gradient obtained; the higher the gradient, the higher the risk of corrosion. The test results can be interpreted based on the following table.

Table 8-2: Half-cell Potential vs Probability of Corrosion

Half–cell potential (mV) relative to Cu-Cu sulphate Ref. electrode	Probability of Corrosion
Less than – 200	90 % No Corrosion
Between – 200 to – 350	50 % (uncertain)
Above – 350	>90 % Possibility of Corrosion activity

8.4.4.3 Significance and Use

- This method may be used to indicate the corrosion activity associated with steel embedded in concrete.
- This method can be applied to members regardless of their size or the depth of concrete cover.
- This method can be used at any time during the life of the concrete member.

8.4.4.4 Influence of Test Conditions

- Dense cover, concrete and higher cover depth will shift the half-cell readings toward a more negative value
- Carbonation of concrete will shift the half-cell readings toward a more negative
- Higher temperature will shift the half-cell readings toward a more negative
- An increase in chloride ion concentration will shift the half-cell readings toward a more negative
- Use of Anodic Corrosion Inhibitors will shift the half-cell readings toward more positive values

8.4.4.5 Reliability & Limitations

The test does not measure the actual corrosion rate or whether corrosion activity has already started, but it indicates the probability of the corrosion activity depending upon the actual surrounding conditions. If this method is used in combination with resistivity measurement, the accuracy is higher. If the concrete surface has dried to the extent that it is dielectric, then pre-wetting of the concrete is essential. However, there are important limitations in terms of data interpretation that need to be taken into consideration.

Table 8-3: Typical ranges of half-cell potentials of rebar in concrete (adapted from RILEM TC-154 EMC: Electrochemical Techniques for Measuring Metallic Corrosion)

Conditions	Potential values (mV/CSE)
Humid, chloride-free concrete	-200 to +100
Wet, chloride-contaminated concrete	-600 to -400
Water-saturated concrete without oxygen	-1000 to -900
Humid, carbonated concrete	-400 to +100
Dry, carbonated concrete	0 to +200
Dry concrete	0 to +200

The effect of the concrete condition (dry or wet), presence of chloride, absence of oxygen at the rebar surface (due to saturation), cover thickness, concrete resistivity, and temperature are all factors that can influence the results by shifting their potential reading towards a more positive or negative value, as shown in Table 8-3. This can make the data interpretation challenging when using the guidelines given in the ASTM C876 (Table 1), especially around the uncertain measurement ranges.

8.4.5 Chloride Testing & Chemical Tests

Chloride testing (and other chemical tests) of concrete can provide extremely useful information regarding the cause of defects or damage to concrete. Corrosion of reinforcement steel cannot occur when the reinforcement steel is encased in concrete. The surface of the reinforcement steel is covered by a stable protective oxide film that is formed in an alkaline environment. Steel is protected by the concrete cover. Steel develops a protective layer on its surface, which is self-generated after cement hydration. As long as that protective oxide film is present, the steel remains intact. When the chloride attacks the concrete, it destroys that protective layer, and corrosion starts taking place, resulting in a subsequent reduction in the strength, serviceability, and aesthetics of the structure. This ingress of chloride ions in concrete is known as Chloride attack. External chloride is the chloride that enters the concrete during post-hardening. When the concrete is directly affected by seawater, external chloride also enters in the form of an airborne salt spray in the structure next to the coast. Then it is necessary to check the trends of origin of chlorides (i.e., external source or contaminated sand or aggregates). It is also necessary to check their movement through the concrete.

Table 8-4: Guidelines for Identification of Corrosion-Prone Locations based on Chemical Analysis

Sl. No.	Chemical Test Results	Interpretations
1	High pH values greater than 11.5 and very low chloride content	No corrosion
2	High pH values and high chloride content greater than threshold values (0.15 % by weight of cement)	Corrosion prone
3	Low pH values and high chloride content (more prone than threshold values of chloride 0.15% by weight of cement.	Increased risk of corrosion

Post-construction contamination can come from chloride salts, used for de-icing, that have dissolved in water and subsequently permeated the concrete from the marine environment. Chloride salts, in the presence of moisture, can cause the accelerated corrosion of the reinforcing steel within the concrete, resulting in an expansive rusting reaction of the reinforcement and subsequent spalling of the concrete. The aim of the chloride testing is to establish the level of chlorides within the concrete structure, either as an overall level or, via the taking of incremental samples, as a profile through the depth of the structure. Chloride content can be determined from broken samples or core samples of concrete. Primarily, the level of chloride near the steel-concrete interface is of prime importance. Chloride present in concrete is fixed (water-insoluble) as well as free (water-soluble).

Though it is the water-soluble chloride ions that are important from a corrosion risk point of view, total acid-soluble (fixed as well as free) chloride contents are determined and compared with limiting values specified for the concrete to assess the risk of corrosion in concrete. The total acid-soluble chloride is determined in accordance with IS: 14959 Part – III – 2001, whereas for assessment of water-soluble chlorides, the test consists of obtaining the water extracts and conducting a standard titration experiment for determining the water-soluble chloride content, and is expressed as water-soluble chloride expressed by weight of concrete or cement. The method gives the average chloride content in the cover region. Further, a chloride profile across the cover thickness will be a more useful measurement, as this can help to make a rough estimate of the chloride content diffusion rate. The total soluble chlorides are determined in accordance with IS: 14959 (Part 2) 2001.

For water-soluble chlorides, obtain the water extracts and carry out standard titration to determine the chloride content. It is expressed by water-soluble chloride and also expressed by the weight of concrete or cement. The water-soluble chloride testing gives the average chloride content in the concrete cover region. Further, a level of chloride across the cover thickness will be a more useful measurement. This can permit the making of a preliminary estimate of the chloride diffusion rate.

One recent development for field testing of chloride content includes the use of chloride ion-sensitive electrodes. This is commercially known as "Rapid Chloride test kit-4". The test consists of obtaining powdered samples by drilling and collecting them from different depths (every 5mm), mixing the sample (of about 15 g weight) with a special chloride extraction liquid, and measuring the electrical potential of the liquid by chloride-ion selective electrodes. With the help of a calibration graph relating electrical potential and chloride content, the chloride content of the samples can be directly determined. Based on the chemical analysis, corrosion-prone locations can be identified as per the guidelines given in Table 8-4.

8.4.6 Carbonation Test

The concrete carbonation test for in-situ concrete is associated with the corrosion of reinforcement steel. Due to the concrete carbonation, reinforcement corrosion often occurs on the structure. The attack of atmospheric CO2 reduces the alkalinity in the concrete. The concrete porosity allows the CO2 to react with the alkalis like calcium, sodium, and potassium hydroxides, which are formed due to cement hydration in concrete. When the atmospheric carbon dioxide (CO2) reacts with hydrated cement minerals (CaOH2). CO2 reacts in the presence of moisture with CaOH2 and produces carbonates (CaCO3). Carbonates slowly penetrate below the exposed surface of the concrete.

Concrete is basically alkaline in nature, having a pH of more than 12.6. Due to concrete carbonation, its alkalinity is reduced, and the corrosion causes an expansion in the volume of the film over the steel bar. This expansion in volume creates cracks in concrete. These cracks further invite CO2. Thus, the vicious cycle continues and ultimately leads to a state of disaster. The carbonation process is also called depassivation. Carbonation penetrates below the exposed surface of concrete extremely slowly. The time required for carbonation can be estimated by knowing the concrete grade and using the following equation:

$$t = (d/k)^2$$

where,

t is the time for carbonation,

d is the concrete cover,

k is the permeability

Typical permeability values are shown in Table 8-5.

Table 8-5: Permeability Values versus Concrete Grade

Grade of Concrete	Permeability $x\ 10^{-9}$ cm/sec
15	17
20	10
25	6
30	5
35	4
40	3.5

In actual practice, CO_2 present in the atmosphere permeates into concrete and carbonates it naturally, reducing the alkalinity of the concrete. The pH value of pore water in the hardened cement paste, which was around pH 13, will be reduced to around pH 9.0. When all the $Ca(OH)_2$ have become carbonated, the pH value becomes carbonated, and it will reduce to about pH 8.3, at which level the passive film on the steel is not stable. The loss of passivity usually occurs around pH 11.

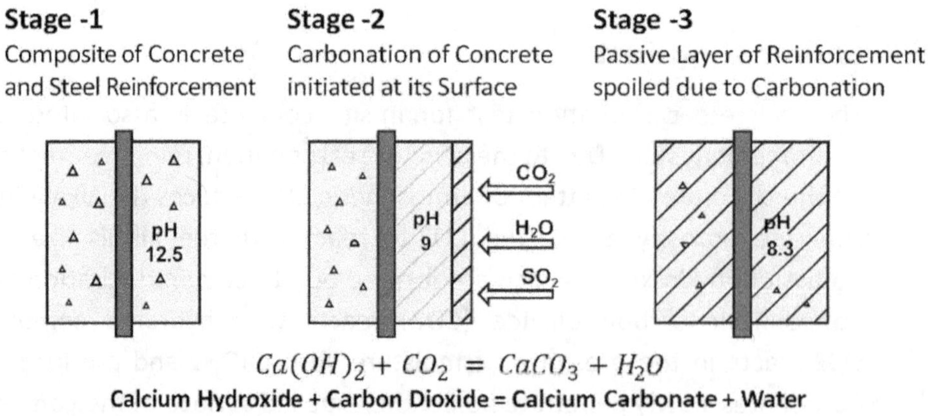

Figure 8.9: The carbonation of concrete

Carbonation is dependent on the relative humidity of the concrete. The highest rates of carbonation occur when the relative humidity is maintained between 50 percent and 75 percent. Below 25 percent relative humidity, the degree of carbonation that takes place is considered insignificant. Above 75 percent relative humidity, moisture in the pores restricts CO2 penetration. Carbonation-induced corrosion often occurs on areas of building facades exposed to rainfall, shaded from sunlight, and having a low concrete cover over the reinforcing steel.

The rate of carbonation can vary, however, and depends on the following factors of the concrete:

i) the level of pore water, i.e., relative humidity;

ii) grade of concrete;

iii) permeability of concrete;

iv) whether the concrete is protected or not;

v) depth of cover; and

vi) time.

8.4.6.1 Equipment

If there is a need to physically measure the extent of carbonation, it can be determined easily by spraying a freshly exposed surface of the concrete with a 1% phenolphthalein solution. Phenolphthalein is an acid-base indicator, which makes it extremely useful for testing for signs of carbonation. The solution will turn bright pink if it interacts with an alkaline and will even show up to a pH of 9.5. As unaffected concrete has a high pH, it will turn pink, but if the concrete is carbonated, it will remain uncoloured. It should be noted that the pink colour indicates that enough Ca(OH)2 is present, but it may have been carbonated to a lesser extent.

8.4.6.2 Procedure

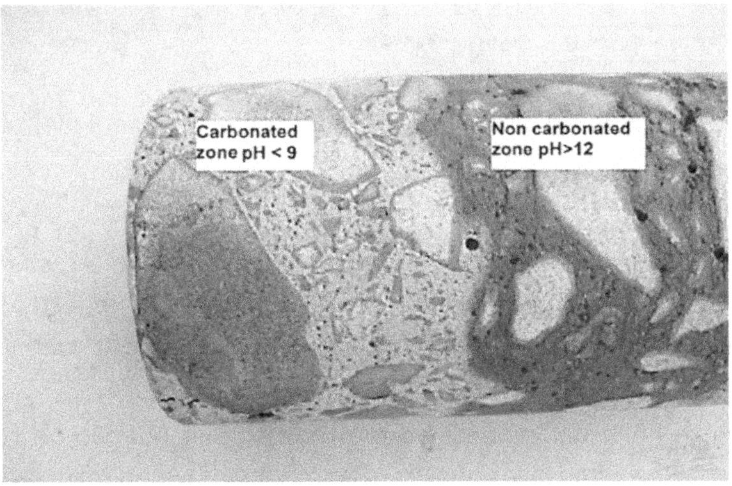

The 1% phenolphthalein solution is made by dissolving 1 gm of phenolphthalein in 90 cc of ethanol. The solution is then made up to 100 cc by adding distilled water. On freshly extracted cores, the core is sprayed with phenolphthalein solution, and the depth of the uncoloured layer (the carbonated layer)

Figure 8.10: Carbonation Test on Concrete Core using phenolphthalein solution

from the external surface is measured to the nearest mm at 4 or 8 positions, and the average is taken. If the test is to be done in a drilled hole, the dust is first removed from the hole using an airbrush, and again, the depth of the uncoloured layer is measured at 4 or 8 positions, and the average is taken.

If the concrete still retains its alkaline characteristic, the colour of the concrete will change to purple. If carbonation has taken place, the pH will have changed to 7 (i.e. neutral condition), and there will be no colour change.

8.4.7 Testing for Sulphate Attack

A sulphate attack generally causes a pattern of micro-cracks in the concrete. As a result, surface scaling and disintegration set in, followed by mass deterioration. When sulphate enters concrete, it combines with hydrated calcium aluminate and/or the calcium hydroxide component of hardened cement paste and begins destroying the paste that holds the concrete together. As sulphate dries, new expansive compounds are formed, often called ettringite. These new crystals occupy empty space, and as they continue to form, they cause the paste to crack, further damaging the concrete. The volume increase and the needle-like shape of ettringite crystals are responsible for creating internal stresses that cause cracking. This is known as a chemical process. A sulphate attack can also be the result of a physical process when water-soluble salts travel through the concrete and get deposited at the surface, causing it to scale and eventually spall.

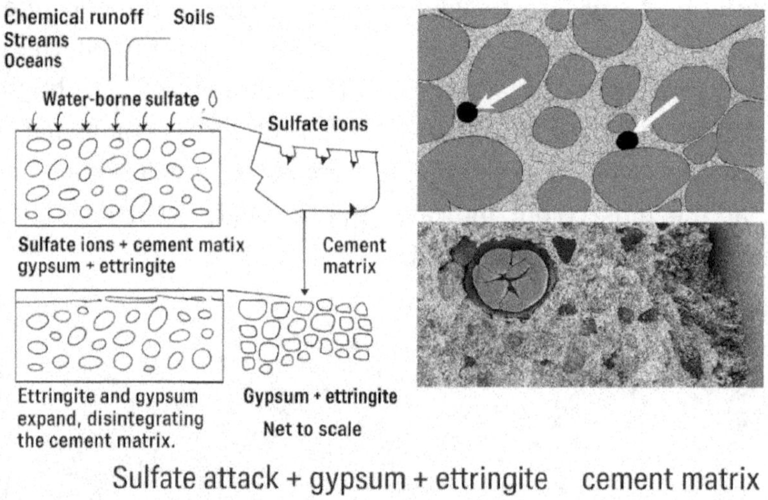

Figure 8.11: Carbonation Test on Concrete

External sources of sulphate are more common and usually are a result of high-sulphate soils and ground waters, or can be the result of atmospheric or industrial water pollution. Seawater is considered to have a "moderate" level of exposure to water-soluble sulphate ions.

Sulphate attack can be visually identified by: spotting salt buildup on concrete surfaces, surface scaling, spalling with traces of salts, de-bonding between the aggregates and paste, and the change of texture of concrete becoming like toothpaste. This happens at a developed stage of the Sulphate Attack. This phenomenon is mainly observed in buried or close-to-the-surface concrete members, such as piers, columns, slabs, foundations, and basements, where sulphate contamination is found in the surrounding soil or water.

8.4.7.1 Testing of Sulphate Content

There are laboratory tests to determine the water-soluble sulphate ion content in the cement. Furthermore, the water and soil surrounding a concrete body can be tested to examine the level of sulphate contamination. The concentration of sulphate is established using gravimetric determination. By adding Barium Chloride, the sulphate is converted to Barium Sulphate precipitates, which are highly insoluble. Therefore, we can collect it, dry it, and weigh it again to figure out the initial concentration of sulphate in the sample.

8.4.7.2 Prevention of Sulphate Attack

There are several methods to prevent or reduce sulphate attacks. Here are some of them:

- Adequate thickness of concrete cover
- Use low water/cement ratio (<0.45)
- Apply good compaction and curing
- Use low-permeability concrete
- Use a cement type that is resistant to sulphate attack
- Remove all salts from the mixed constituents
- Apply a proofing membrane for concrete surfaces in contact with sulphate-contaminated soil/water

8.4.8 Determination of pH Value of Concrete

Although concrete normally begins its life at a very basic pH of about 13, the pH values at exposed surfaces soon drop as reactions take place between carbon dioxide from the atmosphere and alkalis in the concrete. This process is known as carbonation. In due course of time, the front of carbonated concrete, with lesser pH values, advances beneath exposed surfaces.

It is necessary to find out the depth of this front because carbonated concrete can lead to corrosion of reinforcing steel. Carbonation depth can be known by spraying a phenolphthalein indicator solution onto a fractured or cut surface of the concrete and noting the location of a colour change. The depth of the carbonation front can also be estimated by using a drill to collect samples at selected depths.

The Carbonation Depth test procedure is followed as per BS 1881 (Part 201) 1986 clause 2.2. The Structural Condition Survey is a process of assessing the present health of a structure, identifying the effects of actual working loads and forces, and finally assessing the stability & safety of the structure to withstand its remaining life. A structural condition assessment is the process of collecting observations and data and systematically using them to evaluate and assess the condition of an existing structure.

Non-carbonated concrete ↔
Gradual loss of the protective capacity of concrete ↔
Partially carbonated concrete ↔
Completely carbonated concrete →

pH value:	5	6	7	8	9	10	11	12	13	14
RAINBOW-TEST	5	6	7		9		11		13	
Phenolphthalein Test	< 8.5				>9.5					
Thymolphthalein Test	< 9.3						> 10.5			

Figure 8.12: pH Value vs Colour Indication by phenolphthalein solution

Typically, the structural condition assessment process entails the identification and basic inspection of critical members and areas, followed by a detailed inspection of larger portions of the structure, if deemed necessary by the basic inspection. The basic inspection consists of a document review to attempt to determine the original design information and criteria, followed by a visual condition assessment. The visual condition survey involves the close inspection of the structure or portions of the structure to document visual observations. The visual observations that are made vary based on the structure being evaluated and generally include observations of:

- Settlement, heaving, bowing, sagging, warping, or other distortion
- Concrete cracking, scaling, spalling, or delaminations
- Corrosion of metal objects
- Discolouration/staining
- Moisture accumulation
- Construction details and type of construction

The evaluation of the materials of historic structures is one of the most challenging aspects. The challenges are not in simply defining the general nature of the original materials used in the structure, but the specific origin of the material and the transformation of that material over time to the present. This transformation may involve a chemical or physical transformation of the material from years of environmental exposure, fatigue, and one or more incidents or events of damage over the life of the structure. The material may have been transformed or affected by attempts to address deficiencies or damage events by repair or changes in the original design. Any of these transformations changes the material properties or functionality of the existing structure.

8.4.9 Electrical Conductivity Tests

The electrical conductivity of construction materials can be correlated to a variety of engineered properties. These techniques utilise Ohm's Law, correlating resistance to voltage and current. The most common of these methods is a moisture meter. Conductivity moisture meters correlate the electrical resistivity and electrical conductivity. Generally, the greater the moisture in porous materials, the greater the conductivity of the material. Electrical conductivity meters can also be used to evaluate the permeability of construction materials. Four-probe arrays can be used to evaluate the conductivity of porous building materials. This type of instrument is able to utilise the level of connectivity of pores to determine the conductivity of the material. This type of measurement is indirectly related to permeability.

The conductivity principle is also used to detect corrosion of reinforcing steel or embedded metals. Using the half-cell potential technique, the conductivity concept is used to detect the oxidation of embedded steel by completing a circuit between the steel and the surface of the material.

8.5 Observation and Analysis of Cracks

A crack is defined as "an incomplete separation into one or more parts with or without space between". A structural crack is defined as an incomplete separation into one or more parts with or without space between, and it brings into question the structural integrity of an element. Excessive stress or inadequate concrete strength may cause structural cracking.

A cracking survey is an examination of a concrete structure for the purpose of locating, marking, and identifying cracks, and of the relationship of the cracks with other destructive phenomena. In most cases, cracking is the first symptom of concrete distress. Some cracks may occur at an early age and may not be progressive; others may occur at later ages and increase in extent with time; and some may occur following some unusual event. Hence, a cracking survey is significant in evaluating the future serviceability of the structure.

ACI Committee 201 (1980) describes a cracking survey as follows:

- Cracks can be classified by direction, width, and depth; these adjectives used to denote the direction of cracks are longitudinal, transverse, vertical, diagonal, and random.
- By visual observation, cracking patterns can be identified as surface checking, hairline cracking, and D-cracking. A cracking survey shall include descriptions of procedures and devices to be used in measuring the width and depths of cracks.
- For identification purposes, the suggested width ranges are:
    - fine - generally less than 1 mm;
    - medium - between 1 and 2 mm; and
    - wide - over 2 mm.

The crack patterns, width, movement, and growth can be coupled with structural distortions to help understand the causes of multiple issues in historic structures. Cracks should be mapped in as much detail as possible. The width of cracks can be determined by simple crack comparators or a crack microscope, like that shown in Figure 8.12. Other visual methods can identify the discolourations and laitance on materials. These can be used to determine the type of deterioration so that it can be abated and avoided in the future. Conditions associated with cracking should be described. These conditions could include differential movement, spalling, seepage, deposits from leaching, etc. Sometimes cracks can be located by using a hammer, i.e., a hollow sound can indicate the presence of a crack.

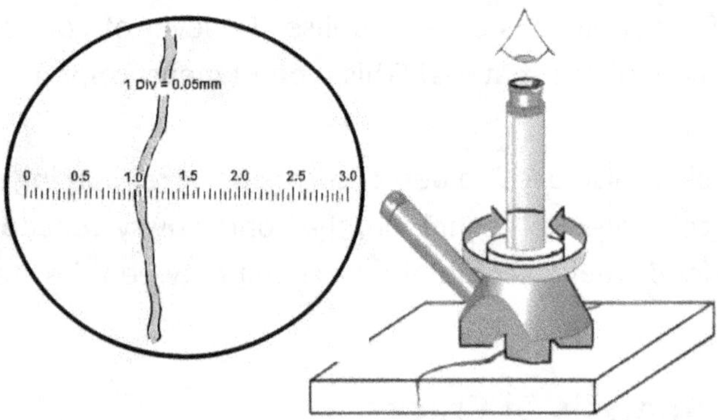

It is often helpful to note crack frequency in assessing the quality of in-situ concrete. Surface mapping of defects can be accomplished using detailed drawings, photographs, or videography.

Figure 8.13: Crack microscope

The importance of photographs of significant distress areas, with a scale or familiar object included, cannot be overemphasised for establishing trends and rates of deterioration. Profiles are recommended when they are feasible to obtain; they show the depth of missing concrete. Some cracking and other distress of concrete in structures can be attributed to excessive stress, design errors, construction errors, or a combination of the three. Critical residual stresses may develop and cause concrete deterioration, which could then compromise the integrity of the structure. The problem is to locate possible areas of high-stress concentrations and determine the stress fields. During a condition survey, it may be desirable to ascertain the in-situ stress at a given location.

Cracks and distortions are often symptoms of more complicated issues. Settlement, structural distress, chemical deterioration, fatigue, overloads, and many other problems are indicated by cracks and distortions.

8.6 Common Surface Defects

The following surface features are common on concrete that has deteriorated. Defects include distortion, cracking, efflorescence, exudation, incrustation, pitting, popout, scaling, peeling, spalling, dummy areas, stalactites, stalagmites, dusting, and corrosion. Additional features include excessive wear due to erosion or cavitation, discolouration, evidence of cement-aggregate reactions, and chemical attack.

8.6.1 Previous repairs

Repairs to older concrete and masonry are not uncommon. The survey team needs to describe any repair in terms of its ability to perform the function for which it was placed. Repairs can include a patch over a form-tie hole or the overlay of the vertical surface of a lock wall. The quality of the patch material (concrete, mortar, epoxy, etc.) should be ascertained; note if the repair is cracked, spalled, etc. If cracked, could the cracking be caused by reflective cracking from the material beneath the repair, or in the repair material? The quality of the bond between the old and repair material should be described. Terms should be used to describe the repair: good (serving intended function); fair (marginal benefit, will last so many more years); poor (has deteriorated, no longer serves intended function).

8.6.2 Cracks due to Carbonation Attack

The alkalinity in concrete is provided by hydroxides of calcium, sodium, and potassium present in the hardened cement gel. Permeation of carbon dioxide into concrete through interconnected pores and its reaction with these hydroxides causes chemical reactions. Because of this, the alkalinity of concrete is reduced. Carbonation of concrete is a problem in areas where the concentration of CO2 is high, and the relative humidity is moderate to high (50-80%).

If the structure is carbonated, the passivating film around reinforcement is neutralised, and the rods are exposed to aggression by the oxygen and humidity that are present in the atmosphere. Under these conditions, a corrosion process of the reinforcement rods initiates, which increases their volume by approximately 6 times their initial volume. The concrete around the reinforcement rods detaches from the rods and may even be completely pushed out. Once the concrete has deteriorated, deterioration to the reinforcement rods will progress even more quickly due to the formation of access points for oxygen and humidity. **Figure 8.14** shows a typical picture of the deterioration of concrete due to carbonation.

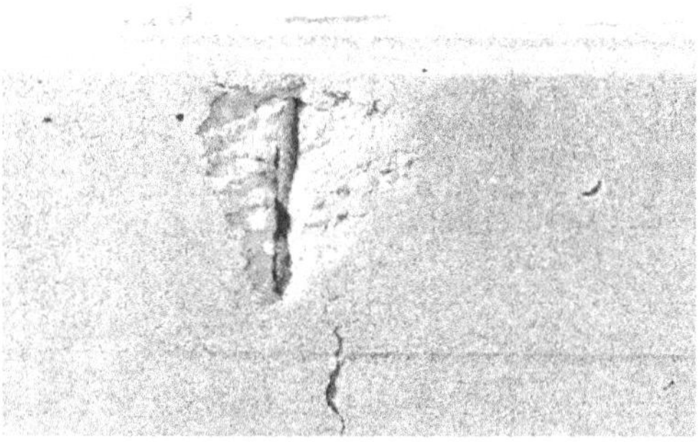

Figure 8.14: A sample picture showing the carbonation effect in concrete

8.6.3 Cracks due to Aggression of Chlorides

Aggression by chlorides on concrete occurs if it is in contact with environments with a high chloride content, such as seawater, etc. Once the chloride penetrates into the concrete and reaches the reinforcement rods, it eliminates the passivating ferrous oxide film on the reinforcement steel so that they are exposed to the corrosion process. Penetration starts on the surface and continues into the concrete. Penetration time depends on:

- The concentration of chlorides that come into contact with the surface of the concrete;
- Permeability of the concrete;
- Percentage of humidity present.

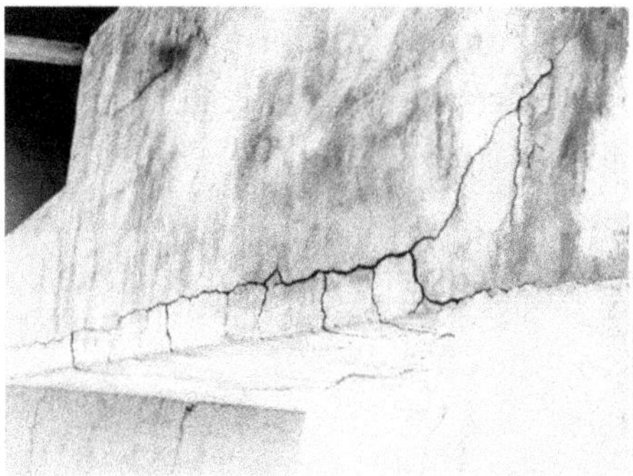

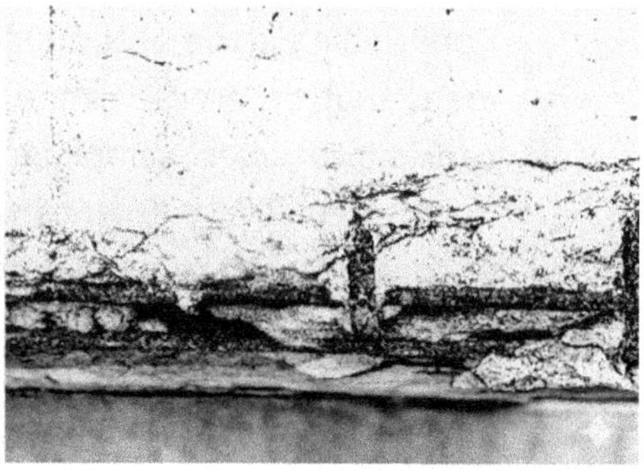

Fig-8.14 (a) **Fig8.14(b)**

Figure 8.15(a): *Crack developed under the soffit of the deck beam due to chlorine-induced corrosion in reinforcement;* ***Figure 8.15(b)****: Spalled concrete due to early induced corrosion in reinforcement and minimum concrete cover to reinforcement*

Corrosion occurs with a combination of two factors, both of which are required for the process: the presence of chlorides, which depassivate the reinforcement rods, and humidity containing oxygen.

The part of marine structures, completely immersed in seawater, will have a higher chloride content, but oxygen will not be able to penetrate. The part of the marine structure, most exposed to chloride deterioration, is the splash zone, the area which is dry and then wet due to the movement of the waves and high and low tides. Once the process has started, wherever the structure is located, corrosion is faster because access routes may be found more easily.

8.6.4 Cracks due to alkali-aggregate reaction

An alkali-aggregate reaction may cause considerable expansion and serious deterioration of concrete structures. Aggregate those that contain reactive silicon, react with alkalis contained in the cement, and due to the porous concrete cover, it reacts with seawater containing sodium chloride (NaCl). This reaction forms a gel that is highly expansive if exposed to humidity, and the gel creates forces that break the concrete around the aggregates. The alkali-aggregates reaction shows up when it starts in the cortex of the concrete, with micro and macro cracks on the surface, or even causing small portions of the concrete above the reactive silicon aggregate to lift (pop-out phenomenon) as shown in **Figure 8.16**.

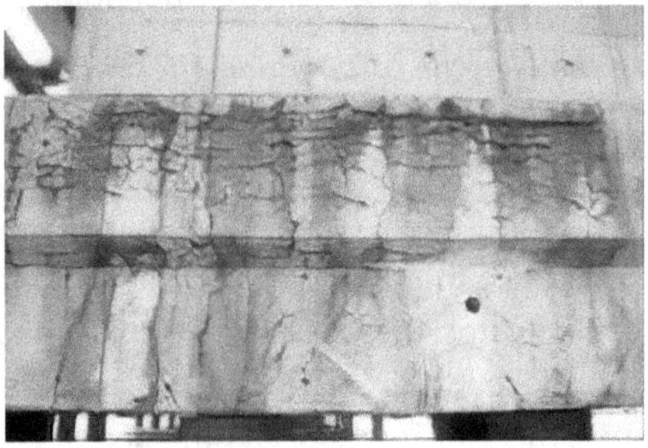

Figure 8.16: A sample picture showing a crack due to the Alkali-Aggregate reaction

8.7 Concrete Core Test

8.7.1.1 Core Drilling

Core drilling is the best method of obtaining information on concrete within a structure in areas that otherwise cannot be observed. Core drilling is expensive and should only be considered when sampling and testing of interior concrete is deemed necessary (ACI Committee 207 1979). It can sometimes be tied in with foundation drilling efforts. Samples of concrete may be recovered by sawing or by breaking pieces off.

The samples should be sufficient in number and size to permit appropriate laboratory examination and testing. For compressive strength, static or dynamic modulus of elasticity, the diameter of the core should be not less than 3.0 times the maximum size of the aggregate. For 6-in. maximum size aggregate concrete, and/or 10-in.-diameter core is generally drilled because of cost, handling, and laboratory loading capabilities.

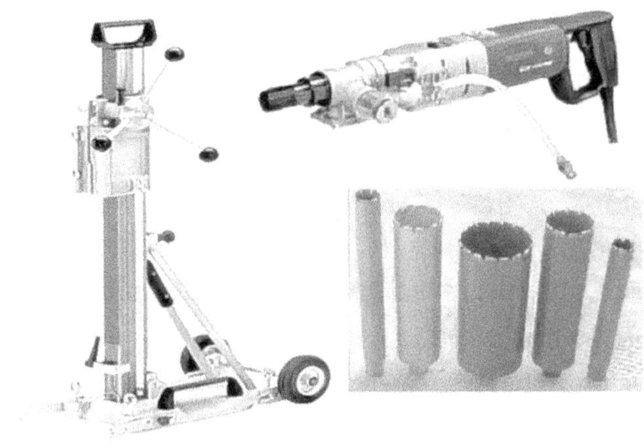

Figure 8.17: A sample picture showing Concrete Core Drilling Apparatus

A word of warning should be given against taking NX-size cores in concrete. When 2- to 6-in. maximum size aggregate concrete is cored, an NX-size (2-1/8-in. diameter) core will generally be recovered in short pieces or as short pieces and broken core. The reason for breakage occurring is because of the ratio of core diameter to aggregate diameter. When this ratio approaches one, there is simply little mortar bonding the concrete together across the diameter of the core.

Thus, the drilling action can easily break the core. When drilling in poor-quality concrete with any size core barrel, the material generally comes out as rubble.

a. Recommended sampling method.

b. Sample size (number of samples).

c. Evaluation of test results on the basis of variability.

d. Evaluation of the quality of concrete.

e. Sampling procedures.

Samples must be properly identified and oriented with permanent markings on the material when feasible. The location of borings must be accurately described and located on photographs or drawings. Cores should be logged by methods similar to those used for geological subsurface exploration.

Logs should show, in addition to general information on the hole, conditions at the surface, depth of obvious deterioration, fractures and conditions on fractured surfaces, unusual deposits, colouring or staining, distribution and size of voids, locations of observed construction joints, and contact with the foundation or other surfaces" (ACI Committee 207 1979). "The concrete should be wrapped and sealed as may be appropriate to preserve the moisture content representative of the structure at the time of sampling, and should be packed so as to be properly protected from freezing or damage in transit or storage, especially if the concrete is very weak" (ASTM C823-75/CRD-C 26-76). Figure 31 illustrates a typical log for a concrete core recovered during a condition survey.

8.7.1.2 In situ stress in concrete

Some cracking and other distress of concrete in structures can be attributed to excessive stress, design errors, construction errors, or a combination of the three. Critical residual stresses may develop and cause concrete deterioration, which could then compromise the integrity of the structure. The problem is to locate possible areas of high-stress concentrations and determine the stress fields. During a condition survey, it may be desirable to ascertain the in situ stress at a given location. The over-coring technique provides a method of accomplishing this end and will yield absolute measurements.

The over-coring technique was originally developed in the study of rock mechanics. However, in the last 10 years, it has also been applied to investigate in situ stress in concrete structures (ACI Committee 207, 1979). The procedure was developed by the Bureau of Mines and documented in their Information Circular No. 8618 dated 1974 (Hooker and Bickel 1974).

8.7.2 Penetration Resistance Methods

8.7.2.1 Introduction

Penetration resistance methods are used to check the hardness of concrete by measuring how deep a metal probe (steel rod or pin) goes into its surface. The idea is simple—the harder the concrete, the less the penetration, and this can be linked to its strength. This method was first discussed by Voellmy in 1954, who tried two approaches. In one, a special hammer called Simbi was used to drill into concrete, and the hole depth was related to cube strength. In the other, steel pins were driven into concrete and their penetration depth was compared with compressive strength. Apart from Voellmy's work, there has been very little research on these methods, and they have not gained much acceptance, especially in Europe, probably because the rebound hammer became popular around the same time. Later, the Windsor probe test was introduced in the USA in the 1960s, followed by pin penetration tests in Canada in the 1980s.

8.7.2.2 Principle

The Windsor probe works on the same basic idea as the rebound hammer—it measures surface hardness. Although probe penetration does not give the exact compressive strength at a point, it does reflect certain properties of the concrete just below the surface. With proper calibration and limits, empirical relationships can be developed between probe penetration depth and concrete strength.

8.7.2.3 Description

The Windsor probe system includes a powder-actuated gun, hardened steel probes, cartridges, and a depth gauge. The standard probe has a diameter of 6.3 mm, a length of about 80 mm, and a pointed tip. Larger probes are available for lightweight concrete. The probe is fixed into the gun and fired into the concrete using a controlled powder charge. The energy level can be adjusted for lower-strength concrete by changing the gun settings.

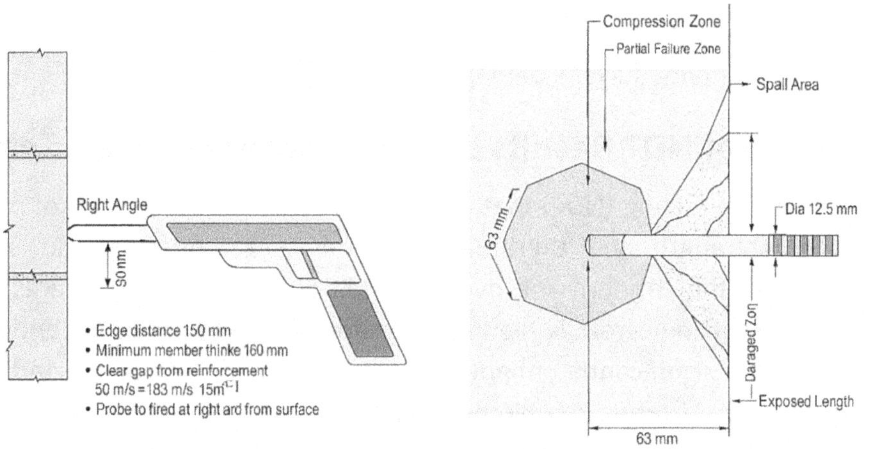

Windsor Probe | Shape of Failure Zone inProbe Penetran Test

Figure 8.18: Representative Image of Windsor Probe Test

8.7.2.4 Correlations between Probe Test Results and Compressive Strength

The usefulness of the probe test depends on how well the penetration depth can be related to concrete compressive strength. Over time, several researchers have developed correlation curves showing the relationship between exposed probe length and 28-day concrete strength. However, these relationships vary depending on concrete type and test conditions.

8.7.2.5 Probe Penetration Test vs. Core Testing

When concrete strength in a structure needs to be checked—especially if test results are doubtful or for older structures—core testing is the most direct method. However, since coring damages the structure, non-destructive or semi-destructive tests like the probe penetration test are often used as quick strength estimation tools.

8.7.2.6 Advantages and Disadvantages of the Probe Penetration Test

The following are the advantages and limitations of a penetration test.

Advantages

- Simple and quick to use
- Portable and rugged equipment
- Requires access to only one surface
- Minimal maintenance
- Fewer influencing factors compared to some other field tests

Limitations

- Needs sufficient distance from edges and reinforcement
- Minimum member thickness required
- Results can be affected by nearby steel bars
- Accuracy is limited, and strength estimates have high uncertainty

- Applicable only up to about 40 MPa strength
- Causes minor surface damage that needs repair

8.8 Validation of NDT Results through Complementary Testing

Non-destructive testing (NDT) methods provide indirect indications of material properties and in-situ strength and therefore have inherent limitations. To achieve a reliable assessment of the strength and overall condition of existing structures, the investigation programme must be carefully planned and executed. The reliability and confidence of the assessment are significantly enhanced when NDT results are supported and corroborated with findings from other complementary investigation methods, including selective core testing. Correlation of NDT results with core test data helps establish representative material properties and enables a more dependable evaluation of the structure.

For example, the assessment of reinforcement corrosion potential should be corroborated using a combination of the following methods:

a. **Half-cell potential test** – to measure the electrical potential and resistance between the reinforcement and the concrete surface, indicating the probability of corrosion activity.

b. **Chloride content analysis in concrete cover** – to determine chloride levels that may initiate or accelerate reinforcement corrosion.

c. **pH value measurement of concrete** – to assess whether the alkaline environment required for reinforcement passivation remains intact.

d. **Concrete cover meter survey** – to assess the depth of concrete cover and the level of protection available against environmental exposure.

Similarly, the health of concrete shall be corroborated using the results of the following tests:

a. **Rebound hammer test** – to provide a rapid and economical preliminary assessment of the surface hardness and relative quality of concrete, typically representative of the near-surface zone up to approximately 100 mm depth.

b. **Ultrasonic Pulse Velocity (UPV) / Impact Echo tests** – to examine the internal quality and uniformity of concrete, detect internal cracks, voids, or defects, and identify deterioration caused by aggressive environmental or chemical exposure.

c. **Concrete core testing** – to directly determine in-situ compressive strength and validate the results obtained from NDT, serving as a benchmark for correlation and confirmation of concrete quality.

The integrated interpretation of the test results provides a comprehensive and reliable assessment of both the reinforcement condition and the overall health of concrete, thereby supporting well-informed decisions related to repair, rehabilitation, or upgradation of existing structures. The selection of investigation techniques shall be based on the accessibility of structural elements for examination and evaluation. In the case of large or open structures, advanced techniques such as Ground Penetrating Radar (GPR) and LiDAR may be used in combination to enhance the effectiveness and coverage of the assessment as discussed in the next Chapter.

8.9 Underwater Investigation of Quay Wall

8.9.1 Visual Inspection by Divers

For the underwater investigation of the quay wall and the seabed profile at the toe, visual inspections conducted by certified divers are commonly employed. This method is inherently ad-hoc and largely dependent on favourable sea conditions, water clarity, and the operational capability of the divers.

In challenging environments, such as turbid or silt-laden waters, the effectiveness of diver-based visual inspections and optical sensors is significantly reduced due to high particle concentrations within the water column, which limit visibility. Additionally, diver inspections are constrained by physical and safety limitations, including restricted working durations at depth and operational risks.

Under such conditions, acoustic survey methods offer a reliable alternative, as their performance is not affected by poor visibility or suspended sediments. Consequently, acoustic techniques provide an effective and innovative solution for in-port underwater surveys, enabling consistent data acquisition where conventional visual inspection methods face practical limitations as discussed below.

8.9.2 Multibeam Echosounder (MBES) Survey Techniques

Multibeam Echosounder (MBES) survey techniques are applied in a customised manner, depending on the quay wall geometry, the required resolution, and the distance between the investigated structure and the sonar transducer.

An MBES system is capable of acquiring high-resolution three-dimensional (3D) point-cloud data of submerged portions of quay wall structures. Similar to conventional sonar systems, MBES transmits acoustic signals in a fan-shaped pattern, and water depth and object position are determined from the travel time of reflected acoustic waves. Unlike other sonar systems, MBES employs beamforming (spatial filtering) to obtain directional information from the returned signals, allowing the collection of a wide swath of depth measurements from a single ping.

By utilizing MBES survey capabilities, detailed information on submerged structural deformation, partial damage, collapse, or instability of the quay wall can be obtained for engineering analysis and decision-making.

A comprehensive condition assessment of the quay wall from the waterfront side should include evaluation of the following aspects:

i) Coping and deck-level structures, including support systems for bollards, fenders, mooring hooks, and associated fittings, together with the effects of superimposed loads on the deck.

ii) Local scour at the toe of the quay wall, potentially caused by natural currents, propeller wash, or bow-thruster action, requiring assessment and definition of appropriate remedial measures such as refilling or stabilization.

iii) Deterioration of construction materials due to continuous exposure to the aggressive marine environment.

iv) Structural deformation or damage caused by vessel impacts, berthing incidents, or other operational accidents.

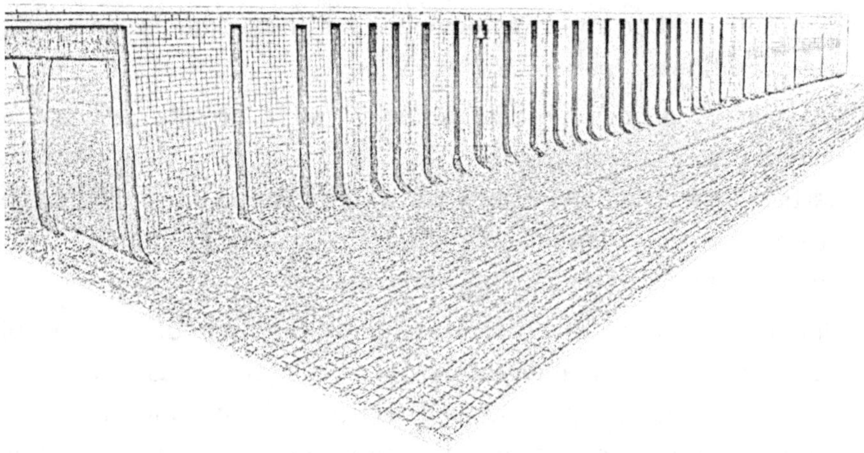

Figure 8.19 : Schematic representation of quay wall geometry and seabed profile derived from Multibeam Echosounder (MBES) survey data.

8.9.2.1 Two-Stage Underwater Inspection Methodology

To ensure the optimal use of MBES data and to achieve a reliable condition and deformation assessment of the quay wall along the waterfront, a two-stage underwater inspection approach is recommended. This methodology significantly reduces survey time and inspection costs while maintaining technical accuracy and coverage.

i) Stage 1 – Preliminary Assessment Using MBES

In the first stage, the MBES system is used to rapidly acquire coarse but accurate 3D point-cloud data of the submerged quay wall. Due to its high coverage rate, MBES mapping provides an efficient overview inspection of underwater structures within a relatively short time frame.

The resulting point-cloud data provides quantitative outputs and can be post-processed for three-dimensional measurements, deformation analysis, alignment checks, and volume calculations. This stage is particularly effective for identifying critical zones requiring further investigation; however, MBES alone may not always provide the resolution necessary for detailed assessment of localized damage or complex subsea features.

ii) Stage 2 – Detailed Inspection Using Scanning Sonar

Based on the results of the preliminary MBES survey, a second-stage inspection is carried out using a scanning sonar system to perform detailed examination of selected areas. Unlike multibeam echosounders, scanning sonar systems offer operational mode flexibility, allowing selection between:

- Profiling mode, providing high-resolution cross-sectional data, and
- Imaging mode, providing detailed acoustic images of structural surfaces.

Scanning sonar imaging delivers high-resolution qualitative information, enabling precise visual interpretation, identification of localized defects, and two-dimensional measurements of cracks, offsets, erosion, or other structural anomalies.

The combined two-stage methodology integrates the wide-area efficiency of MBES with the high-resolution diagnostic capability of scanning sonar, resulting in a comprehensive, technically robust, and cost-effective underwater inspection and condition assessment of quay wall structures.

Modernising Maritime Infrastructure

Modernising Maritime Infrastructure

Geotechnical Studies for Brownfield Port Upgradation

Chapter Summary

Investigations and studies for redevelopment in a brownfield port environment are challenging due to ageing infrastructure, complex subsurface conditions, and existing utilities and structures. Geophysical surveys are non-invasive techniques that safely detect underground features—such as old utilities or contaminated zones—without drilling, enhancing both accuracy and cost-efficiency. With flexible planning and the integration of geophysical and geotechnical methods, a more comprehensive understanding of subsurface conditions can be achieved. Geophysical surveys identify hidden features quickly, while geotechnical tests provide precise soil data. Together, they support better planning, reduce costs, and improve safety in complex redevelopment projects.

Learning Objectives

- Key Challenges in Brownfield Port Redevelopment
- Purpose of Subsurface Investigations
- Enhancing Subsurface Analysis through Geophysical-Geotechnical Integration
- Geotechnical Investigation: Common Challenges
- Applying Empirical Approaches to Streamline Site Studies
- Versatile Applications of Ground Penetrating Radar (GPR)

9.1 Introduction

A brownfield port facility refers to a previously developed maritime site that is currently in need of redevelopment. These locations often have a history of industrial and commercial activity, leading to complex site conditions such as environmental contamination, ageing infrastructure, and regulatory limitations. The presence of existing subsurface and surface infrastructure—including underground utilities, roadways, drainage systems, and structural remnants—further complicates site evaluation and planning.

9.1.1 Challenges in Redeveloping Brownfield Port Facilities

Detailed investigations and studies for redeveloping brownfield port facilities involve addressing a range of challenges across environmental, technical, regulatory, and operational aspects:

- **Subsurface Uncertainties**: These sites often harbour hidden or undocumented infrastructure such as old pipelines, buried foundations, trenches, waste disposal areas, unstable fill materials, and legacy utility networks. These concealed elements can significantly hinder invasive ground investigation works and pose safety risks during drilling, excavation, and redevelopment activities.
- **Structural Degradation Assessment**: Existing structures within the port may have deteriorated over time due to environmental exposure or lack of maintenance. A thorough structural evaluation is essential to determine whether retrofitting is feasible or if demolition is the more viable option.
- **Regulatory and Legal Constraints**: In active port zones, subsurface investigations with borings and soil sampling activities must be carefully aligned with the regulations and legal frameworks of authorities. This includes obtaining necessary permits and compliance.
- **Operational Continuity**: Ports often remain operational during redevelopment, necessitating meticulous planning to avoid disruptions. Coordination with port authorities and stakeholders is crucial to ensure that construction activities do not interfere with cargo handling, logistics, or maritime operations.
- **Site Accessibility**: Physical constraints within the port—such as limited space, ongoing operations, and infrastructure congestion—can impede access for essential activities like drilling, sampling, and excavation. These limitations require innovative engineering solutions and adaptive construction strategies.

In complex port redevelopment projects, flexible planning combined with geophysical surveys can be very helpful for geotechnical data collection. Geophysical methods allow engineers to study underground conditions without drilling or digging, making the process safer and more affordable. These techniques help detect important hidden features like old pipes, voids, or contaminated soil that could affect construction. To get more accurate results, experts often use a combination of methods—such as ground-penetrating radar, electrical resistivity, and seismic testing—especially in busy and unpredictable brownfield port areas.

9.2 Geophysical-Geotechnical Integration for Enhanced Subsurface Analysis

Geophysical exploration involves the collection of data from the Earth's surface to infer what lies beneath, without the need for extensive digging or drilling. This is achieved through various methods such as seismic, magnetic, electrical, and gravitational measurements. The data gathered provides insights into the subsurface conditions, including the structure and composition of geological layers, the presence of discontinuities, and the location of groundwater.

Integrating geophysical and geotechnical investigations offers a more complete and reliable understanding of subsurface conditions, which is critical for safe and cost-effective project planning and execution. Geophysical methods—such as ground-penetrating radar, seismic surveys, and electrical resistivity—are non-invasive and can quickly scan large areas to detect anomalies like buried objects, voids, or changes in soil composition. These insights help identify potential problem zones before physical drilling begins.

Geotechnical investigations, on the other hand, provide direct measurements of soil and rock properties through techniques like borehole drilling, standard penetration tests (SPT), and lab analysis. While geotechnical data is highly accurate at specific points, it can be limited in spatial coverage.

By combining both approaches, engineers can use geophysical data to guide and optimize the placement of geotechnical tests, reducing the number of boreholes needed and improving cost-efficiency. This integration also helps in validating geophysical interpretations with physical data, increasing confidence in the overall ground model.

In complex environments—such as brownfield sites, ports, or urban areas—this combined approach enhances risk management, supports better design decisions, and minimizes unexpected delays during construction.

The investigation methods may be divided into the following types or groups:

(a) geophysical methods

(b) in situ tests with borings and soundings

(c) soil and rock sampling for lab tests

The methods most commonly used are described in the following sections. Different methods may provide similar information. The selection of a method will, to a large extent, be determined by availability and costs. Each method has certain advantages and limitations, which should be taken into consideration.

9.3 Geotechnical Investigations

The design of marine and waterfront structures involves a multitude of design parameters and boundary conditions of various disciplines. Each discipline contributes its own set of critical information regarding these parameters and conditions, which are essential for the structural integrity and functionality of quay walls.

Among the other site studies, a comprehensive geotechnical investigation is crucial for the planning and design of marine facilities and quay structures. This investigation should aim to understand the site's subsurface conditions, including the sequence and nature of soil and rock strata, groundwater conditions, and the physical and mechanical properties of the soil. Moreover, understanding the depth and breadth of the soil's influence on the construction allows engineers to anticipate and mitigate potential complexities that could arise during the design and construction phase at the site, and also the depth of the investigation should cover the entire zone expected to be influenced by the construction to provide a complete understanding of the site conditions.

The ultimate goal of a geotechnical investigation is to gather sufficient sub-soil information to manage all potential risks and enable the design for a safe, efficient, and cost-effective structure. This activity also assists the engineers in exploring different structural options and making informed decisions throughout the planning and pre-construction process, and during the execution too. In this process, the collaboration between structural and geotechnical engineers is also very important to bring their expertise and knowledge of local conditions for interpreting the data accurately to navigate risks and perform the design of a structure that is both safe and cost-effective, ensuring the longevity and integrity of the construction.

9.3.1 Objectives of Site Investigations

Before carrying out a design either in a green-field or brown-field environment, it is imperative to understand the ground conditions and anticipate how soils or rocks will react to construction activities. For any construction activities, it is necessary to assess the load-bearing resistance and characteristics of deformation of the ground against various active loads. By evaluating the load-bearing resistance and deformation characteristics against various active loads, engineers can predict and mitigate potential issues that may compromise the structure's stability and overall functional requirements throughout the design life. The objectives of these investigations are well captured by Peck's 1962 insights, which highlight the importance of three key areas of knowledge:

i) **Knowledge of Precedents**: Familiarity with previous cases in similar ground conditions can prevent unexpected issues during design and construction phases.

ii) **Working Knowledge of Geology**: Understanding geological aspects helps predict possible variations in ground conditions, which is vital for planning and executing borings.

iii) **Knowledge of Soil Mechanics**: Proficiency in soil or rock mechanics is essential to ensure that the ground behaves as expected during and after the construction, minimising the risk of structural failures or other issues.

These elements form the foundation of subsurface engineering, enabling engineers to design and construct safely and efficiently, minimising risks and ensuring the integrity of the structures.

9.3.2 Setting up the Investigation Plan

Setting up an investigation plan is a crucial step in the design process, as it helps to determine the design parameters and boundary conditions efficiently. This plan should include a clear justification for the investigation, detailing why it is necessary for each discipline involved. It should outline the objectives, the methods to be used, the expected outcomes, and how the results will influence the overall design. By doing so, it ensures that the investigation is aligned with the project's goals and can provide valuable insights that inform the design decisions. Finally, a well-structured investigation plan can save time and resources and contribute to the success of the project. The various considerations for setting up an investigation are outlined below;

i) **Objective Formulation:**

Clearly define what is the aim to achieve with the investigation. This could be the assessment of the sub-soil strata to understand the following;

- The nature and sequence of strata
- The groundwater conditions at the site;
- The physical properties of underlying strata, such as colour, texture and composition of soil;
- The mechanical properties of strata, such as strength and compressibility, and other specific information, such as the chemical composition of the groundwater, and the characteristics of foundations, including assessing the method of construction, environmental impact, or other specific goals relevant to the project.

ii) **Investigation Methodology**

Investigation methodology shall outline the strategy and techniques to be employed during the investigation. This typically includes:

- **Geotechnical investigations**, such as drilling boreholes to collect soil samples at various depths.
- **Laboratory testing** of these samples to determine properties like grain size distribution, Atterberg limits, shear strength, and compressibility.
- Depending on project needs, **geophysical surveys** (e.g., Electrical Resistivity Tomography or Seismic Refraction) may be used to assess subsurface profiles over larger areas.
- **Remote sensing techniques**, such as satellite imagery or drone-based LiDAR, can be applied for terrain analysis or mapping inaccessible regions.

Example: For a proposed quay wall, boreholes may be strategically positioned along the planned wall alignment, covering both the structural footprint and zones expected to experience significant stress. To complement this, seismic refraction surveys can be conducted to verify the continuity and characteristics of subsurface layers between boreholes and in adjacent areas, ensuring a comprehensive understanding of the ground conditions.

iii) **Phasing and Motivation**

This part details the timeline and logical sequence of investigation activities, explaining the purpose of each phase and its contribution to the overall objectives.

- **Phase 1:** Conduct a desktop study and site reconnaissance to collect and review existing information from previous investigations, published literature, and publicly available data sources. This initial phase helps develop a preliminary understanding of the site conditions and supports the identification of critical zones for targeted field investigations..

- **Phase 2:** Field investigations (e.g., drilling, sampling, geophysical surveys).
- **Phase 3:** Laboratory testing and data analysis.
- **Phase 4:** Reporting and interpretation to support design decisions.

Example: Phase 1 helps avoid redundant drilling by identifying previously studied areas, while Phases 2 and 3 focus on collecting new site-specific data essential for accurate foundation design and engineering analysis.

iv) Expected Outcomes

Describe the anticipated results from the investigation, which may include:

- Detailed soil profiles and stratigraphy.
- Engineering parameters for design (e.g., bearing capacity, settlement estimates).
- Geotechnical baseline reports to guide construction planning.
- Identification of potential geohazards like liquefaction zones or expansive soils.

Example: The investigation may reveal the presence of compressible subsurface layers along the proposed quay wall alignment, which could adversely affect structural stability. In such scenarios, suitable structural alternatives and construction techniques may be devised prior to construction to ensure sufficient bearing capacity and to mitigate potential settlement issues.

v) Justification for Investigation Type

Explain why specific investigation methods were chosen, and how they align with project requirements.

- Borehole drilling is selected for direct sampling and accurate depth profiling.
- Seismic surveys are justified for non-invasive mapping of subsurface layers over large areas.
- Laboratory tests are chosen based on expected soil types and design needs.
- Modelling studies (e.g., slope stability or settlement analysis) are proposed to simulate site behaviour under load.

Example: In sites with variable subsoil conditions, SPT is preferred over CPT as it provides discrete resistance readings at specific depths, better captures layer changes, and allows sample recovery for visual classification and lab testing—making it more suitable for heterogeneous ground profiles.

vi) Cost Analysis

Offer a detailed breakdown of the projected expenses associated with the investigation. This could cover;

- Personnel costs: Geotechnical engineers, field technicians, lab staff.
- Equipment rental: Drilling rigs, geophysical instruments, sampling tools.
- Laboratory testing fees: Charges per test type.
- Reporting and data processing: Compilation, interpretation, and documentation.

- Contingency: Allowance for unforeseen site conditions or additional testing.

9.3.3 Importance of Geo-technical Investigations

Thus, the soil investigation is a critical step in construction and should be prioritized accordingly. A thorough and careful investigation can prevent many potential problems by providing essential data for a good analysis of soil conditions. This ensures that the foundation of any structure is built on solid ground, both literally and figuratively. However, it often gets overlooked or rushed, which leads to problems later on. When soil investigation is skipped or done poorly, major issues can show up later in the project. Foundations can end up being under-designed because the ground conditions weren't properly understood, which may lead to cracks, uneven settlement, or even structural failure. Unexpected soil types—like soft clay or high groundwater—can be encountered, causing delays and extra costs. Safety can be put at risk, and approvals might get held up due to missing or incomplete data. If soil properties aren't clearly identified, wrong design choices can be made, affecting how well the structure performs over time.

The examples below testify to this.

i) **Soil Investigation shortcuts lead to costly surprises.**

Shortcuts in geotechnical exploration should be avoided because reliable design depends on having accurate information about the ground. Incomplete or rushed investigations can miss weak soil layers, groundwater issues, or other hidden conditions, leading to unsafe design decisions and costly problems during construction. Proper, thorough site investigation helps prevent structural failures, reduces risk, and ensures safer and more economical project outcomes.

The narration provided below highlights how important solid soil investigation is for construction. It gives a real-life example where trying to cut costs on soil testing ended up being way more expensive. A quay structure ran into trouble after completion—the anchoring system wasn't stable because the backfill materials acted unpredictably. Turns out, there was a layer of sensitive marine clay that got missed since the borehole tests were spaced too far apart and only a few lab tests were done. By the time they figured it out, it was too late—the quay wall had collapsed into the sea. The lesson? Don't skimp on soil investigations. Spending a bit more up front can save you from major issues later.

ii) **When determining soil parameters, take the design into account**

In geotechnical engineering, the accurate determination of soil parameters is crucial for designing safe and efficient structures. Standard tests, such as triaxial compression tests, provide valuable data for soil of cohesive nature, but they must be interpreted within the context of the specific design requirements. Understanding the expected failure mechanisms and stress-deformation behaviour from the design phase is essential. This knowledge guides the selection of appropriate tests and parameters that reflect the actual conditions the structure will face. For instance, under a caisson, a direct shear test might be more indicative of the soil behaviour than a triaxial test, as it could better simulate the stress distribution and potential failure planes. Therefore, integrating design considerations into soil parameter determination is not just beneficial but necessary to ensure the reliability and safety of geotechnical solutions.

a) ***Poor geotechnical exploration is one of the biggest hazards in civil engineering.*** Adequate geotechnical exploration forms the foundation of every civil engineering project. When it is poorly conducted—or neglected altogether—it becomes one of the greatest hazards because every structure ultimately depends on the behaviour of the ground beneath it. Unlike materials such as concrete or steel, which can be manufactured and controlled to specification, soil and rock conditions vary widely and often unpredictably. Failing to properly investigate them introduces risks that can compromise safety, performance, cost, and even human life.

b) The ground is highly unpredictable.

- Soil is a natural, uneven material. Without proper investigation, hidden issues like weak layers, high groundwater, voids, swelling clays, or loose sands remain unnoticed—until they cause major problems.

c) Poor exploration leads to wrong design assumptions.

- When the soil profile is unclear, designers are forced to assume uniform strength or stable groundwater levels, which often isn't true. Designs based on guesswork are unsafe, and many failures could be prevented with proper geotechnical studies.

d) Geotechnical uncertainty directly affects stability.

- Inadequate investigation can lead to incorrect estimates of bearing capacity, settlement, or shear strength—causing anything from tilting buildings to quay wall and embankment collapses. In most cases, the soil didn’t behave unexpectedly; it simply wasn’t understood well enough.

e) Fixing soil problems later is extremely expensive.

- Once construction is complete, correcting foundation issues often requires underpinning, ground improvement, grouting, or reconstruction—plus shutdowns. A small investment in proper soil investigation upfront avoids massive costs later.

f) Safety is ultimately at stake.

- Poor geotechnical work threatens structural integrity, worker and public safety, and the environment. It’s not just a technical oversight—it’s a serious safety risk.

Thus, inadequate subsurface investigation is not simply a technical deficiency—it is a major safety hazard. Without reliable subsurface data, even the best engineering designs may fail, whereas thorough ground investigation drastically reduces risk, enhances safety, improves constructability, and saves cost.

9.4 Stages of Site Investigation

In essence, a well-conducted geotechnical investigation is a step-by-step process that builds a comprehensive understanding of the subsurface conditions, which is essential for any construction project to be built on solid ground, both literally and figuratively. As described above, the expertise of geotechnical and structural engineers is paramount throughout this process, as their judgment and experience guide the investigation, ensuring that the project rests on a solid foundation, both literally and figuratively. The investigation may involve the following steps;

9.4.1 Desk Study and Site Walkover Survey

The desk study and site walkover survey are indeed crucial steps in ground investigation, serving as the foundation for understanding the subsurface conditions of a site. The process of investigation begins with a desk study and site walk-over survey, which provides an initial understanding of the site's history, current condition, and potential challenges. This phase often involves reviewing historical data, maps, and records, as well as a physical inspection of the site. By gathering and analysing all the available existing information on a site, including topographical and geological maps, records of previous site investigations, and current and previous land use (information on constructions, buried electrical cables, trenches, foundations, landfills, along with the possibility of contaminated land), professionals can form a preliminary assessment of the site's characteristics. This information is invaluable for identifying potential hazards like contamination or unstable ground.

9.4.2 Preliminary Investigation

A preliminary investigation is carried out, involving more in-depth research and testing to assess the soil, groundwater conditions, and other environmental factors.

During this phase, a series of strategically placed borings is performed to collect soil samples at various depths. These samples are essential for identifying the characteristics of the soil and any layers that may impact the structural integrity of future constructions, such as shallow foundations, piles, raft or retaining walls, etc. The data gathered from these borings enable engineers to conduct a thorough qualitative analysis, which includes assessing the soil's behaviour under different conditions, identifying potential problematic strata, and evaluating the basic soil parameters necessary for design. In areas prone to under-seepage, it's vital to determine the pervious layers' thickness to design measures that mitigate this risk. The spacing and depth of borings are tailored to the site's specific geological context, ensuring a comprehensive understanding of the subsurface environment.

9.4.3 Detailed Site Investigation

The detailed investigation delves deeper, with extensive testing and analysis to fully characterise the site conditions. This might include drilling boreholes, conducting in-situ tests, extraction of soil samples and conducting tests on soil samples in the laboratory aimed at understanding the subsurface conditions to a degree that allows for the accurate design and construction of a structure. The process begins with the development of a detailed sub-soil profile, which includes the quantification of material parameters through a series of borings. These borings, typically two to five times the number of preliminary borings, are strategically placed based on the geologic conditions and the characteristics of the proposed structure.

The goal is to confirm both the changes in subsurface conditions and the continuity of consistent conditions. This is achieved by obtaining undisturbed and disturbed soil samples for laboratory testing and performing in situ tests. With the data gathered, a final site characterisation is conducted, resulting in an idealised ground profile. This profile is not only a simplified representation of the subsurface but also includes empirical correlations to aid in the design process. It serves as the foundation for stability analysis and design, ensuring that the structure will be built on a thoroughly understood and well-characterised ground.

9.4.4 Supplementary Investigation and Construction Control

In the field of geotechnical engineering, the complexity of sub-soil conditions can indeed pose significant challenges. While a handful of boreholes may offer a glimpse into the underground environment, they are often insufficient for comprehensive analysis. Ultimately, such thorough exploration safeguards against delays and additional costs, underlining the importance of a detailed and proactive approach to construction control.

Basic borehole data gives a rough idea, but you usually need extra checks to spot hidden loose or hard soil patches that can mess with stability or pile installation. Tools like seismic surveys help spot underground variations that regular tests miss. Getting a full picture of the soil early on helps design better foundations, smooth out construction, and avoid delays. Bottom line: smart, flexible planning in Geotech work keeps things on track and reduces risks.

In a project spread over a large area, by integrating geophysical exploration early in the project, alongside minimal direct subsurface exploration like boreholes, a more comprehensive understanding of the site can be developed. This approach not only aids in identifying key geological features but also offers a cost-effective alternative to extensive coring or excavation. The information obtained is instrumental in making informed decisions for construction, mining, and environmental projects, ensuring safety, efficiency, and economic viability.

9.5 Geotechnical Design Considerations

Geotechnical analysis is a comprehensive process addressing the behaviour of soil against the imposed loads during the design phase, especially when it comes to understanding the complex dynamics between soil and structures. It helps engineers to determine if the ground can support the proposed structures, assessing various factors, like the type of soil, its ability to bear load, and the presence and behaviour of groundwater, ensuring that potential issues can be addressed proactively rather than reactively. Geotechnical and structural engineers need to make well-informed decisions by providing a deep understanding of the various geotechnical aspects, including the time-dependent and stress-dependent behaviour of soil that could affect a project.

In the process of upgradation of quay structures, a comprehensive approach that includes both geotechnical investigation and condition assessment of the Structure. provides crucial data on the stability of geo-media, and concurrently, the condition assessment evaluates the current state of the structure, focusing on the demands placed on it for future requirements. This dual approach ensures that any upgradation of a structure requires a thorough understanding of both the existing conditions of the structure and the geotechnical characteristics of the site.

Further, it's critical to select construction materials that are compatible with the site's conditions to maintain structural integrity. For instance, when choosing steel pile sections for the quay, it's important to ensure that they are drivable and adaptive to the site, without affecting the surrounding structure due to the vibrations imposed on the soil during the driving of piles.

The goal shall be to create a harmonious collaboration between the geotechnics and the structure to establish a resilient system. This integrated process helps in minimising the risk of hazards during the installation or post-construction stages, thereby promoting safety and longevity of the structure. It's a strategic and methodical approach that aligns the structural upgrades with the unique challenges of the site, ensuring that the structure can withstand future demands. In essence, geotechnical studies are not just a formality—they're a smart engineering practice that leads to safer, stronger, and smarter infrastructure.

9.5.1 Strength of Soil

The shear strength of soil is a critical factor in geotechnical engineering, determining the soil's resistance to deformation under shear stress. It is influenced by the soil's cohesion, friction between particles, and the normal effective stress acting upon it. This property is essential for calculating the bearing capacity of soils, which in turn affects the stability of foundations. The stability of foundations is paramount for the safety and longevity of structures, as it ensures resistance to various forces such as wind, seismic activity, and lateral loads. Proper foundation design involves assessing the shear strength of soil through tests like the Direct Shear Test, Triaxial Compression Test, and others, to prevent excessive movement or failure of the structure. Engineers can enhance soil shear strength using methods like mechanical compaction or chemical stabilisation, and by incorporating reinforcing materials like geogrids. Understanding these aspects is vital for the design of stable and durable foundations that can support the intended loads and withstand environmental challenges.

9.5.2 Depth of Investigation and Influence Foundation Stress

The depth of investigation for a construction site is a critical factor that ensures the stability and safety of the structure. It involves a thorough exploration of the soil and rock layers that will be affected by the construction to determine their bearing capacity and other physical characteristics. Moreover, for areas prone to liquefaction, a minimum depth of 15 meters is investigated to evaluate the risk and inform appropriate construction strategies. These guidelines are vital for the engineering design and for predicting and mitigating potential issues that may arise during and after construction.

It is essential to explore the soil or rock to a depth where all the changes due to construction will have an impact. This means that the investigation must reach into the zone where the soil or rock will bear the load of the construction.

- Investigation depth should be great enough to fully penetrate soft, highly compressible soils (e.g. peat, organic silt, soft fine-grained soils) into competent material of suitable bearing capacity (e.g., stiff to hard cohesive soil, compact dense cohesionless soil, or bedrock).
- In case of heavily loaded piles, the depth of investigation should extend below the anticipated pile or shaft tip elevation a minimum of 6 m. For shafts supported on or extending into rock, a minimum of 3 m of rock core, or a length of rock core equal to at least five times the shaft diameter for isolated shafts or two times the maximum shaft group dimension, whichever is greater, shall be extended below the anticipated shaft tip elevation to determine the physical characteristics of rock within the zone of foundation influence.

This thorough investigation is necessary to prevent any future problems with the structure, which can be catastrophic and costly to fix. Thus, the depth of investigation is therefore not just a technical requirement; it is a crucial step in risk management for any construction project. It ensures that the structure will stand the test of time and provide a safe and secure environment.

9.5.3 Horizontal Spread of Geotechnical Exploration

The horizontal spread of investigation points is a strategic decision based on the variability of subsoil conditions and the sensitivity of structures to differential settlements. For instance, in areas with residual soil deposits, where lateral variations can be abrupt, a closer spacing of 20m to 30m for boreholes or test locations is recommended. Conversely, in more uniform natural alluvial deposits, a wider spacing of 60m to 100m may suffice, although closer inspection near river courses might be necessary. Marine deposits typically require a 30 to 50m spacing, while structures that are particularly sensitive to differential settlements may need a more cautious approach with spacings of 20 to 30m. Ultimately, geotechnical investigation is an interplay between field investigations, such as Standard Penetration Tests, Vane Shear Tests, and Permeability tests, and laboratory analyses. These methods complement each other to provide a comprehensive understanding of the ground conditions, ensuring that the foundations of our infrastructure are as safe and robust as possible.

9.6 Field Investigation

9.6.1 Standard Penetration Test (SPT)

The Standard Penetration Test, known as the SPT, is commonly for subsurface investigations and geotechnical designs. It is one of the most broadly used tests worldwide to characterise in-situ soil strength. While other in-situ tests are available, CPT, CPTU and dilatometer, to mention a few, only the SPT test enables the drill crew to retrieve soil samples. The SPT test is made by dropping a free-falling hammer weighing 65 Kg onto the drill rods from a height of 750mm to achieve the penetration of a standard sample tube 300 mm into the soil. The sampler is driven using the hammer for 450 mm, recording the number of blows for every 150 mm. The number of blows for the last 300 mm drive shall be reported as the N value.

The advantage of the SPT tests is that'

- The drillers can collect samples for further classification and laboratory testing.
- This is a simple and economical test
- SPT N values are empirically correlated with geotechnical design parameters such as soil density, consistency, friction angles, undrained shear strength, Young's modulus, shear modulus, settlement of shallow and deep foundations in sand, bearing capacity values, and to provide an index of soil liquefaction resistance of cohesionless soils.
- It can also be used to determine the unconfined compressive strength of cohesive soils.
- SPT allows for the collection of soil samples, which can be analysed further for detailed soil classification and testing.
- It's a cost-effective method that provides empirical correlations with various soil properties and geotechnical design parameters.
- SPT data can inform the design and safety of structures by assessing soil density, consistency, bearing capacity, and potential for liquefaction.

The Standard Penetration Test (SPT) is indeed a fundamental procedure in geotechnical engineering, offering a practical and economical means to assess soil characteristics. The test's ability to provide immediate results on-site is invaluable, particularly in preliminary stages of construction projects where quick decision-making is crucial. The N-value, obtained from the SPT, is a direct indicator of the soil's relative density and consistency, which are vital parameters for the design and analysis of foundations.

The simplicity of the SPT also allows for widespread application, making it a default choice in many geotechnical investigations. However, as with any field test, the accuracy and reliability of the SPT can be affected by various factors. The type of hammer, the energy delivered to the sampler, and even the length of the rods can alter the N-values. It is also recognised that different operators may achieve different results due to variations in their technique and experience.

These variables necessitate careful standardisation and calibration of equipment, as well as training for operators to minimise discrepancies. The use of safety hammers and automatic trip hammers has been encouraged to enhance the uniformity of the energy transfer. Additionally, corrections are often applied to the raw N-values to account for overburden pressure, borehole diameter, and rod length, among other factors. In soils where the SPT is less effective, such as cohesive soils with low permeability, the results may not be representative of the true soil conditions. In these cases, alternative methods like the CPT, which measures soil resistance to penetration, or the dilatometer test, which assesses soil deformability, can provide more accurate data.

The SPT's role in evaluating liquefaction potential is particularly noteworthy. The test has been instrumental in developing empirical correlations to predict soil behaviour during earthquakes, which is critical for constructing buildings in seismically active regions.

Despite its limitations, the SPT remains a staple in geotechnical practice due to its historical significance, ease of use, and the extensive data available for correlation with various soil properties. It serves as a bridge between past and present geotechnical methodologies, linking empirical experience with modern analytical techniques. As the field of geotechnical engineering evolves, the SPT continues to adapt, with ongoing research aimed at refining the test procedure and enhancing its interpretative frameworks. This enduring test not only connects us to the foundational aspects of soil mechanics but also propels us towards a future of innovative engineering solutions.

9.7 Laboratory Testing Program

Prior to planning the laboratory-testing program, it is again necessary to review the design needs of the project. Specifying unnecessary laboratory tests will add time and cost to the project and consume samples. The laboratory test may include the results of moisture content, Atterberg limits, grain size distribution, unit weight, void ratio, shear strength and compressibility.

Once the field program is complete, engineers must analyse the data to pinpoint critical areas where the laboratory test results could significantly impact the design. Identifying these areas is essential for conducting targeted performance testing, which can reveal issues like unexpected clay deposits that may necessitate additional consolidation testing. Such findings are vital for making informed decisions about the project, whether it involves relocating structures, redesigning them, or considering alternative foundation solutions to address potential settlement issues.

Selecting the right samples for laboratory analysis is another critical step. The goal is to choose samples that accurately represent the site conditions. In heterogeneous areas, a larger number of samples may be needed to capture the variability and ensure that the test results provide a comprehensive understanding of the site. Conversely, in more homogeneous areas, fewer samples might suffice.

The laboratory testing program should focus on both representative and critical specimens to ensure that the data collected will inform the project effectively. This data is instrumental in creating detailed subsurface cross sections, which, along with in-situ testing results and laboratory index test results, serve as invaluable tools for the engineering team. These tools aid in visualising the subsurface conditions and assessing the suitability of the proposed design, ensuring that the project is built on a solid foundation of empirical data and sound engineering principles. In essence, the laboratory testing program is a foundational component of the engineering process, providing the Insights needed to navigate the complexities of construction and design with confidence and precision.

9.8 Challenges in Geotechnical Investigation

Geotechnical investigation is indeed a field that presents a number of challenges, primarily due to the intricate nature of soil as a material. Unlike more uniform materials like steel or concrete, soil's heterogeneity and the presence of multiple phases make it a complex subject for study. Therefore, obtaining more reasonable and accurate geotechnical information is a challenging task for geotechnical engineers. The most common challenges are;

i) **Maintaining the Quality of Field Investigation**

Ensuring the quality of field investigations is crucial, as this stage lays the groundwork for all subsequent analysis and construction. The expertise of the drilling team, the oversight of the geotechnical engineer, and the quality of the equipment are all pivotal factors in this process.

ii) **Obtaining Undisturbed Samples**

The process of obtaining undisturbed soil samples is crucial for accurate soil analysis in geotechnical engineering. In cohesive soils, which can be either very soft or hard, maintaining the soil's original structure and properties during sampling is difficult. The challenge is amplified in cohesionless soils, such as sands and gravels, where traditional sampling methods often result in some disturbance. Advanced samplers like the Sherbrooke sampler and the stationary piston sampler are designed to minimise disturbance during the sampling process. Despite these advancements, the quest for perfecting undisturbed soil sampling methods continues, as current techniques may still alter the soil's natural state to some degree.

iii) **Quality of Laboratory Test**

Laboratory testing is another area where accuracy is paramount, as the results directly influence design decisions. Operator skill and precision are essential to ensure that test results are reliable and reflective of the soil's behaviour in its natural state.

Overall, geotechnical investigation requires a delicate balance of field expertise, meticulous laboratory work, and a deep understanding of soil mechanics. It's a field that's as demanding as it is fascinating, with each project presenting its own unique set of challenges and learning opportunities. Geotechnical engineers must continuously adapt and innovate to overcome these obstacles, ensuring the safety and stability of the structures they help to build.

9.8.1 Empirical Correlations for Simplified Geotechnical Assessment

Empirical correlations play an important role in geotechnical engineering by using past data and practical experience to estimate soil properties. These correlations connect simple soil index properties with more complex mechanical and physical behaviour, particularly for cohesive soils. Parameters such as the Atterberg limits relate soil water content to consistency and provide useful insight into shear strength and compressibility. Similarly, the plasticity index helps indicate expected soil strength, while the liquid limit gives an idea of consolidation potential. Such correlations help engineers take informed decisions during design and construction.

It is important to note that these correlations are based on extensive research and real-world application rather than theory alone. Institutions such as the US Army, US Navy, and various international bodies have developed and validated empirical relationships through design manuals and field studies. References like the Canadian Foundation Engineering Manual highlight the widespread acceptance and reliability of these methods, making them essential tools in everyday geotechnical practice.

In practice, these correlations allow engineers to make educated guesses about soil behaviour under various conditions, which is particularly useful when laboratory or in situ testing is not feasible. They enable a qualitative evaluation of soil parameters, providing a preliminary understanding that guides further investigation and testing.

The empirical correlations between soil consistency limits and geotechnical parameters offer a practical approach to understanding the behaviour of cohesive soil masses. These correlations are grounded in the observation that the natural moisture content (W_M) relative to the liquid (W_L) and plastic limits (W_P) can indicate the consolidation state of the soil. For instance,

If W_M is close to W_L,	***soil is normally consolidated***
If W_M is close to Wp,	***soil is some- to heavily over-consolidated***
If W_M is intermediate,	***soil is somewhat over-consolidated***
If W_M is greater than W_L	***soil is on verge of being a viscous liquid***

Generally, the presence of platy clay minerals influences the soil's shear strength, typically reducing it due to their low frictional resistance. As these minerals increase in the soil, both the liquid limit (W_L) and the plasticity index (I_P) tend to rise, suggesting a decrease in the angle of internal friction (ϕ'). This relationship is crucial for predicting the soil's behaviour under stress, particularly its shear strength. Empirical correlations have been developed to link the plasticity index with the drained peak strength parameters, such as the angle of internal friction for over-consolidated conditions (ϕ'_{oc}) and the cohesion for over-consolidated conditions (c'_{oc}). These correlations are derived from extensive triaxial compression tests on undisturbed over-consolidated clays with varying degrees of plasticity.

Understanding these correlations allows for a preliminary assessment of the soil's shear strength and consolidation characteristics, as well as its stress history, using simple index properties like the liquid limit (L_L), plastic limit (P_L), plasticity index (PI), and void ratio. This can be particularly cost-effective, providing significant insights with minimal financial investment. However, while these correlations offer valuable qualitative insights, they are not substitutes for quantitative analysis. For detailed and precise geotechnical evaluations, additional methods and tests are necessary to obtain quantitative values of sub-soil parameters. These might include more advanced laboratory tests or in-situ testing methods, which can provide a more comprehensive understanding of the soil's mechanical properties and behaviour under various loading conditions.

9.8.2 Assessment of Time-Dependent Behaviours in Cohesive Soils

The time-dependent behaviour of cohesive soils, particularly fine-grained ones, is a critical aspect of geotechnical engineering. These soils exhibit unique characteristics under varying stress conditions due to their low permeability. In the short term, they tend to show undrained behaviour, where the shear strength is a key parameter.

Additionally, factors like void ratio and the sensitivity of the clay can further influence the shear strength, with higher void ratios typically leading to lower strength. In construction, the disturbance of the soil's structure can lead to significant strength loss. The concept of undrained condition under rapid loading represents a short-term scenario, while the increase of undrained shear strength with depth, expressed as the ratio of $Su/\sigma v$, correlates with the soil's plasticity and over-consolidation ratio, providing insights into the soil's behaviour under different loading conditions.

i) Time-Dependent Behaviour of Cohesive Soils

Cohesive soils, such as clay, exhibit time-dependent behaviour due to their low permeability, which affects how they respond to stress changes. When fine-grained soils are subjected to stress changes develop excess pore pressures because of their low permeability. Thus, their behaviour is time-dependent due to their low permeability, resulting in different behaviour under short-term and long-term loading conditions. For short-term loadings, the undrained shear strength of soil is applied.

ii) Short-*Term* vs. *Long-Term Loading*

Under short-term loading, cohesive soils show undrained shear strength, while long-term loading can alter the physio-mechanical behaviour of the soil, leading to softening and strain-softening. During long-term loading, the physico-mechanical behaviour of clay gets altered. When over-consolidated hard or stiff clay strata are subjected to stress change and associated shear and/or volumetric plastic strains (or both), it causes breaking or de-structuring of inter-particle bonds of clay and loss of the extra strength achieved due to over-consolidation. The phenomenon is termed softening and strain-softening of clay. Since an undrained condition may be expected to occur under "fast" loading in the field, it represents a "short-term" condition.

iii) Brittle *to Ductile Transition*

Over time, the behaviour of cohesive soils can change from brittle to ductile, affecting mechanical parameters like modulus of elasticity, cohesion values, and water content.

Over time, the physico-mechanical properties of the soil can change significantly, especially in over-consolidated clays. Stress changes can lead to a breakdown of inter-particle bonds, resulting in a softening of the clay and a reduction in its strength. This process, from brittle to ductile transformation, affects the mechanical parameters of the soil, such as elasticity, cohesion, and water content. It's essential to consider these changes when assessing the long-term stability of structures and during the design process, where the residual shear strength of clay is a vital factor.

iv) ***Loss of Cohesion***

Long-term stress changes can cause the de-structuring of inter-particle bonds in over-consolidated clay, leading to a loss of cohesion and a decrease in the basic friction angle.

This is a time-dependent process, and the behaviour of soil changes from brittle to ductile, implying a variation in mechanical parameters (such as Modulus of elasticity, cohesion values, and water content) of soil. This deterioration is usually thought to be responsible for the partial or complete vanishing of the cohesion, and the final process causes the decrease in the basic friction angle from the critical to the residual value. It is imperative to evaluate the following parameters;

- Residual Shear Strength: It is crucial to evaluate the residual shear strength of clay for design purposes, as it represents the critical shear parameters after the soil has undergone softening.
- Void Ratio and Shear Strength: The second factor, a higher void ratio, generally means lower shear strength. Furthermore, in some (sensitive) clays, the loose structure of the clay may be disturbed by construction operations, leading to a much lower strength and even a liquid state.
- Undrained Conditions: In the field, fast loading can result in undrained conditions, which are considered short-term scenarios.

v) ***Effective Stress and Shear Strength***

The undrained shear strength (Su) of cohesive soils increases with depth, and is often represented by the ratio Su/σv, where σv is the effective vertical stress. This ratio is roughly correlated with the plasticity index and over-consolidation ratio of the soil.

The drained shear strength of normally consolidated clays is comparable to that of loose sands, with both having an effective cohesion (c') of zero, but the angle of internal friction (ϕ) is typically lower in clays. An empirical relationship exists between the effective angle of internal friction (ϕ') and the plasticity index in normally consolidated clays, which can be visualised in graphical form as shown in Figure 9.1.

Over-consolidated clays share similarities in drained shear strength with dense sands, characterised by a peak strength where c' is nonzero and a residual shear strength where c' equals zero.

In geotechnical engineering, it is standard practice to analyse both undrained and drained conditions separately when dealing with problems involving clay unless one condition is clearly more applicable. The more critical of the two conditions, undrained or drained, will dictate the design approach, with total stresses considered for undrained shear strength and effective stresses for drained conditions. This leads to the categorisation of analyses into total stress analysis (for undrained conditions) and effective stress analysis (for drained conditions).

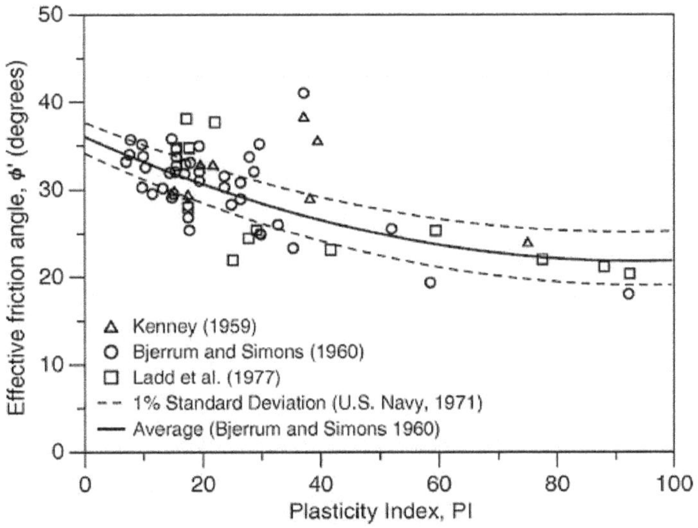

Figure 9.1: Effective angle of friction versus PI for normally consolidated clays

- Materials like clay shales, which show a significant reduction in shear strength upon shearing, necessitate that sliding analyses for structures founded on them include checks for residual shear strengths.
- The behaviour of over-consolidated clays in terms of consolidation settlement around structures is influenced by their historical effective stress, showing shear strength closer to the pre-consolidation pressure rather than the current stress level. When subjected to stresses exceeding the pre-consolidation pressure, over-consolidated clays tend to behave similarly to normally consolidated clays.

9.9 Geophysical Methods for Ground Exploration

Geophysical methods are suitable for initial surveys offshore because they can provide a significant amount of information at a modest cost. Such methods do not provide accurate data on soil and rock conditions, and need to be calibrated using information obtained from boring, sampling and/or in situ tests.

The commonly used methods are Ground-Penetrating Radar (GPR), seismic reflection and seismic refraction.

9.9.1 Ground-Penetrating Radar (GPR)

GPR uses high-frequency electromagnetic waves to detect subsurface anomalies based on dielectric contrasts. Very effective in identifying buried trenches, undocumented foundations, electrical conduits, utility lines, and zones of disturbed soil or fill, which is discussed in detail in section-9.10.

9.9.2 Seismic reflection Method (Acoustic Profiling)

Acoustic profiling is performed using equipment towed by a vessel at low speed (2–5 knots), working according to the same principle as an echo sounder used for depth recording, but operating at a different frequency that allows penetration into the seabed. A high-energy acoustic pulse passes through the water and into the soil (Figure 9.2). The pulse is reflected back by the seabed and by subsoil interfaces corresponding to changes in density and sonic velocity. A profile of the return times from the various reflecting horizons is continuously recorded and transformed into electric signals that are visualised by printing a 2D diagram (echogram), which needs to be interpreted by a specialist. Different types of equipment are available, with different characteristics and capabilities.

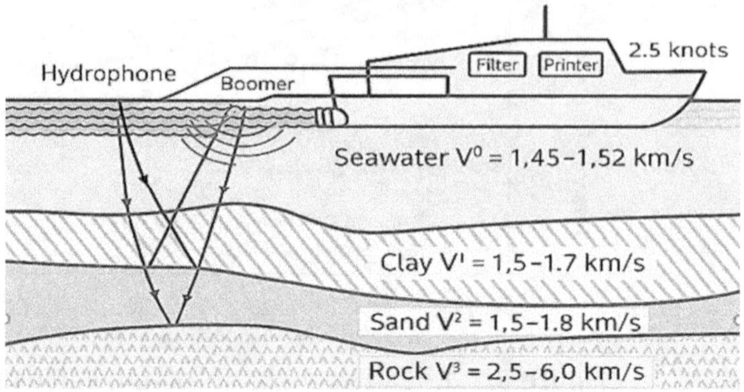

Figure 9.2: Schematic Diagram of Seismic Reflection Method

The main difference is the frequency, which determines the maximum penetration depth and the resolution accuracy. Low-frequency pulses achieve deep penetration but have a low resolution. The main types are in Table 9.1.

Table 9-1: Seismic Reflection Survey with various frequencies

Type	**Frequency Range**	**Penetration Depth**	**Resolution**	**Notes**
Pinger	3–10 kHz	Up to 25 m	~0.5 m	Shallow profiling
Boomer	0.5–2.5 kHz	Up to 100 m	~1 m	Moderate depth
Spraker	0.05–1.0 kHz	Up to 500 m	~5 m	Deep profiling

- Penetration Validity: Effective in clay, silt, and sand.
- Advantages: Fast, cost-effective, covers large areas (~20 km²/day).
- Limitations: Low accuracy; requires calibration with borehole data.

9.9.3 Seismic Refraction Method

Refraction Seismology or Refraction investigations are based on the times of arrival of the initial ground movement generated by a source recorded at a variety of distances. Later arriving complications in the recorded ground motion are discarded. Thus, the data set derived from refraction experiments consists of a series of times versus distances. These are then interpreted in terms of the depths to subsurface interfaces and the speeds at which motion travels through the subsurface within each layer. These speeds are controlled by a set of physical constants, called ***elastic parameters*** that describe the material

- Principle: Measures arrival times of shock waves at geophones.

- Source: Explosives or hammer blows.
- Application: Determines depth to bedrock and soil boundaries.
- Accuracy: Higher than reflection but more costly and limited in area.
- Use Case: Preliminary investigations and supplement to acoustic surveys.
- Focus: Initial ground motion arrival times.
- Output: Depth and velocity profiles of subsurface layers.
- Industry Use: Preferred in environmental and engineering fields.

Seismic Refraction is best suited for mapping the depth to bedrock or identifying layered structures with increasing velocity.

- It is used to describe a subsurface that has simple, horizontal layering.
- To determine the layer thicknesses and P-wave velocities.
- It can have moderate depth coverage at a moderate budget.
- It shall be avoided if the subsurface has low-velocity-over-high-velocity layers or has complex geometry.

Limitations:

- **Velocity Layering Assumption**: It assumes that seismic velocity increases with depth. If a low-velocity layer overlies a high-velocity layer (e.g., soft clay over rock), refraction methods may fail or misinterpret the structure.
- **Complex Geometries**: In areas with dipping layers, faults, or irregular interfaces, refraction data can be difficult to interpret accurately.
- **Limited Vertical Resolution**: It provides good horizontal layering information but lacks the vertical resolution needed for detailed stratigraphic analysis.

Depth Limitations: Although it can reach moderate depths, it's not suitable for deep investigations compared to reflection methods.

9.9.4 Seismic Reflection Method

In reflection experiments, analysis is concentrated on energy arriving after the initial ground motion. Specifically, the analysis concentrates on ground movement that has been ***reflected*** off of subsurface interfaces. In this sense, reflection seismology is a very sophisticated version of the echo sounding used in submarines, ships, and radar systems.

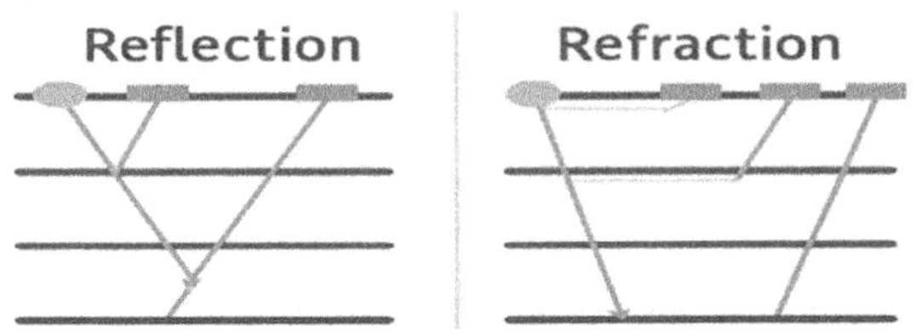

Figure 9.3: Schematic Diagram of Reflection and Refraction Seismology

Each of these techniques has specific advantages and disadvantages when compared to the others and when compared to other geophysical techniques. For these reasons, different industries apply these techniques to differing degrees.

- Focus: Analyses reflected energy after initial ground motion.

- Application: Detailed subsurface imaging using amplitude and waveform.
- Industry Use: Dominantly used in oil & gas exploration.

Seismic Reflection is best suited for High-resolution imaging of complex subsurface structures.

- It is used when a detailed stratigraphy, fault mapping, or resource exploration is required.
- The project requires deep penetration and precise layer delineation.
- Budget allows for advanced acquisition and processing.

It shall be avoided if the site is small, shallow, or has budget constraints.

Limitations:

- **Cost and Complexity**: Requires sophisticated equipment, skilled personnel, and extensive data processing, making it expensive and time-consuming.
- **Not Suitable for Small or Shallow Sites**: For shallow investigations or small-scale projects, the method may be overkill and inefficient.
- **Sensitive to Noise and Ground Conditions**: Urban noise, poor coupling, and heterogeneous near-surface conditions can degrade data quality.
- **Requires High Data Quality**: Poor acquisition or processing can lead to misleading interpretations, especially in complex terrains

9.9.5 MASW (Multichannel Analysis of Surface Waves)

Seismic surveys utilise different wave types to investigate subsurface conditions. Refraction surveys analyse first-arrival refracted waves to estimate layer velocities and depths, with advanced methods like seismic refraction tomography offering improved resolution.

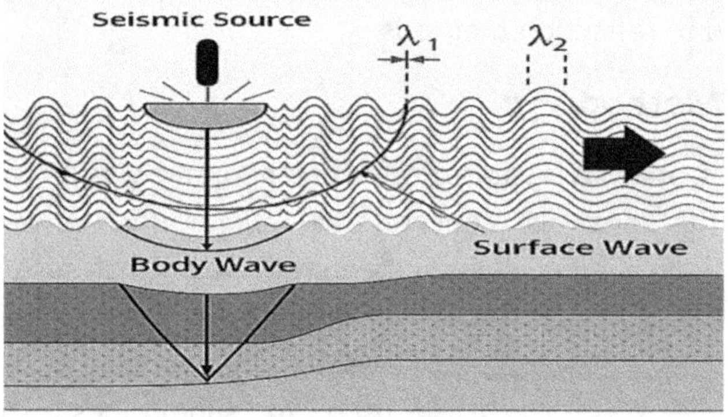

Reflection surveys use waves reflected from interfaces with contrasting elastic properties, employing the CDP (common-depth-point) method to produce detailed cross-sectional images, traditionally used in resource exploration and now adapted for shallow geotechnical studies.

Figure 9.4: Geophysical Investigation by Multichannel Analysis of Surface Waves

Surface waves, particularly Rayleigh and Love waves, are dispersive and sensitive to shear wave velocity (Vs) variations with depth. The MASW (Multichannel Analysis of Surface Waves) method, developed in the early 2000s, leverages this property to efficiently and cost-effectively derive Vs profiles, offering simple field acquisition, easy processing, and broad area coverage.

The **MASW (Multichannel Analysis of Surface Waves)** method is a widely used geophysical technique for determining shear wave velocity (Vs) profiles of near-surface materials. It analyses the dispersion of surface waves, primarily Rayleigh waves, which are sensitive to Vs variations with depth.

Table 9-2: Details of Various Geophysical Investigations

Method	Wave Type	Output	Depth Range	Cost	Accuracy	Field Complexity
MASW	Surface (Rayleigh)	Vs profile	Up to ~30 m	Low	Moderate	Low
Refraction	Body (P-wave)	Layer depth & velocity	Up to ~100 m	Moderate	Moderate	Moderate
Reflection	Body (P-wave)	Subsurface imaging	Deep	High	High	High
Down-hole	Body (P/S-wave)	Direct Vs/Vp measurement	~30–50 m	High	High	High

MASW offers several key advantages: simple and robust field data acquisition, as surface waves carry the strongest energy; easy and accessible data processing, even for non-experts; efficient coverage of large areas in short timeframes; and high cost-effectiveness compared to conventional body-wave methods like S-wave refraction or down-hole surveys. These benefits make MASW particularly suitable for geotechnical site characterisation, seismic hazard assessment, and soil stiffness profiling.

Choosing the right seismic method depends on the project goals, site conditions, required depth, resolution, and budget. Here's a guide to help decide when to use each method:

- Thus, MASW is best applicable for Shallow subsurface investigations (up to ~30 m), especially for shear wave velocity (Vs) profiling. It is used when:
- Recommended when:

- A non-invasive, cost-effective, and rapid technique is preferred.
- Assessment of soil stiffness or liquefaction potential is required.
- The site is located in a noisy or urban environment.

Limitations:

- **Depth Restriction**: MASW typically investigates only the upper ~30 meters. For deeper geotechnical or geological studies, it lacks sufficient penetration.
- **Resolution Constraints**: While effective for general Vs profiling, MASW may not provide the fine resolution needed for detailed stratigraphy or small-scale features.
- **Limited in Complex Geologies**: In areas with strong lateral heterogeneity or abrupt changes in subsurface properties, MASW may produce ambiguous results.

- **Not Ideal for High-Resolution Imaging**: It's primarily designed for velocity profiling, not for imaging detailed subsurface structures like faults or thin layers.

9.9.5.1 Down-hole / Cross-hole Surveys

This method is based on direct measurement of Vs and Vp at specific locations.

- It is used when High accuracy and site-specific data are needed.
- Boreholes are already planned or available.

It shall be avoided if broad area coverage is required or want to avoid invasive methods.

Limitations:

- **Invasive** Nature: Requires boreholes, which can be costly and logistically challenging, especially in remote or environmentally sensitive areas.
- **Limited Spatial Coverage**: Provides point-specific data; not suitable for mapping large areas or continuous profiles.
- **Dependent on Borehole Quality**: Accuracy depends on borehole alignment, casing, and coupling. Poor borehole conditions can compromise results.
- **Time and Cost Intensive**: Drilling, equipment setup, and data acquisition can be resource-heavy, especially for multiple locations

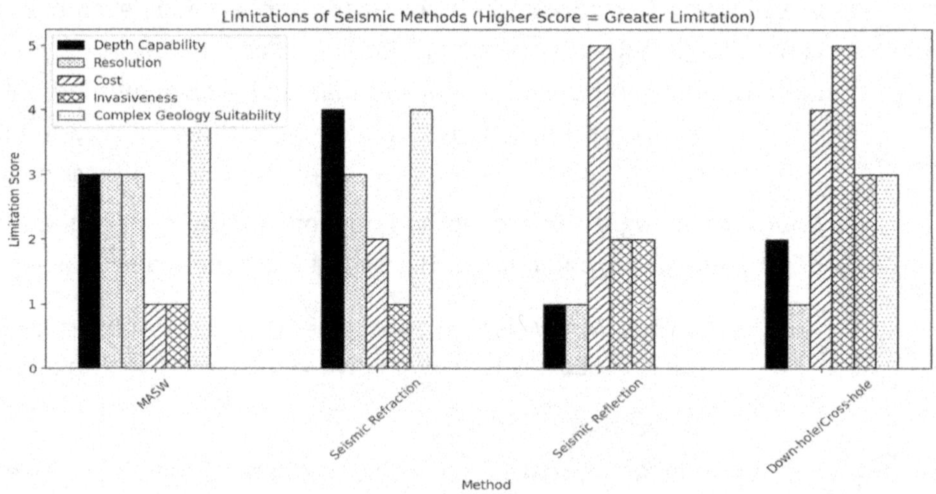

Figure 9.5: Graphical presentation of limitations of various Seismic Methods

9.10 Ground-Penetrating-Radar Technique

Traditional inspection methods like visual checks, hammer sounding, chain dragging, or even digging test pits and coring aren't enough to accurately detect what's happening below the surface. That's why alternative techniques are needed.

GPR is one of the most effective and flexible non-destructive tools for routine underground inspections. It works by sending electromagnetic waves through different materials and analysing how they travel.

GPR is reliable, cost-effective, and doesn't damage the surface. It's great for spotting hidden defects, voids, or buried objects. Originally used in geology and geophysics, GPR is now widely applied in civil engineering too. Its versatility and wide range of uses make it stand out from other testing methods. GPR uses radar signals between 10 MHz and 3,000 MHz and follows the ASTM D6432 standard.

9.10.1 Working Principles of GPR

GPR sends out radar signals into the ground, and when these signals hit something with a different material (like a pipe, void, or buried object), they bounce back. The strength of the reflected signal depends on how different the materials are. As the GPR device moves along the surface, it creates a 2D image (called a radargram or B-scan) showing what's underneath.

If we know how fast the signal travels through the ground and measure how long it takes to bounce back, we can figure out how deep the object is. With advanced tools, these signals can even be turned into 3D images, giving a clearer view of underground structures.

The GPR system includes a transmitter and receiver antenna connected to a processing unit. It sends out radar pulses at specific frequencies. Higher frequencies give better detail but only scan shallow depths, while lower frequencies can reach deeper but with less detail. So, the choice of frequency depends on whether you need high resolution or deeper scanning.

9.10.2 Where GPR Is Used

GPR is super handy for scanning below the surface in areas like soil, pavement layers, concrete slabs, and RCC decks. It's great for finding things like rebar, live electrical conduits, and utility ducts.

Figure 9.6: GPR Survey Method

Some common uses include:

- Finding and mapping steel reinforcements like rebars and tendons
- Estimating the thickness of concrete structures
- Spotting defects or voids like delaminated zones
- Detecting metal objects inside non-conductive materials like concrete

One of GPR's biggest advantages is that it can collect data quickly and continuously using mobile units, without needing direct contact with the surface. Plus, it doesn't interrupt traffic or operations, which saves time and money.

However, GPR has some limits. It doesn't work well in areas with high-conductivity materials like wet clay, and mixed ground conditions can mess with the signals. Also, reading the radar images (called radargrams) isn't always straightforward—it takes skill and experience to interpret them correctly.

Different antennas are used depending on what you're looking for. Low-frequency antennas go deeper but give less detail, while high-frequency ones show more detail but only scan shallow depths. Scanning the same spot with multiple frequencies helps get the most complete data.

9.10.3 Impact of Frequency on Cost in GPR Surveys

The choice of frequency in Ground Penetrating Radar (GPR) doesn't just affect depth and resolution—it also influences survey cost.

i) High-Frequency Antennas (e.g., 1000–2000 MHz) are used for shallow, high-detail scans (like locating rebar or cracks in concrete), which *is more expensive per square* meter because, they require slower scanning speeds to capture fine details and often need multiple passes to cover the area thoroughly. *Therefore, data processing is more complex and time-consuming*

ii) Low-Frequency Antennas (e.g., 100–400 MHz) are generally cheaper per square meter because, they cover larger areas faster, data is simpler and quicker to process and fewer passes are needed.

iii) Multi-Frequency or Dual Surveys is the most expensive option, but provides maximum data, combining both shallow and deep scans which is useful for complex sites like ports or brownfields where both surface and deep features matter.

- ***Higher frequency = better detail, higher cost***
- ***Lower frequency = deeper reach, lower cost***
- ***Dual-frequency = best results, highest cost***

iv) Choose the Right Frequency for the Job

- Avoid over-specifying resolution: Use low-frequency antennas (e.g., 250 MHz) for deep scans where fine detail isn't needed.
- Use high-frequency antennas (e.g., 1000 MHz+) only when shallow, high-resolution data is essential—like detecting rebar or cracks in concrete.

Avoid using high-frequency antennas for large-area surveys unless necessary—they take longer and cost more.

v) Limit Survey Area Strategically

- Focus on critical zones identified through visual inspection or historical data.
- Use GPR to verify suspected problem areas rather than scanning the entire site.

vi) Combine GPR with Other Methods

- Use geophysical data to guide geotechnical drilling, reducing the number of boreholes needed.
- This hybrid approach improves accuracy and lowers overall investigation costs.

vii) Use Mobile or Vehicle-Mounted Systems

- For large sites like roads, ports, or airfields, vehicle-mounted GPR units can scan faster and more efficiently than handheld systems.

9.10.4 Hybrid survey Methods with GPR

Hybrid survey methods that combine GPR with other techniques to improve subsurface investigations while keeping costs and risks under control:

- GPR + Geotechnical Boreholes

Working principles: Use GPR to scan large areas and identify zones of interest, then drill boreholes only in those specific spots. It reduces the number of boreholes needed, saving time and money while improving targeting accuracy.

- GPR + Electrical Resistivity Tomography (ERT)

Working principles: GPR detects shallow features with high resolution, while ERT maps deeper layers and moisture content. The combined data offers a full-depth profile—ideal for detecting voids, buried utilities, and groundwater zones.

- GPR + Seismic Refraction

Working principles: GPR maps shallow structures; seismic refraction identifies deeper rock layers and stiffness. The combined data is useful in identifying bedrock depth and designing the foundation and slope stability analysis.

- GPR + Magnetometry

Working principles: GPR detects non-metallic and metallic objects, whereas magnetometry focuses on ferrous materials like buried tanks or pipes. Thus, it enhances the detection of buried metallic objects or hazards in brownfield or port / industrial sites.

- GPR + Drone-Based Photogrammetry or LiDAR

Working principles: GPR scans below ground, and drones map the surface features and topography. The combined surface and subsurface data are ideal for comprehensive site modelling, such as large infrastructure projects.

Before specifying geophysical investigations, the following factors should be considered to ensure their effective use:

1. **Type of site** – Whether the location is a port, industrial area, roadway, dam, or something else should be identified.
2. **Main investigation goals** – The purpose of the survey, such as locating buried utilities, structural issues, soil layers, voids, or contamination, should be clearly defined.
3. **Depth of interest** – The depth range to be scanned (shallow: 0–1 m, medium: 1–5 m, or deep: over 5 m) should be determined.
4. **Site conditions** – Ground characteristics like dryness, moisture, clay content, rockiness, or mixed conditions should be assessed.
5. **Operational constraints** – Any limitations on site access or working hours should be taken into account.

9.11 Other Applications of GPR

The use of applications of GPR in civil engineering apparently began in the mid-70s and early 80s, mainly focused on the detection of hidden utilities and pavement inspection. From these first studies, the number of applications and developments of the methodology increased remarkably. Nowadays, it is being successfully employed for a variety of tasks, and a brief overview of the civil engineering applications is presented here.

9.11.1 GPR surveys in pavements

GPR surveys in pavements (rigid or flexible) are frequently applied as a complementary technique for the assessment of roads and airport pavements. GPR images provide valuable information about the continuity of layers and their thickness, which could be combined with falling weight deflectometer (FWD) tests. In addition, GPR inspection allows for the analysis of the pavement layer's condition and locating defects, which is crucial in preserving the structure and its service life. In road assessment, the main analyses are focused on measuring layer thickness and detecting voids or delamination.

It is also applied in the assessment of **rigid pavements**, with one of the main objectives being the identification of metal bars and layer thickness, and measuring the rebar cover depth. The assessment to detect damage, such as cracking and voids, is another application in rigid pavements. These studies are mainly focused on the detection of damage areas, but in some cases, the origin of the damage is also investigated, including the effect of roots in pavements and the impact of weather and load conditions.

9.11.2 In the study of quay walls

In the study of quay walls, GPR focuses on the detection of rebar in concrete retaining walls, detecting joints and dowels, and the location of damage and water infiltration. In some cases, the GPR analysis objective is the detection of the contact between the wall and the natural soil or between the wall and further restorations of scouring holes, etc.

Other applications are the **detection and the analysis of the foundations**. Gravity retaining walls, which are massive and require a significant gravity load to be stable under the soil's horizontal pressures. These structures can be built with concrete, masonry, stone, or precast concrete boxes filled with coarse material structures. Those walls support high loads, and the assessment of their state is important in maintaining or repairing parts of the structure.

The main advantage is that they enable faster data collection by increasing the extension of the investigated area per time unit, and they make it easier for the operator to produce 3D images.

However, the studies about the relation between the GPR images and the material parameters allowed for the application of the technique in the assessment, focused on particular aspects such as the material density for quality controls [28], control of layer thickness, detection of voids, measuring the depth to damage, and the detection of contact between rehabilitated (new) and old pavement.

As a consequence of ageing, degradation of materials, changes in the loads (due to variations in the water content or in consequence of the construction of other structures) or changes in the temperature and moisture, the walls can suffer different levels of damage, leading even to the collapse of the structure. The detection of the pathologies reduces the possibility of structural failure. The assessment of those walls requires the study of the structure and the analysis of the interaction between the structure and the soil.

The inspection of the structures for retained soil requires the assessment of the wall that determines the decisions about the structure (repair, maintenance, change or frequency of future inspections). The assessment involves the visual inspection and simple sonic and thermographic tests. Depending on the results, more invasive tests could be needed.

Non-destructive inspection techniques are currently applied to obtain valuable information about the state of the wall. GPR is a promising technique in this field.

Only a few applications in masonry retaining walls can be found in the literature Those applications are based on detecting the contact with the natural soil, estimating the thickness of the structure, and changes due to old repair works and the inner structure including the foundations [119]. Although in many cases, GPR data is acquired with the antenna on the surface of the wall, in some cases, the analysis of the wall acquiring data from its top part (Figure 6) could provide information about the number of stone rows and the contact between ashlars.

However, the detection of the contact between the natural soil and the wall in order to identify possible voids, changes in humidity or defects on the surface of the wall, is a common application in all types of walls. In masonry walls, the common objectives of surveying are the analysis of layers, the contact between blocks, and the assessment of foundations. Conversely, in concrete walls, the main goal focuses on the detection of the rebar and possible defects that occurred during the construction of the wall.

The use of NDT technology becomes essential for the health assessment of these former structures in order to preserve their historical character and structural integrity. In this context, the GPR method has been demonstrated as effective for the diagnosis of masonry arch structures, in which the detection of the following characteristics was possible:

- Unknown geometries remaining in the interior of the bridge, such as hidden arches and ancient profiles (shape) of the structure.
- Evidence of restorations and/or reconstructions in stonework.
- Existence of cavities and fractures/cracking in masonry.
- Moisture in masonry.
- Foundation condition.
- Filling distribution in masonry.

For a satisfactory result from GPR, the ground-coupled antenna is placed on that platform, assuring the maximum contact between the antenna and the pavement surface, and the maximum transmission of the wave into the medium.

9.11.3 Soil Subgrade Assessment and the Detection of Bedrock

Soil Subgrade Assessment and the Detection of Bedrock. Some of the first studies in pavements were focused on the analysis of the soil subgrade. This layer is highly sensitive to the environmental changes, being also affected by the traffic loads. It may involve following three applications:

- Analysis of road alignments, studying the ground materials and the depth to the bedrock and to the water table, reaching in some cases a depth of about 5 m.
- Assessment of existing roads in order to detect filling materials, voids or weak zones under the pavement.
- Quality control of restorations.

In practical, it is observed that the air-coupled antennas are more appropriate in detecting interfaces between layers and with the subgrade; whereas the ground-coupled antennas are more appropriate in determining anomalous zones that could be associated with damage.

9.11.4 Benefits of Drone Integrated GPR Surveys in Port

i) **Rapid Surface Mapping** - Drones equipped with photogrammetry or LiDAR can quickly capture high-resolution surface data across large and complex port areas. This help in creating accurate topographic maps, 3D models, and site layouts before subsurface investigations begin.

ii) **Enhanced Planning for GPR Surveys** - Drone data helps identify obstacles, access routes, and zones of interest, allowing GPR teams to plan more efficient and targeted scans. It reduces time spent on-site and minimizes disruption to port operations.

iii) **Improved Safety** - Drones can survey hard-to-reach or hazardous areas (e.g., near water, under cranes, or congested zones) without putting personnel at risk. It is ideal for initial reconnaissance in active or restricted port environments.

iv) **Integration with Subsurface Data -** Surface models from drones can be combined with GPR and other geophysical data to build comprehensive 3D site models. This integration improves the interpretation of subsurface features in relation to surface structures and utilities.

v) **Real-Time Monitoring and Documentation -** Drones can provide live video feeds and progress tracking during construction or redevelopment and it is useful for documentation, stakeholder reporting, and quality control.

Part-III
Engineering Approach for Port Infrastructure Upgradation

Advanced Materials for Repair and Upgradation

Chapter Summary

The development of advanced materials has significantly improved the performance, durability, and sustainability of structural repair and rehabilitation. Polymer-based compounds, in particular, offer tailored solutions for strengthening and upgrading existing infrastructure. However, adopting these innovative materials comes with challenges. Key barriers include high initial costs, complex regulatory requirements, limited industry experience, environmental considerations, and the need for compatibility with existing structures. To overcome these challenges, bridging the knowledge gap is essential. Promoting collaborative learning, sharing practical case studies, and providing targeted training can build confidence among professionals and support informed decision-making. A proactive and technically sound approach—aligned with regulatory standards—can enable the construction industry to deliver safer, more resilient, and cost-effective infrastructure that meets evolving demands.

Learning Objectives

- Gain an understanding of various synthetic construction materials and their specific applications in structural repair and strengthening.
- Recognize the unique features and limitations of these materials in different contexts.
- Learn the essential precautions to be followed during material application.
- Develop the ability to select appropriate repair materials based on structural needs and performance criteria.

10.1 Introduction

The rapid advancement in construction technology and chemical engineering has led to the development of a wide range of polymer-based and synthetic materials specifically designed for structural repair and enhancement. These materials offer superior performance characteristics such as high tensile strength, corrosion resistance, and adaptability to complex structural geometries.

Common examples of polymer-based materials used in structural repair include:

Epoxy Resins – Ideal for crack injection, bonding, and coating due to their excellent adhesion and chemical resistance.

Polyurethane Grouts – Effective for sealing water ingress and filling voids, particularly in underground or marine structures.

Fibre-Reinforced Polymers (FRP) – Including Carbon FRP (CFRP) and Glass FRP (GFRP), used to strengthen beams, columns, and slabs.

Acrylic Polymers – Applied in surface treatments and protective coatings to improve durability.

Vinyl Ester Resins – Known for their resistance to acids and solvents, making them suitable for industrial environments.

Selecting the right material based on structural requirements and environmental conditions can effectively address deficiencies and extend service life in a cost-efficient manner. However, successful application demands a well-informed approach—understanding material properties, ensuring proper training, and maintaining thorough documentation to support regulatory approvals. This ensures compliance with safety, durability, and environmental standards, contributing to resilient and sustainable infrastructure.

Despite these advancements, the construction industry often shows caution in adopting new materials and methods. This hesitation is largely due to several key challenges:

i) Lack of Knowledge and Experience

The reluctance to adopt new technologies and materials often stems from limited knowledge and practical experience. This can slow down innovation and hinder progress. Overcoming this resistance requires targeted efforts to share relevant information and provide hands-on exposure. Practical training helps professionals become more confident in using new materials.

Additionally, case studies and success stories from projects that have effectively implemented innovative materials can serve as powerful tools. They not only demonstrate the practicality and benefits—such as improved safety and sustainability—but also help shift mindsets. Sharing such examples encourages professionals to explore and embrace novel solutions.

Bridging this knowledge gap is essential for advancing the industry and making full use of modern technologies. A proactive approach to learning builds a strong foundation for innovation and adaptability. Ultimately, the goal is to foster an environment where knowledge and experience with new materials are not only accessible but actively pursued. This leads to more informed decision-making based on evidence and best practices, rather than familiarity, resulting in safer, more sustainable, and cost-effective construction outcomes.

ii) Compatibility with Existing Infrastructure

When upgrading or strengthening existing structures, it is crucial to ensure that new materials are compatible with the original or existing infrastructure. Materials must work in harmony to maintain structural integrity. Incompatibility—such as differing rates of expansion, contraction, or chemical reactivity—can lead to failures.

Therefore, engineers must carefully assess the properties of both existing and new materials, considering their interaction over the structure's expected lifespan. This holistic evaluation is vital for a successful and sustainable upgrade.

iii) Environmental Impact

Evaluating the environmental impact of new materials, especially in marine environments, is a critical step. Materials must be tested for durability under stressors like continuous saline exposure. A full lifecycle assessment—from manufacturing to disposal—is necessary to understand their ecological footprint.

This comprehensive understanding supports informed decisions about material selection, application, and maintenance, ensuring that sustainability remains a core priority.

Thus, by integrating advanced materials with informed engineering practices and addressing key challenges such as knowledge gaps, compatibility, and environmental impact, the construction industry can move toward more resilient, sustainable, and long-lasting infrastructure solutions.

10.2 Composite Fibre Reinforcement

Wrapping with composite fibre reinforcement, particularly carbon fibre, is a highly effective technique for structural repair and strengthening. This method entails applying a composite material—typically a carbon or glass fibre fabric combined with epoxy resin—onto the surface of existing structures. Due to its high tensile strength and lightweight properties, carbon fibre is an ideal choice for enhancing structural integrity without significantly increasing weight, which is essential for preserving the balance of forces within the structure. Notably, carbon fibre offers a superior strength-to-weight ratio compared to conventional materials like carbon steel and can last up to 50 years, depending on environmental conditions. Additionally, the flexibility of carbon fibre wraps enables them to conform to various shapes and sizes, making them suitable for a wide range of structural forms. The ease of installation, combined with cost-effectiveness and minimal disruption during repairs, makes composite fibre wrapping a preferred solution for reinforcing both distressed and non-distressed structures, including those located in seismically active regions.

10.2.1 Core Components of the Composite Fibre Wrapping System

- **Fabric - Glass Fibre Woven Roving (GFRP):** Available in thicknesses ranging from 200 gsm to 800 gsm or higher.
- **Isophthalic Polyester Resin:** A general-purpose polymer resin used for bonding.

Accelerator (Enhances the curing process of the resin)

Catalyst (Initiates polymerisation and hardening of the composite material)

These three materials (Accelerator, Catalyst and Hardner) are added to initiate polymerisation and hardening

Hardener (Facilitates the final setting and strength development of the resin system.)

Fibre Reinforced Polymer (FRP) or Glass Fibre Woven Roving (GFRP) is a "cast-in-place" system using two primary components:

Fibres are usually made of materials like glass, carbon, or aramid, and a polymer matrix, typically an **epoxy resin**. The installation process begins with the delivery of fibres to the site in the form of dry, flexible fabrics. These fabrics are then carefully wrapped around the structure where reinforcement is needed. Initially, the fabrics are saturated with the uncured epoxy resin. As the epoxy cures, it hardens, forming a rigid FRP composite that intimately conforms to the shape of the structure. The cured FRP composite not only strengthens the structure but also enhances its durability to extends its service life.

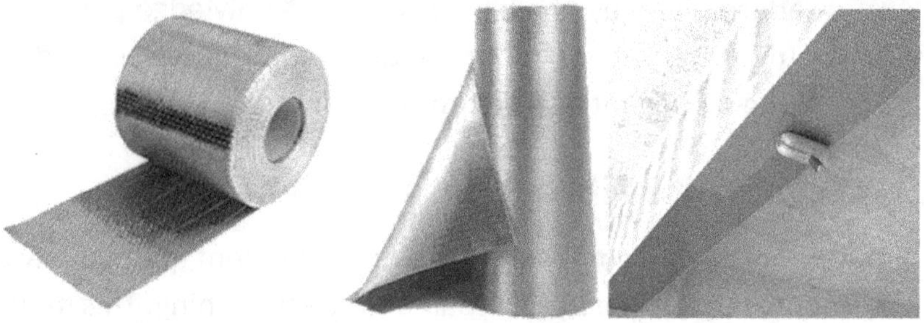

Figure-10.1: Fibre Reinforced Polymer (FRP) or Glass Fibre Woven Roving (GFRP)

There are three types of fabrics available [as shown in Figure 10.1], and selection is based on the right fibre for the right purpose basis.

i) FRP Laminates

- FRP Laminates are composed of carbon fibre reinforced with epoxy resin, providing both strength and flexibility. The installation process involves bonding these laminates to concrete surfaces using a specialised epoxy adhesive to ensure a secure and durable fit. Standard laminate dimensions—such as 50/1.4 and 100/1.4—indicate their width and thickness in millimetres, allowing for precise reinforcement tailored to structural requirements.
- A major advantage of FRP Laminates is their versatility. They can be applied to various structural elements, including wooden and concrete beams, columns, slabs, and walls, making them suitable for both retrofitting existing structures and enhancing new constructions. Supplied in rolls or pre-cut lengths, FRP Laminates offer flexibility and minimise material waste—especially beneficial for custom or small-scale applications requiring specific lengths.

ii) Near Surface Mounted (NSM) FRP

The Near Surface Mounted (NSM) technique using FRP is an advanced method for improving structural integrity. It involves embedding thin carbon fibre laminates into grooves cut into the substrate, offering an effective solution for both flexural and shear strengthening.

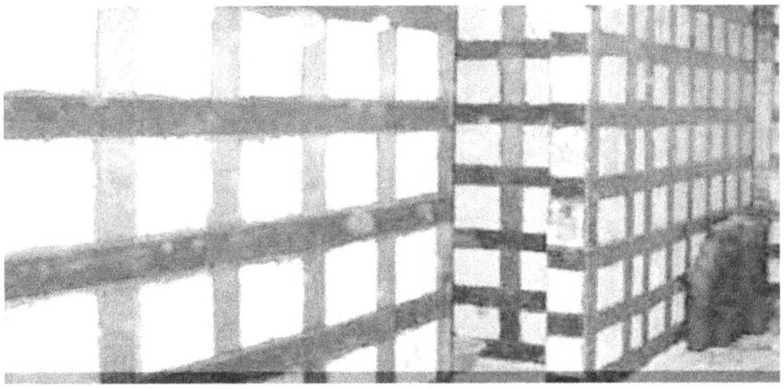

Figure-10.2: Carbon Fibre Laminâtes for NSM

This approach is particularly beneficial due to the non-corrosive nature of the materials, allowing for shallower grooves and reduced material cover compared to conventional steel reinforcement. Additionally, the adaptability of carbon fibre laminates enables customisation to meet specific project requirements.

10.3 Fibre Reinforced Concrete (FRC)

Concrete is inherently weak in tension and exhibits brittle behaviour. The concept of incorporating fibres aims to enhance its construction properties. While continuous reinforcement in concrete (i.e., reinforced concrete) improves strength and ductility, it offers limited control over micro-crack development. Alternatively, introducing fibres in discrete form into plain or reinforced concrete presents a more effective solution. The addition of fibres transforms concrete into a more homogeneous and isotropic material. Upon cracking, the randomly distributed fibres become active, arresting crack initiation and propagation, thereby improving both strength and ductility.

The different types of fibres commonly used in FRC are polypropylene fibres, steel fibres, and macro synthetic fibres.

- **Polypropylene Fibres** are made from synthetic polymers. They are Monofilament (single-strand) fibres, have Lightweight, Good resistance to alkalis and chemicals.
- **Glass Fibres** are straight fibres with a diameter range from 0.005 to 0.015mm (may be bonded together to form elements with diameters of 0.13 to 1.3mm).
- **Steel Fibres** are typically made from low-carbon steel. They can be Straight, crimped, twisted, hooked, ringed, and paddled ends. Generally, its diameter ranges from 0.25 to 0.76mm. Steel fibres have high tensile strength but are susceptible to corrosion in aggressive environments
- **Macro Synthetic Fibres** are made from synthetic polymers (e.g., polypropylene, polyethene, or nylon). They are Multifilament (multiple strands bundled together) having properties similar to polypropylene fibres and moderate tensile strength.

The choice of fibre type depends on the specific project requirements, cost considerations, and desired performance. Each type has its advantages and limitations; therefore, it shall be selected appropriately based on the application.

FRC serves as secondary reinforcement, effectively reducing shrinkage and controlling cracking in concrete. It is user-friendly, requiring no changes to the concrete mix design or special equipment modifications. It is also suitable for both pumping and shotcreting applications.

Figure-10.3 : Various types of micro-filament fibres

Performance Benefits

When fibres are incorporated, repair concrete formulations show:

- Reduced plastic and hardened shrinkage
- Improved impact strength
- Increased fatigue resistance and toughness
- Enhanced long-term durability and structural performance
- Effective secondary and temperature crack control through a synergistic fibre system with long-length design

10.3.1 Dosage Rate Guidelines

i) **Low Dosage (~0.1% by volume)**

- Polypropylene fibres primarily enhance crack control and inhibit crack initiation.

ii) Moderate Dosage

- Improves concrete durability by minimising shrinkage and temperature-induced cracks.
- Enhances resistance to chloride ion penetration, sulphate corrosion, and carbonation.

iii) High Dosage

- While crack control benefits plateau, concrete toughness increases significantly.
- High-dosage FRC demonstrates superior post-cracking performance and residual strength.

Therefore, the specific dosage should be adjusted based on the concrete mix design, fibre type, and other project-specific requirements.

When the above fibres are added, the repair concrete formulations reduce plastic and hardened shrinkage, improve the impact strength, increase the fatigue resistance and concrete toughness hence providing long term durability, structural enhancement and effective secondary/temperature crack control by incorporating a unique synergistic fibre secondary/temperature crack control by incorporating a unique synergistic fibre system of long term of long length design.

A. Recommended Dosage

In compliance with ASTM C-1116, fibres are typically added in proportions ranging from 0.2% to 2% of the concrete volume during mixing. For structural repair applications, high-strength concrete is conventionally designed with fibre dosages between **2 kg/m³ and 99 kg/m³**, depending on the specific requirements.

This fibre-reinforced concrete is particularly effective for:

- Jacketing of concrete piles and columns
- Strengthening of foundations
- Rehabilitation of deteriorated jetty decks

To enhance structural performance, steel reinforcement is often used in combination with polypropylene fibres, providing improved load-bearing capacity and crack resistance

The 4 C's of Polypropylene Fibre Reinforcement

i) **Configuration**

A synergistic blend of twisted-bundle fibres, non-fibrillating monofilaments, and polypropylene fibres designed to enhance structural performance.

ii) **Chemistry**

Manufactured exclusively from 100% virgin polypropylene to ensure consistency and durability.

iii) **Contents**

Fibre dosage must be tailored to meet specific design requirements for optimal effectiveness.

iv) **Correct Length**

Standard fibre length is 2.25 inches (54 mm), suitable for most structural applications.

B. Standard Specification

All fibre-reinforced concrete products must conform to ASTM C-1116 to ensure quality and performance standards

10.4 Ferrocement Lining

Ferrocement is a highly adaptable construction material, consisting of a mortar matrix reinforced with multiple layers of mesh and metal. Known for its thin sections and excellent strength-to-weight ratio, this composite typically includes ordinary Portland cement, well-graded sand, and specialized admixtures that enhance plasticity, bonding, and sealing characteristics. The application process often utilizes orbital vibrators to achieve a compact and uniform layer, which is critical for the structural integrity of ferrocement components. Installation does not require highly skilled labour; with basic training, general workers can effectively carry out the construction, making ferrocement a practical solution for diverse applications.

Ferrocement is essentially a thin reinforced concrete formed from cement mortar and closely spaced layers of continuous, small-diameter mesh. Its components also include woven wire hot-dip galvanised mesh, serving as skeletal reinforcement, along with chemical admixtures that enhance mortar performance.

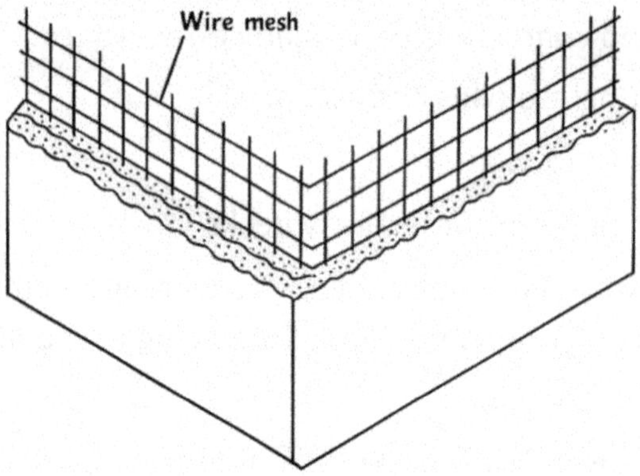

Figure-10.4 : Ferrocement Lining

i) Advantages of Ferrocement

- High strength under various stress conditions—compression, tension, shock, impact, and explosion.
- Resistance to mild chemicals, making it suitable for diverse environmental conditions.
- Enables a high-quality surface finish.
- Excellent crack-arresting capability due to dispersed reinforcement.
- Highly cost-effective—approximately 1/6th the cost of FRP jacketing.
- Non-combustible and does not emit harmful fumes, unlike FRP.
- Simple construction process; can be executed by general labour with minimal training.

10.4.1 Composite Matrix Components

- **Ordinary Portland Cement:** Serves as the primary binding agent.
- **Well-Graded Zone II Sand:** Medium, silt-free sand for optimal consistency.
- **Special Admixtures:**
  - *Plasticisers* to enhance workability
  - *Bonding agents* to improve adhesion

- *Pore sealants* to reduce porosity
- **Water:** Essential for mixing and curing.

Application Technique

Special orbital vibrators are used to ensure uniform distribution and compaction of the ferrocement, which is critical for maintaining the integrity of thin layers.

10.4.2 Ferrocement Lining Procedure (for RCC Column Jacketing)

i) Surface Preparation

a. Remove Existing Plaster

Strip off all existing plaster from the column surface to expose the concrete substrate.

b. Surface Cleaning and Roughening

Mechanically hack the surface to create a rough texture, enhancing the bond between the ferrocement and the substrate. Clean thoroughly using a wire brush to remove dust, loose particles, and contaminants.

c. Additional Surface Removal (for aged columns)

In the case of very old columns, it is recommended to remove approximately ½ inch of cover concrete to expose sound material and improve bonding.

ii) Anchor Installation

a. Drilling Anti-Carbo

Drill holes at pre-determined, structurally designed locations to accommodate anchor pins.

b. Cleaning and Surface Preparation

Clear all drilled holes of debris and dust. Wash the entire surface using a high-pressure water jet to ensure cleanliness and proper adhesion.

c. Fixing Anchor Pins

Use polymer-modified mortar to securely fix anchor pins into the drilled holes. Ensure that the pins are properly projected to serve as skeletal reinforcement for the ferrocement mesh.

iii) Jacketing Thickness

a. Uniform Application

Apply the ferrocement layer with a consistent thickness of 6 mm across the entire surface.

Use orbital vibrators during application to ensure even distribution, proper compaction, and elimination of voids—critical for the performance of thin ferrocement sections.

10.5 High-Performance Concrete (HPC)

High-Performance Concrete (HPC) is a breakthrough in modern civil engineering, engineered to meet the most demanding structural requirements. With its exceptional strength and durability, HPC is the preferred material for constructing robust and resilient structures. Its high modulus of elasticity enables it to bear heavy loads and withstand high traffic conditions, making it ideal for critical infrastructure applications.

The superior performance of HPC is attributed to its carefully optimised mix design, which typically includes:

- Low water-to-cement ratios (ranging from 0.20 to 0.45)
- Plasticisers to improve fluidity without compromising strength

HPC is not only strong but also intelligently designed to resist harsh environmental conditions, chemical attacks, and mechanical wear over time. It is widely used in critical applications of structure, demonstrating innovation in materials science.

Typical Properties of HPC May Include:

- Very high compressive and early strength
- High modulus of elasticity
- Excellent abrasion resistance
- Long-term durability in severe environments
- Low permeability and high resistance to chemical penetration
- Enhanced toughness and impact resistance
- Easy placement and compaction without segregation

High-performance concretes are produced using carefully selected, high-quality materials and optimised mix designs. These concretes are batched, mixed, placed, compacted, and cured in accordance with the highest industry standards. Typically, they feature a low water-to-cementitious materials ratio, ranging from 0.20 to 0.45. Plasticisers are commonly incorporated to enhance fluidity and workability without compromising strength.

A sample of a mix design for HPC is presented below for understanding purposes.

Table-10.1: Mixture Proportions and Properties of Commercially Available High Strength Concrete (Burg and Ost 1994)

Units per Cu.m.	*Mix Number*					
	1	*2*	*3*	*4*	*5*	*6*
Cement, Type I, Kg	*564*	*475*	*487*	*564*	*475*	*327*
Silica fume, Kg	-	*24*	*47*	*89*	*74*	*27*
Fly ash, Kg	-	*59*	-	-	*104*	*87*
Coarse aggregate SSD (12.5 mm crushed limestone), Kg	*1068*	*1068*	*1068*	*1068*	*1068*	*1121*
Fine aggregate SSD, Kg	*647*	*659*	*676*	*593*	*593*	*742*

HRWR Type F, Liters	*11.6*	*11.6*	*11.22*	*20.11*	*16.44*	*6.3*
HRWR Type G, Liters	--	-	-	-	-	*3.24*
Retarder Type D, Liters	*1.12*	*1.05*	*0.97*	*1.5*	*1.5*	-
Water to cementing materials ratio	*0.28*	*0.29*	*0.29*	*0.23*	*0.23*	*0.32*
Fresh concrete properties						
Slump, mm	*197*	*248*	*216*	*254*	*235*	*203*
Density, Kg/m³	*2451*	*2453*	*2433*	*2486*	*2459*	*2454*
Air content, %	*1.6*	*0.7*	*1.3*	*1.1*	*1.4*	*1.2*
Concrete temp, °C	*24*	*24*	*18*	*17*	*17*	*23*
Compressive Strength, 100 x 200-mm moist-cured cylinders						
3 days, MPa	*57*	*54*	*55*	*72*	*53*	*43*
7 days, MPa	*67*	*71*	*71*	*92*	*77*	*63*
28 days, MPa	*79*	*92*	*90*	*117*	*100*	*85*
56 days, MPa	*84*	*94*	*95*	*122*	*116*	-
91 days, MPa	*88*	*105*	*96*	*124*	*120*	*92*
182 days, MPa	*97*	*105*	*97*	*128*	*120*	-
426 days, MPa	*103*	*118*	*100*	*133*	*119*	-
1065 days, MPa	*115*	*122*	*115*	*150*	*132*	-
Modulus of elasticity in compression, 100 x 200-mm moist-cured cylinders						
91 days, GPa	*50.6*	*49.9*	*50.1*	*56.5*	*53>4*	*47.9*
Drying shrinkage, 75 by 75 x 285-mm prisms						
7 days, millionths	*193*	*123*	*100*	*87*	*137*	-
28 days, millionths	*400*	*287*	*240*	*203*	*233*	-
90 days, millionths	*573*	*447*	*383*	*320*	*340*	-
369 days, millionths	*690*	*577*	*520*	*453*	*467*	-
1075 days, millionths	*753*	*677*	*603*	*527*	*523*	-

10.6 Self-Compacting Concrete (SCC)

Self-Compacting Concrete (SCC) marks a major advancement in the construction industry, offering an efficient method of concrete placement without the need for mechanical vibration. This innovative material flows effortlessly, navigating through congested reinforcement and filling complex formworks with ease.

Its unique composition—characterised by a high powder content, reduced coarse aggregate, and the inclusion of specific admixtures such as superplasticisers and viscosity-modifying agents—contributes to its exceptional performance. The carefully balanced mix ensures both fluidity and stability, preventing segregation and maintaining uniform consistency and strength throughout the structure.

Additionally, SCC's low water-to-cement ratio promotes higher early strength, enabling faster demoulding and accelerating construction timelines. Beyond simplifying the placement process, Self-Compacting Concrete enhances both the durability and aesthetic quality of concrete structures

Table-10.2 : Sample Mix Proportioning of SCC for M30, M35 and M80

Sr. No.	Ingredient per m^3	Kg/m^3			Remarks
		M30	M35	M80	
1	Cement (OPC 53 Gr.)	225	330	400	
2	Fly Ash	225	150	160	
3	Micro silica	Nil	Nil	40	
4	Corse Aggregate (CA) – 20mm – 10mm	354 : 107 : 355	455 : 764 : 309	284 : 711 : 427	
5	Fine Aggregate (FA)	288 River 684 crushed	917	970	Crushed sand
6	Water (W/C ratio)	165	163(0.34)	135 (0.225)	
7	PC Admixture As % of Binder by weight	0.8% Type G	0.8%	1.2%	
8	VMA	0.60%	0.20%	0.20%	
9	Ratio FA/CA	1.37	1.2	1.364	

10.7 Grouting for Structural Repair & Strengthening

Grouting is a versatile technique employed by engineers for the maintenance, repair, and strengthening of ageing structures, honeycombed concrete, and for leak-proofing applications in walls, tunnels, and similar elements.

Grout is a high-strength binding material in liquid form that can be poured or pumped into cracks, voids behind concrete, or interstices within other materials. Upon setting, it forms a solid, cohesive mass that fills the entire space while maintaining its original volume. Hardened grout must meet key performance criteria:

i. Strength

ii. Durability

iii. Ability to protect embedded reinforcement

10.7.1 Key Considerations for Fresh Grout

Due to the limited accessibility of grouting regions, the properties of fresh grout are critical to ensure effective application and long-term performance. Key factors to consider include:

i) Grout Consistency

- Flow characteristics
- Bleeding behaviour before setting
- Strength development in the hardened state

ii) **Nature of Voids or Cracks:** Assess the size, shape, and distribution of voids or cracks to determine appropriate grout type and method.

iii) **Selection of Grouting Material**: Choose materials compatible with structural and environmental conditions.

iv) **Grouting Pressure:** Apply pressure suitable for the type of voids and the surrounding substrate to avoid damage or incomplete filling.

v) **Equipment Type:** Use appropriate grouting tools and machinery to ensure controlled and uniform application.

vi) **Grouting Hole Spacing:** Plan the layout carefully, including:

- Accurate positioning of nipples or packers

10.7.2 Grouting Method

i) **Manual Grouting (Gravity Flow)**

This method relies on hydrostatic head to enable grout flow. Self-flowing grouts require minimal head pressure and are self-levelling and self-compacting, making them easy to apply. However, this technique is suitable only for open areas and is not effective for grouting cracks.

ii) **Pressure Grouting**

This is the most commonly used method in structural rehabilitation and strengthening works. It allows grout to be injected under controlled pressure, making it ideal for filling cracks, voids, and inaccessible zones.

10.7.3 Types of Grouts

Two main types of grout compositions are commonly used:

i) **Cementitious Grouts**

These grouts exhibit improved flow characteristics through the addition of chemical admixtures and stabilisers. They offer good injectability and favourable rheological properties. Cementitious grouts are easy to handle, minimise material wastage, and are cost-effective.

ii) **Chemical Grouts**

Composed of polymeric materials such as epoxies and polyurethanes, chemical grouts provide excellent injectability along with outstanding mechanical and bonding properties. They exhibit low shrinkage upon setting and superior resistance to heat and chemical exposure. However, they are significantly more expensive—typically six times or more than cementitious grouts—and should not be applied to wet surfaces. Manufacturer-specific procedures and mix proportions must be strictly followed, as injectable epoxies vary by grade.

10.8 Micro-silica in Concrete

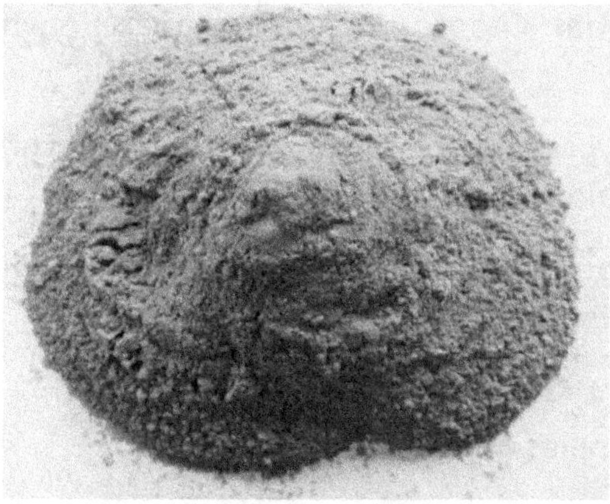

Microsilica is a byproduct of silicon metal, where these metals are produced in submerged electric arc furnaces. The physical characteristics of micro silica fume are quite different from those of other concrete components. But the chemical makeup is rather similar. Microsilica is an extremely fine particulate, 100 times finer than cement particles. This superfine material works wonders in concrete, serving both as a filler and enhancing the strength of the mix

Figure-10.6: Micro-Silica Additive for Concrete

It's fascinating how these tiny particles can make such a big impact, filling in the gaps and reacting chemically to create a denser, more robust material. Typically added in proportions of 7 to 10 percent, microsilica can create exceptionally strong concrete, although it does increase the need for water in the mix. It's a delicate balance, but when done right, the results are solid as, well as concrete.

10.8.1 How Microsilica Works in Concrete

Pozzolanic reaction, and microfiller effect. When water is added to Portland cement, hydration occurs, forming two products, as shown below:

Cement + Water -----------> Calcium Silicate Hydrate +Calcium Hydroxide

Calcium Hydroxide + Micro-silica + Water -----------> Calcium Silicate Hydrate

$$Ca(OH)2 + SiO2 + H2O \text{ -----------> } CSH$$

This additional CSH formation improves bonding within the concrete matrix and helps reduce permeability, while the reduction in calcium hydroxide improves concrete durability.

The extreme fineness of micro silica allows it to fill the microscopic voids between cement particles. The microfiller effect is credited with greatly reducing permeability and improving the paste-to-aggregate bond of microsilica concrete compared to conventional concrete.

Depending on the dosage of micro silica fume, the fresh concrete may be more cohesive and sticky than conventional concrete. Due to this cohesiveness, slumps 25 to 50 mm (1 to 2 in.) higher than normal should be used for similar types of placement.

Despite the increased cohesiveness, microsilica powder fume concrete produces a very creamy paste and is very pumpable. One of the biggest differences in using microsilica concrete occurs during finishing.

10.9 Development of Advanced Reinforced Concrete

10.9.1 Stainless Steel Reinforced Concrete

Stainless steel, an alloy known for its corrosion-resistant properties, is widely used in various industries, especially in construction, due to its durability and strength. While it is more resistant to corrosion compared to carbon steel, it is not completely immune to it. Thus, stainless steel reinforcement is a critical component in the longevity and durability of concrete structures, particularly in corrosive environments such as coastal areas or places with high chloride exposure. The defining characteristic of stainless steel is its minimum chromium content of 10.5%, which imparts remarkable resistance to corrosion. This resistance is primarily due to the formation of a passive, self-repairing chromium-rich oxide layer on the surface.

The performance of stainless steel can be further enhanced by adding other elements like nickel, which is evident in the austenitic and ferritic varieties of stainless steel. Austenitic stainless steel, with higher nickel content, offers superior ductility, formability, weldability, and overall corrosion resistance compared to ferritic stainless steel, which has less nickel and a different microstructure.

Certain conditions can lead to various forms of corrosion, such as crevice, intergranular, and galvanic corrosion. The presence of halide ions, especially chlorides, can be particularly detrimental as they may disrupt the protective oxide layer that guards the steel against corrosion.

Despite these vulnerabilities, stainless steel remains a superior choice for reinforcement, especially in environments prone to atmospheric corrosion. Its compatibility with current construction practices is a significant advantage, as highlighted by the British Highway Agency, which emphasises that no special skills are required for its use in reinforced concrete structures.

Despite its high resistance to corrosion, stainless steel is not impervious to all forms of degradation. Issues such as crevice corrosion, intergranular corrosion, and galvanic corrosion can still arise under certain conditions. For instance, the presence of halide ions, particularly chlorides, can compromise the integrity of the protective oxide layer, leading to corrosion.

However, the overall performance of stainless steel in resisting atmospheric corrosion is significantly better than that of carbon steel, making it a preferred choice for reinforcing materials in harsh environments. The British Highway Agency's statement underscores the compatibility of stainless steel with existing construction practices, requiring no additional skills for its implementation in reinforced concrete structures.

In summary, while stainless steel reinforcement offers numerous advantages in terms of durability and corrosion resistance, it is essential to consider the specific environmental conditions and potential corrosion mechanisms when selecting the appropriate type of stainless steel for a given application.

Based on a report by the British Highway Agency (2009), no additional skills are required for the design and use of stainless steel reinforcement compared to normal carbon steel reinforcement, and the detailing and construction processes are unchanged. Stainless-steel-reinforced concrete is applied especially to structures and substructures in maritime environments that are exposed to seawater. The use of stainless steel reinforcement enables durability requirements to be relaxed, but its significance is best demonstrated by the considerable reduction of maintenance costs over the service life, even though the price of stainless steel is 5.6 and 7.8 times more than carbon steel reinforcement for bars less than 20 mm and over 25 mm, respectively.

Austenitic and duplex stainless steel can offer high corrosion resistance. The benefits of using stainless steel reinforcement include:

- Cover requirements can be reduced to 30 mm
- allowable crack width can be increased to 0.3 mm
- very low corrosion rates (at least 1000 times lower than for carbon steel reinforcement)
- low maintenance requirements over the whole service life.

Generally, the cost of austenitic stainless steel is about 8–11 times more than that of unalloyed steels, and duplex is about 12 times as much. It may be prohibitively expensive to replace all carbon steel reinforcement with stainless steel. One option would be to replace the carbon steel reinforcement of those concrete structures at a high risk of chloride-induced corrosion.

10.10 Reinforcement with Fibre-reinforced Polymer

Fibre-reinforced polymer (FRP) bars, particularly glass fibre reinforced polymer (GFRP) bars, have become a popular choice in the construction industry for their numerous advantages, including resistance to corrosion, high strength-to-weight ratio, and cost-effectiveness. Their application in various reinforced concrete structures has demonstrated satisfactory performance, even in harsh environmental conditions. However, concerns regarding the long-term durability of FRP, especially GFRP, within the alkaline environment of concrete have been raised. Studies have indicated a potential reduction in the strength of GFRP's components when exposed to wet, alkaline conditions commonly found in service. For instance, research by Robert et al. (2009) suggested a predicted decrease in the strength of GFRP rebars by 20% to 25% over a period of 100 to 200 years. Moreover, accelerated durability tests have shown that GFRP can suffer significant strength loss under certain conditions, such as elevated temperatures and rapid wet-dry cycles.

The thermal compatibility between FRP bars and concrete is another critical factor, with the strength of the interfacial bond being a vital aspect of the overall performance of FRP-reinforced concrete structures. The difference in thermal expansion coefficients between concrete and FRP materials can affect the integrity of these bonds. To mitigate these issues, especially in hot climates, careful design considerations are necessary. For example, the concrete cover should be increased to 1.5–2 times the diameter of the FRP reinforcement, and the use of FRP should be avoided in temperatures exceeding 50°C.

In conclusion, while FRP bars offer significant benefits for reinforced concrete structures, it is essential to address the challenges related to their durability and thermal compatibility to ensure their effective and safe use in long-term applications. Ongoing research and development in this field are crucial for optimising the material properties and design methodologies to overcome these concerns.

10.10.1 Concrete–Polymer Composites

Concrete-polymer composites are an innovative class of materials that enhance the properties of traditional concrete by incorporating polymers. These composites offer improved workability, durability, and a range of mechanical strengths, making them suitable for various construction and repair applications. Their rapid curing time and resistance to environmental factors also contribute to their popularity in the construction industry. Despite their higher cost and the need to handle certain ingredients carefully due to safety concerns, their benefits in terms of performance and longevity often justify the investment. As with any construction material, it's crucial to balance these factors to meet the specific needs of a project.

10.10.2 High-performance and Ultra-high-performance Concrete

High-performance concrete (HPC) and ultra-high-performance concrete (UHPC) represent significant advancements in the construction industry. HPC, known for its low water-to-binder ratio, offers enhanced durability and strength, making it ideal for structures exposed to harsh conditions. HPC that has a water/binder ratio of 0.3–0.4 is usually more durable than ordinary concrete, not only because it is less porous but also because its capillary and pore networks are somewhat disconnected. Its dense structure reduces permeability, thus limiting damage from environmental factors. UHPC takes it a step further by incorporating fibre-reinforced composites, pushing the boundaries of strength and durability even more. These materials not only extend the lifespan of structures but also allow for new design possibilities, thanks to their superior mechanical properties and resistance to environmental degradation. The evolution from HPC to UHPC marks a notable shift towards more resilient and long-lasting construction materials.

Compared to HPC, UHPC shows advanced mechanical and durability properties due to:

- internal fibre reinforcement to ensure non-brittle behaviour
- high binder content and special aggregates
- very low water content with adequate rheological properties through a combination of optimised granular packing and the addition of high-range water reduction admixtures.

Durability is strongly affected by the curing treatment. Notably, the use of steam treatments significantly decreases the ability of chloride ions to penetrate the concrete and increases abrasion resistance.

10.11 Epoxy Injection Grout

Epoxy injection grout is a specialised solution developed for the repair of concrete structures. It is a two-component system that is water-insensitive, making it particularly effective for sealing moisture-affected cracks in marine environments. Its low viscosity ensures excellent flowability, allowing it to penetrate even the finest cracks to form a strong, durable seal.

This grout exhibits high chemical resistance, enhancing its durability in a variety of conditions, including underwater applications. It is capable of sealing voids and cracks even under negative water pressure, a common challenge in submerged structures.

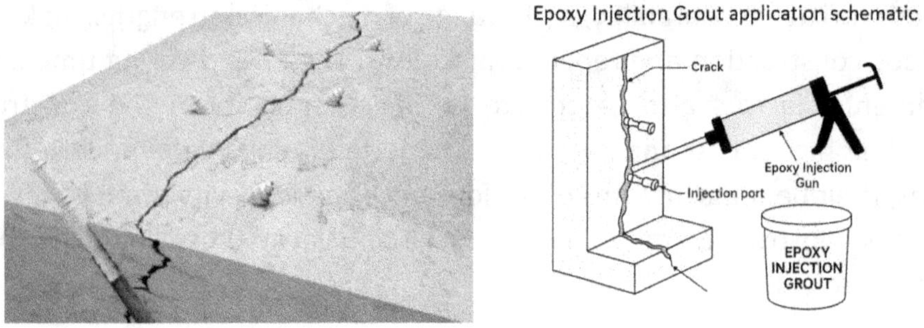

Figure-10.6: Schematic Arrangement of Epoxy Injection Grout

Application Process

- Identify and mark the cracks.
- Drill holes at specified intervals.
- Install injection nozzles.
- Inject the grout using a low-pressure system to ensure deep penetration and uniform distribution.

This method is essential for the repair and rehabilitation of underwater structures, deck slabs, RCC walls, and other critical structural elements, offering a long-lasting repair solution. Before injection, it is crucial to thoroughly mix the epoxy grout to achieve a homogeneous consistency, which is vital for the strength and durability of the repair. When properly applied, epoxy injection grout can restore structural integrity and significantly extend the service life of concrete structures.

10.12 Free Flow Cementitious Grout

Free-Flow Cementitious Grout, also known as Self-Flow and Self-Compacting Grout, is an advanced grouting material designed for a wide range of construction and repair applications. It is characterised by high strength, non-shrink properties, and excellent flowability, making it ideal for use in foundations, rail fixing, and structural strengthening.

This grout complies with ASTM C-1107 standards for hydraulic cement grouts, ensuring consistent quality and performance. The Self-Flow and Self-Compacting Grout is supplied in a ready-to-use form, requiring only the addition of water on-site, which simplifies the application process. Its formulation includes hydraulic setting binders, graded fillers, and expansion control additives to prevent shrinkage, making it highly suitable for critical grouting operations demanding durability and reliability. It is especially effective for structural repairs and encasing steel sections, providing long-term support and stability.

10.12.1 Features & Benefits of Free-Flow Cementitious Grout

i) **Non-Shrink with Controlled Expansion**

The grout is engineered to expand slightly during setting, compensating for shrinkage and ensuring full contact with surrounding surfaces. This makes it ideal for jacketing applications, where it enhances the structural integrity and provides additional support to existing elements.

ii) **High Flowability and Pourability**

Its self-levelling and free-flowing nature allows for easy application in the field, enabling the grout to fill voids, gaps, and recesses thoroughly without requiring extensive manual intervention. This ensures efficient placement, especially in congested or complex geometries.

iii) **Bleeding and Segregation Resistance**

Even in a highly fluid state, the grout maintains its homogeneous composition, preventing separation of components and water bleeding. This results in consistent strength and durability across the entire grouted area.

iv) **High Load-Bearing Capacity**

Designed to withstand both static and dynamic loads, the grout provides reliable performance in structural applications such as base plates, machine foundations, and column jacketing.

v) **Chloride-Free and Non-Corrosive**

The formulation excludes chlorides and other corrosive substances, helping to preserve embedded steel reinforcement and extend the service life of the structure.

vi) **Self-Compacting Nature**

The grout naturally flows into every corner, crevice, and void, eliminating the need for vibration or manual compaction. This ensures a dense, void-free fill, critical for structural stability and long-term performance.

10.12.2 Method of Application

i) **Mixer Preparation**

Wet the inside of a mechanically powered, padded/drum-type grout mixer before starting the first batch.

ii) Mixing Procedure

- Begin with a clean, damp mixer free of excess water.
- For a standard 50 kg pack, add 6.5 to 7.5 litres of water at 27°C to achieve a flowable consistency.
- Start by adding 85–90% of the total water, then gradually incorporate the grout powder.
- Mix for 1–2 minutes, then add the remaining water to reach the desired consistency.
- Continue mixing for 2–3 minutes until the grout is smooth and uniform.

i) Placement

Pour the grout from one side only to ensure even distribution. Use watertight shuttering to contain the material and prevent leakage.

ii) Precautions

- For large pours, conduct the grouting in stages using a high head or grouting pump to avoid air entrapment.
- Do not add extra water under any circumstances.
- Maintain a gravity head of 6 to 8 inches during pouring to ensure proper flow.

10.13 Micro Concrete

Micro concrete is a highly versatile material used in construction for a wide range of repair and strengthening applications. It is particularly effective for jacketing of columns, slab extensions, and localised patch repairs. Supplied as a ready-to-use dry powder, micro concrete simplifies the repair process by requiring only the addition of clean water on-site to produce a free-flowing, non-shrink repair mortar.

Its formulation includes cementitious binders and special additives, making it ideal for repairing damaged reinforced concrete elements, especially in areas with restricted access or where traditional vibration methods are impractical. The mix is customised based on project requirements, with a maximum aggregate size of 5 mm, and incorporates polymers and superplasticisers to achieve the desired performance characteristics. This ensures not only ease of application but also durability and long-term structural integrity.

Figure-10.7: Schematic Arrangement for Micro-concrete Preparation

Also referred to as micro-cement, this material is designed for thin-layer applications. It typically consists of high-quality cement, fine aggregates (such as silica sand or quartz flour), water-based resins, polymers, and performance-enhancing additives. These components are carefully selected to ensure strong adhesion, flexibility, and a smooth, aesthetic finish. Polymers improve crack resistance and flexibility, while additives like plasticisers and setting agents enhance workability and speed up curing. The result is a quick-drying, impact- and water-resistant surface, suitable for both indoor and outdoor use.

Micro concrete offers a reliable, efficient, and durable solution for structural repairs and surface refinishing, making it a preferred choice in modern construction and rehabilitation projects.

Features & Benefits of Micro Concrete

i) Flowable Nature

Micro concrete's high flowability allows it to be easily pumped or poured into confined or inaccessible areas, making it ideal for complex structural repairs where conventional concrete placement is challenging.

ii) Gaseous Expansion System

The inclusion of a controlled expansion mechanism helps counteract shrinkage and settlement during the plastic stage, ensuring dimensional stability and maintaining the integrity of the repair.

iii) High Strength and Low Permeability

Once cured, micro concrete forms a dense, durable matrix with high ultimate compressive strength and low permeability, which is essential for long-term structural performance and protection against environmental degradation.

iv) Rapid Strength Gain

Its ability to achieve early strength development enables quick reinstatement of repaired sections, reducing downtime and accelerating construction schedules.

v) Chloride-Free Composition

Free from chlorides and other corrosive agents, micro concrete is safe for use in marine and corrosive environments, helping to preserve embedded reinforcement and extend the service life of structures.

vi) Versatility and Reliability

Suitable for a wide range of repair, rehabilitation, and strengthening applications, micro concrete offers a dependable solution for restoring structural integrity with minimal disruption.

Areas of Application

i) Micro concrete is suitable for the repair of damaged reinforced concrete elements such as beams, columns, and walls, particularly in locations where access is limited and mechanical compaction is not feasible.

ii) It is designed for thin-layer applications, typically between 2 mm to 3 mm, making it ideal for surface renovations without significantly affecting the structure's height or weight.

iii) For applications where the thickness exceeds 100 mm, it is recommended to add pre-calculated aggregates to maintain the material's integrity and performance.

iv) Micro concrete also offers rapid drying, allowing treated surfaces to become functional shortly after application, making it a practical solution for fast-paced construction and repair environments

10.14 Epoxy Zinc Primer

Epoxy Zinc Primer is a specialised coating designed to provide cathodic protection to steel surfaces and reinforcing bars (re-bars), particularly in corrosive environments. It functions by allowing the zinc particles within the primer to act as a sacrificial layer, corroding in place of the steel and thereby preserving its integrity. This is especially beneficial in marine and moisture-prone environments, where corrosion risks are elevated.

The primer is a two-component system, consisting of metallic zinc and epoxy resin, which together form a durable, grey-coloured protective coating. It not only resists corrosion but also enhances the bond between steel and subsequent repair materials. Widely used across industries—including marine construction—this primer is valued for its robust anti-corrosive properties and ease of application.

i) Features & Benefits

- Two-Component System: Can be applied using a brush or spray, offering flexibility for various project requirements.
- Quick Drying: Becomes touch-dry within 15 to 45 minutes, enabling faster progress on site.
- Excellent Base Coat: Serves as a reliable foundation for epoxy or polyurethane top coatings, ensuring a strong and durable finish.
- High Zinc Content: Provides sacrificial protection, where zinc corrodes preferentially, safeguarding the steel beneath.
- Corrosion Resistance: Ideal for marine and aggressive environments, extending the life of steel structures.

ii) **Areas of Application**

a) Coating of reinforcing bars (re-bars)

b) Protection of steel surfaces in concrete structures

iii) **Method of Application**

- **Surface Preparation:** Clean the steel surface thoroughly to remove rust, scale, and loose particles—preferably by grit or sandblasting.
- **Timely Application:** Apply the primer within three hours of surface preparation to prevent re-oxidation.
- **Primer Application:** Use a suitable brush to apply a uniform coat, ensuring coverage on all exposed surfaces, especially the backs of re-bars.
- **Second Coat (if needed):** After the first coat dries, apply a second coat to cover any missed areas or discontinuities.
- **Concrete Overlay:** Do not leave the primed surface exposed for extended periods. Apply the concrete repair material once the primer is fully dry to ensure optimal bonding and protection.

iv) **Precautions & Limitations**

- Temperature Restrictions: Do not apply when ambient temperature is below 10°C or above 40°C.
- Flammability: The primer is flammable and must be kept away from sources of ignition.

10.15 Epoxy Repair Mortar

Epoxy Repair Mortar is a high-performance, three-component system designed for efficient and durable structural repairs. Its fast application and curing process make it ideal for time-sensitive projects. With excellent chemical and abrasion resistance, it is particularly suitable for use in harsh environments, such as marine structures, where long-term durability is essential.

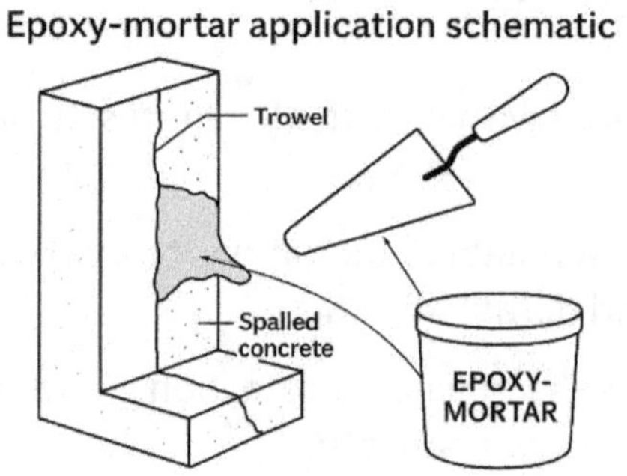

Epoxy mortar is effective for crack filling, surface levelling, and high-strength patch repairs, offering a reliable solution for extending the service life of concrete structures. Its ease of application and rapid strength gain help minimise downtime, making it a preferred choice for maintenance operations.

Figure-10.6: Schematic details of Epoxy-mortar Application

Features & Benefits

i) **Universal Application**

Epoxy repair mortar is suitable for a wide range of industries, making it a versatile solution for structural maintenance and rehabilitation. Its adaptability allows it to be used in various environments—from industrial floors to marine structures—where reliable and durable repairs are essential.

ii) **High Load-Bearing Capacity**

Engineered to withstand substantial mechanical stress, epoxy mortar delivers excellent performance under both static and dynamic loads. This makes it ideal for repairing structural components that are subject to heavy usage or operational vibrations.

iii) **Abrasion and Chemical Resistance**

The formulation offers superior resistance to physical wear and chemical exposure, ensuring long-term durability even in aggressive environments. It performs effectively in areas exposed to chemicals, oils, solvents, and marine conditions, preserving the integrity of the repaired surface.

iv) **Rapid Strength Development**

The mortar gains strength quickly after application, allowing for early reinstatement of the repaired area. This minimizes downtime and supports fast-paced maintenance schedules, making it an efficient choice for time-critical repairs.

v) Chemical resistance

Easily applied as a universal repair mortar in all industries and has excellent load bearing, abrasion resistance and chemical resistance properties. Give a non-slip Surface and excellent early strength. Chemical resistance are as follows:

Medium	**Rating**
Water	Excellent
Steam	Good
30 % caustic soda	Excellent

Liquid ammonia	Excellent
50 % sulphuric acid	Fair
15 % sulphuric acid	Excellent
Conc. hydrochloric acid	Poor
15 % hydrochloric acid	Excellent
10 % nitric acid	Good
40 % phosphoric acid	Excellent
Xylylene	Excellent
Methanol	Poor
50 % ethanol	Good
Chlorobenzene	Fair
Carbon tetrachloride	Fair
Fruit juice	Excellent

vi) **Non-Slip Surface Finish**

Once cured, the mortar provides a textured, non-slip surface, enhancing safety in operational zones such as walkways, platforms, and industrial workspaces. This feature is particularly beneficial in wet or high-traffic areas.

Method of Application

- **Surface Preparation:** Ensure the substrate is clean, dry, and free from contaminants such as oil or grease. Use wire brushing or grit blasting for thorough cleaning, followed by the application of a suitable epoxy primer.
- **Primer Mixing & Application:** Mix the hardener into the base component of the epoxy primer until homogeneous (minimum 3 minutes using a slow-speed drill). Apply a thin coat adjusted to the surface's absorption and roughness. Recoat if dry patches appear after the first layer dries.
- **Mortar Mixing:** Mix the hardener into the base component, then gradually add the filler using a paddle mixer attached to a heavy-duty, low-speed drill. Ensure a consistent, lump-free mix.
- **Mortar Application:** Apply the repair mortar over the tacky primer, working in lanes for effective levelling and trowelling. Finish with a steel trowel. Overcoating can be done 24 hours after application.

10.16 Curing Compound

Curing Compound is a wax-based emulsion specifically formulated to aid in the curing of repaired concrete surfaces. It functions by forming a temporary membrane over the concrete, which helps retain moisture essential for proper hydration and strength development. This method is significantly more efficient than traditional wet curing, which typically requires continuous moisture for up to 28 days.

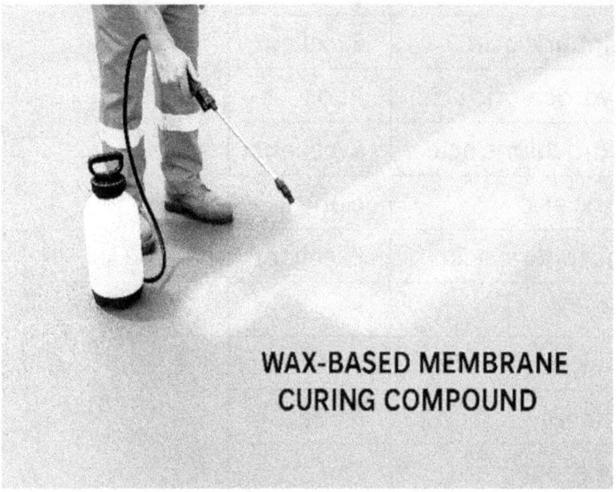

The compound contains white pigments that serve a dual purpose: they reflect sunlight, reducing surface temperature and minimising the risk of thermal cracking. Products conforming to ASTM C-309-89, Type 2 standards ensure reliable performance in terms of water retention and reflectance, making them particularly suitable for curing repaired concrete in areas with limited access.

Figure-10.7: Application of Waz-based Membrane Curing compound

Features & Benefits

- **Advanced Moisture Retention:** Forms a moisture-retaining film that promotes optimal cement hydration, eliminating the need for continuous water curing.
- **Environmentally Friendly:** Typically water-based and free from harmful fumes, making it safe for use in enclosed or sensitive environments.
- **Ease of Application:** Can be applied using a brush or spray, making it suitable for both small-scale and large-scale projects.
- **Reflective Properties:** The white pigment helps maintain cooler concrete temperatures, reducing the likelihood of surface cracking due to heat.
- **Natural Degradation:** The membrane breaks down over time, allowing for subsequent finishing work without requiring manual removal.
- **Versatile Use:** Effective for both indoor and outdoor applications, especially in areas where traditional curing methods are impractical.

Method of Application

- **Timing:** Apply the curing compound immediately after final trowelling, once the surface is free from excess water.
- **Technique:** Use a sprayer or brush to apply a uniform coat, forming a non-penetrating film that prevents rapid moisture loss.
- **Vertical Surfaces:** For vertical elements, apply the compound as soon as formwork is removed to protect the freshly exposed concrete.

10.17 Epoxy Bonding Agent

Epoxy Bonding Agent is a high-performance, two-component, solvent-free epoxy-based adhesive designed to create a strong and durable bond between different construction materials. It plays a critical role in structural repairs and modifications, particularly in bonding new concrete to existing concrete surfaces, ensuring seamless integration and long-term stability.

The solvent-free formulation enhances safety for both users and the environment by minimising the release of volatile organic compounds (VOCs). This makes it suitable for use in enclosed spaces and environmentally sensitive areas. With various reliable formulations available—such as CICO BOND EPO for high-strength applications and BASF Master-Brace 1414 for versatile bonding needs—epoxy bonding agents are widely used across industrial, commercial, and residential projects.

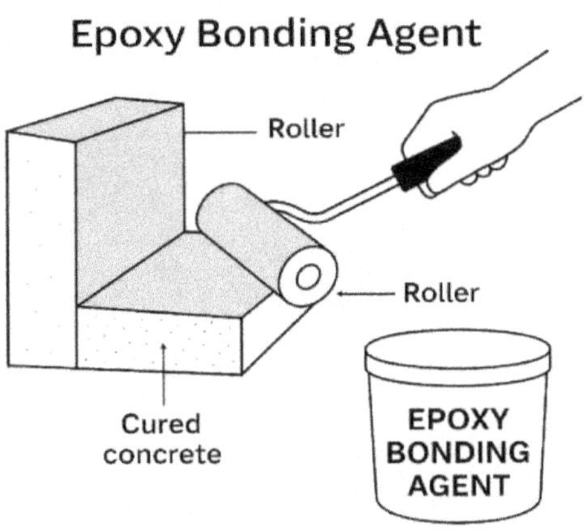

Figure-10.8: Application of Epoxy-Bonding Agent

Mixing & Application Procedure

Mixing

- Pour the hardener component into the base resin.
- Mix thoroughly using a low-speed mechanical mixer (150–200 rpm) for 3 to 4 minutes until a uniform colour and consistency is achieved.

Surface Preparation: Ensure the substrate is clean, dry, and free from dust, oil, grease, or any contaminants that may hinder adhesion.

Application: Apply the mixed epoxy bonding agent uniformly with a brush over the prepared surface.

- The overlay of concrete or mortar must be placed while the bonding agent is still tacky to ensure proper adhesion.
- If the bonding agent has dried before overlay placement, apply a second coat to restore tackiness before proceeding.

Modernising Maritime Infrastructure

Method of Repair, Rehabilitation for Service Life Enhancement

Chapter Summary

Many port structures become obsolete due to the rapid evolution in vessel dimensions and types or deteriorate prematurely—well before their intended design life—owing to environmental degradation or excessive loading. Upgrading, rehabilitating, or strengthening such infrastructure demands a structured methodology: assessing the current condition, identifying underlying causes, and implementing appropriate repair solutions. The longevity and performance of marine structures depend on the use of proper materials, adopting suitable repair techniques, and thorough condition evaluations. This chapter presents strategic approaches for preventive maintenance, corrective measures, retrofitting, and renovation, with a strong emphasis on effective repair and rehabilitation practices to ensure durable restoration.

Learning Objectives

- Concrete defects and deterioration weaken structures—knowing their causes is key to effective repair.
- Obsolescence assessment guides whether to repair, upgrade, or replace a structure, ensuring safety and adaptability.
- Renovation helps structures meet current standards, balancing old design with modern needs to extend their life and keep them safe.
- Rehabilitation of structures requires thorough condition assessment and a mix of preventive and corrective actions to address distress and ensure durability.

11.1 Introduction

In marine environments, the durability and safety of reinforced concrete (RCC) structures are major concerns due to aggressive exposure conditions. Chlorides present in seawater accelerate corrosion of steel reinforcement, which is a leading cause of structural deterioration and can result in serious damage or failure. Contrary to common belief, RCC structures do require regular maintenance. In marine conditions, the loss of concrete alkalinity initiates steel corrosion, leading to cracking, reduced strength and stiffness, and spalling of the protective concrete cover. If not addressed in time, these effects can severely distress the structure. Therefore, an effective repair and rehabilitation strategy is essential. Before discussing such strategies, the key types of structural deficiencies are outlined below.

11.2 Defects and deterioration in concrete structures

Defects and deterioration are different problems but both affect the strength and life of concrete structures. Defects are visible flaws that happen mainly due to construction issues, such as shrinkage or temperature cracks, settlement cracks, misalignment, honeycombing, or surface wear. Deterioration, however, is the gradual decline of the structure over time due to chemical and physical processes like steel corrosion, chemical attack, or overloading. Both can reduce strength and safety and often lead to expensive repairs. Understanding these issues is essential for the proper repair and rehabilitation of concrete structures.

11.2.1 Cracks - Common Defects in RCC Structure

Cracking in RCC structures is a common occurrence, and these are often required to look into their causes and to carry out suitable repairs and remedial measures. For effective repair and remedial measures, it is important to have a proper understanding of the causes of cracking and deterioration. For investigating the cause, it is necessary to observe the location, shape, size, depth, behaviour, and other characteristics of the cracks. Understanding the Causes and Remediation of Cracks in Structures. Cracks in structural elements are a common issue that can arise from a variety of causes, necessitating careful investigation to determine the appropriate repair and remedial actions. It's crucial to comprehend the underlying reasons for cracking and deterioration to ensure effective intervention.

Causes of cracks may be divided into two categories

1. Structural cracks
2. Non-structural cracks

i) Structural Cracks

cracks are those whose inherent cause lies in either incorrect design or faulty construction, or overloading, and are the ones that can endanger the safety of a given structure. These cracks may pose a threat to the structure's integrity and safety.

They often result from: a) Incorrect design; b) Faulty construction; c) Overloading;

ii) Non-Structural Cracks

Non-structural cracks are those that have an underlying origin due to moisture or thermal fluctuations, elastic deformation, creep, chemical reaction, or a reason related to the foundation soil, such as its movement or settlement, or unhindered vegetation. Non-structural cracks are mostly due to internally induced stresses in building materials, and these generally do not directly result in structural weakening. These do not typically compromise the strength of the structure but are still a concern.

They can be caused by -

- Moisture changes;
- Thermal fluctuations;

- Elastic deformation;
- Creep;
- Chemical reactions;
- Soil movement or settlement;
- Vegetation growth

The non-structural cracks can be repaired, provided the reasons for the cracks are identified, and suitable remedial measures are taken to prevent their reoccurrence.

11.2.2 Investigation for Cracks

To address these issues effectively, a thorough examination of the cracks is required, focusing on their location, shape, size, depth, and behaviour. This analysis helps in understanding the severity and cause of the cracks, which is essential for determining the most suitable repair techniques. By understanding the various factors that contribute to cracking, engineers and construction professionals can better prevent and manage these issues, ensuring the longevity and safety of structures. Regular maintenance and monitoring are also key to early detection and remediation of potential problems.

Limitations of Crack Width

Table-11.1: The limitations on crack width are imposed based on exposure conditions.

S. No.	Members in	Permissible value of crack width in RCC structures
1	Structures containing chemically hazardous materials	0.1 mm
2	structures containing chemically	0.2 mm
3	Moderate atmosphere	0.3 mm

11.2.3 Obsolescence and Deterioration in Structure

Obsolescence refers to the condition of being out of date. Obsolescence may be defined as reduced functional and structural adequacy of the structure compared to the best and most suitable structure that could be constructed at present. It manifests itself in many ways, including the following:

i) an increase in operating loads due to which the old structure is not capable of meeting the current demands,

ii) reduced safety factors due to increased loads on the structure due to current operating demands,

iii) a lowering of safety standards because the old structure is unable to meet current safety standards.

Here are some key points about obsolescence:

i) Functional Obsolescence

Functional obsolescence refers to the situation where a structure or design becomes outdated and can no longer fulfil the requirements for which it was originally designed. This could be due to a variety of reasons, such as degradation and deterioration of the construction material of the structure or wear and tear over time, or evolving functional needs of the space. Taking the example of a jetty, if the structure is designed to withstand certain loads and conditions. Over time, the materials can deteriorate due to exposure to the elements, such as chloride-induced corrosion, leading to a reduction in the structure's load-bearing capacity. When a jetty can no longer support the loads it was designed for, it's considered functionally obsolete. This doesn't necessarily mean it's unsafe, but it does mean that it can't be used as intended without risk or without modifications.

In such cases, the extent of the obsolescence must be assessed to decide whether the structure shall be repaired, upgraded, or replaced. This decision is crucial for ensuring safety, functionality, and cost-effectiveness. It's a delicate balance between preserving the past and embracing the future, and it's a challenge that engineers face regularly as they work to maintain and improve our built environment. Thus, functional obsolescence is a reminder of the ever-changing nature of our needs and the importance of adaptability in design.

11.2.3.1 Obsolescence due to Technological Advances

Technological advancements indeed have a profound impact on infrastructure, leading to what is known as obsolescence. This phenomenon occurs when existing structures or systems become outdated due to the emergence of new technologies. For example, a jetty structure designed decades ago may no longer be suitable for today's shipping industry due to a significant increase in ship sizes. This mismatch can affect the safety of maritime operations, necessitating upgrades or strengthening with some vital replacements like fenders and bollards to accommodate larger vessels.

If the jetty structure and port basin are not upgraded, it will still serve its purpose for smaller vessels, and it may not be deemed obsolete for all users. However, for a port aiming to remain competitive and handle larger cargo ships, an upgrade or redesign might be necessary to avoid obsolescence. Thus, the concept of technological obsolescence is not just about the physical inability of structures to meet current demands; it also encompasses a broader range of factors. These include economic considerations, user needs, and the competitive landscape in determining whether a technology is considered obsolete. For planning of port facilities, a comprehensive understanding of technological obsolescence is essential in planning and decision-making, ensuring that items are not devalued prematurely due to technological progress. Strategies to manage technology obsolescence include staying informed about new developments, planning for future compatibility, and conceptualising the adaptable and upgradable systems.

11.2.3.2 Deterioration of Structure

As discussed above, deterioration is very different from obsolescence. It is defined as the reduced structural adequacy of the structure compared to that of the original design when the structure was new. To reduce the deterioration of the structure, it involves maintenance and repair operations to restore the old structure to its prior condition. Removing cracked or delaminated concrete, protecting the exposed steel reinforcement, and replacing the concrete with a new, improved material are typical tasks that must be performed to tackle the deterioration. These operations contrast with reconstruction or enhancement operations to combat obsolescence. While the two operations frequently overlap, it is important to draw some distinction when considering the alternatives available for overcoming structural deficiency and functional obsolescence.

11.2.3.3 Alternatives for Reversing Deterioration

When it comes to reversing the deterioration or degradation of structures, there are several strategies that can be employed, as briefly outlined below;

i) **Protection or application of protective coating** - involves applying coatings or treatments on the outer surface of the structure that shield the material from environmental factors that cause degradation.

ii) **Patching or patch repairing** is a more direct approach where damaged areas are repaired, often with new material that bonds with the old.

iii) **Rehabilitation** may include strengthening or upgrading the structure to improve its performance and extend its lifespan.

iv) **Renovation** often encompasses a broader scope, potentially involving both aesthetic and functional updates to meet current standards or requirements.

In the field of port engineering, these methods are crucial for maintaining the integrity and safety of infrastructure. In the marine environment, concrete structures are also susceptible to various types of deterioration, such as corrosion of embedded metals, stresses due to overloading, and chemical attacks. Each of these issues requires a specific approach to mitigate and reverse the damage.

Furthermore, the challenges and solutions related to the deterioration of structures are vast and varied, but with the right approach, it is possible to restore and even enhance the longevity and performance of these critical assets.

i) Protective Coating on Reinforced Concrete

Physical barriers, such as waterproof overlays or membranes, halt or slow the process of deterioration. Electrochemical methods can also be used to remove chlorides or upset the required chemical actions. These alternatives are viable if deterioration has not progressed beyond certain limits.

Protection from chloride ingress is an alternative for new structures and structures where chloride contamination has not yet reached critical limits. Under these conditions, every effort should be made to use the best technologies available in design, construction, and operation to prevent contamination and extend life. Barriers are most effective when introduced during the maintenance operations as noted below.

To mitigate chloride ingress, various physical barriers can be applied to the concrete surface. These barriers prevent or slow down the movement of chloride ions into the concrete, which can lead to corrosion and structural weakness. Applying physical barriers is a common approach to slow down or prevent the movement of chloride ions. These barriers can include various coatings like epoxy or cement-based paint, which act as a sealant to protect the concrete surface.

ii) **Waterproof Overlays**

These are thin layers of waterproof material (such as polymer-modified cementitious overlays or epoxy-based coatings) applied directly to the concrete surface. waterproofing is a critical consideration, especially when it comes to protecting structures from the damaging effects of water and chloride ions. Waterproof overlays and membranes serve as the first line of defence in this battle, each with its unique properties and applications. Overlays, often composed of polymer-modified cementitious materials or epoxy-based coatings, are applied as a thin layer directly onto the concrete surface. This creates a robust barrier that not only prevents water from seeping through but also blocks chloride ions, which can lead to corrosion of steel reinforcement within the concrete.

iii) **Waterproof membranes**

Membranes, on the other hand, offer a different approach. Made from materials like bitumen, PVC, or EPDM, these flexible sheets are adhered to the concrete, forming a waterproof shield. This shield is impermeable to moisture and chlorides, thus safeguarding the concrete beneath. The choice between using an overlay or a membrane depends on various factors, including the severity of environmental exposure, the specific durability requirements of the structure, and, of course, budget considerations.

While both methods are effective at reducing chloride ingress, they each have their own set of advantages and limitations. Overlays are typically more cost-effective and easier to apply, making them suitable for a wide range of applications. They are particularly beneficial in scenarios where the concrete surface is exposed to heavy traffic or chemical exposure, as they provide an additional layer of protection that can be easily renewed or repaired. Membranes, with their flexible nature, are ideal for situations where the concrete structure is subject to movement or settling, which could cause rigid overlays to crack. They are also preferred in cases where a seamless barrier is required, such as under the decks of wharf and in inter-tidal zones of piles and walls. However, membranes can be more labour-intensive to install and may require more frequent maintenance or replacement over time.

iv) **Protection against Carbonation**

Carbonation is a natural process that can significantly affect the integrity and longevity of concrete structures. To mitigate this, various protective strategies can be employed. Anti-carbonation coatings are specially formulated to interact with carbon dioxide, creating a reaction that neutralizes its potential to lower the pH of the concrete. These coatings are designed to be permeable to water vapor, allowing the concrete to 'breathe' while providing an effective shield against CO_2 ingress. To protect the concrete against carbonation, the concrete surface can be coated with a protective layer. Here are some common methods:

i) **Anti-Carbonation Coatings**: These coatings contain alkaline materials (such as silicates) that react with CO_2, neutralizing its effect. They form a barrier against carbonation.

ii) **Paints and Sealers**: Acrylic or elastomeric paints and sealers can be applied to the concrete surface. They provide a protective layer that reduces CO_2 penetration.

Paints and sealers offer another line of defence. Acrylic-based paints are favoured for their durability and ease of application, while elastomeric sealers provide a flexible, waterproof barrier that accommodates the natural movement of the concrete without cracking. Both options serve to limit the amount of carbon dioxide that can reach the concrete's surface.

Quality control during application is crucial; even the best protective systems can fail if not properly applied. Regular inspection and maintenance are also necessary to ensure the continued performance of the protective layer, as damage or wear over time can lead to exposure of the underlying concrete.

v) Patch Repairing

Patching is defined as the process whereby small localized areas of deterioration in the structure are repaired by removing the deteriorated concrete, cleaning any corroded reinforcing steel, and replacing the concrete with a substitute material. The objective is to reinstate the original characteristics of the structure as far as possible. There is no expectation of increased life, and patching does not enhance the original functional characteristics of the structure.

Patching is essentially a quick, short-term procedure that may need to be repeated on several occasions. The timing of the work is not fixed, and it can be delayed for fairly extensive periods determined only by safety, serviceability, and the fact that delayed work costs more than work undertaken at the most opportune time. The various materials and methods for patch repairing are outlined in section……

11.2.3.4 Renovation

Renovation aims to update the structure to meet current standards and regulations. Both processes are essential for maintaining the functionality and extending the service life of structures, especially in the face of challenges like chloride-induced corrosion. It's a thorough balance between preserving the original design and adapting to modern requirements, ensuring that the structures remain safe, functional, and relevant for future use. The expectation of increased life span, without necessarily enhancing the original functional characteristics, underscores the pragmatic approach of these engineering efforts. It's about respecting the past while securing the future.

11.2.4 Rehabilitation

Rehabilitation is more extensive than patching. Large areas of chloride-contaminated or deteriorated concrete are systematically removed, the corroded steel is cleaned, and the area is repaired with materials similar to those originally used. These are critical aspects of civil engineering that ensure the safety and longevity of a structure. While rehabilitation focuses on restoring the structural integrity by removing and repairing the damaged areas.

11.2.4.1 Rehabilitation strategies

Rehabilitation strategies for RCC in marine structures involve a combination of preventive and corrective measures. Corrective measures become necessary when signs of distress appear. These may include retrofitting or strengthening the structure, or undertaking repairs. Such interventions are essentially required where the structure has developed distress due to factors like corrosion, overloading, or mechanical damage, or when it needs to be modified to meet revised loading requirements or codal specifications, such as changes in earthquake zoning or usage. Preventive measures include the application of protective coatings on concrete surfaces to preserve the initial alkaline environment of the concrete and prevent the ingress of aggressive substances. However, before adopting a repair and rehabilitation measure to restore the strength of the structure, assessment of the condition of the structure and the type of distress is important.

i) Assessment of the condition of Structure

Assessment of the condition of RCC structural elements is an essential aspect to formulate and implement cost-effective repairs. In summary, the maintenance and repair of RCC structures in marine environments needs a comprehensive and holistic study that considers the root causes of deterioration and distress in structures, corroborating the interactions of various factors. Most often, visual inspection by experienced structural experts and non-destructive tests are employed by experience structural experts to conduct the condition survey of structures, which are outlined in Chapter-8.

ii) Setting Methodology for Repair and Rehabilitation

Concrete repair is a critical process that involves the restoration of concrete to its original integrity and function. It encompasses a wide range of activities, from fixing minor defects to addressing significant structural damage caused by environmental factors or mechanical stress. Effective repairs require;

- A thorough understanding of the underlying causes of damage,
- the selection of appropriate repair materials and methods, and careful execution to ensure that the repaired area adjoins seamlessly with the surrounding concrete and restores the intended characteristics of the structure.

The general requirements for quality repair need a systematic approach to concrete restoration, ensuring that each step, from surface preparation to the application of repair materials, is carried out with due precision and care.

iii) Finalisation of Procedures

For restoration and rehabilitation works of concrete structures, correct methods and techniques shall be emphasised to ensure their effectiveness. Poor procedures and workmanship can lead to substandard repairs. For a successful repair, it is essential to remove all deteriorated concrete, reinforcements and replace them carefully by following standard procedures, ensuring proper anchorage.

The repair and rehabilitation work shall be adopted in a structured way, and a general approach followed for the repair work are as follows;

- Determine the cause(s) of damage
- Evaluate the extent of damage
- Evaluate the need to repair
- Select the material for repair
- Select the repair method
- Prepare the old concrete for repair
- Apply the repair method
- Cure the repair properly

iv) Determine the Cause(s) of Damage

Determining the cause of damage in concrete structures is crucial for effective repair and prevention of further deterioration. The process involves a thorough investigation to identify whether the damage is due to a one-time event or ongoing issues. One-time events like application of severe impact loads, earthquakes, or structural overloads, which may not require action beyond the initial repair, if they are unlikely to recur. However, if the damage is happening from continuous or repeated causes, it is essential to address the root cause of the problem to prevent future damage. Continuous causes of damage often involve a combination of factors like deterioration of concrete due to carbonation, chloride and sulphate attack, alkali-silica reaction, including improper design or workmanship followed during construction or use of low-quality materials. These factors can significantly reduce the concrete's durability and make it more vulnerable to other types of damage.

To ensure the longevity of concrete repairs, it is necessary to use methods and materials that are resistant to the identified causes of damage. This might include designing for better load distribution, selecting higher-quality materials resistant to environmental stresses, or employing more rigorous construction practices.

In summary, a successful concrete repair strategy must start with a correct diagnosis of the damage causes. If the damage is due to ongoing issues, remediation efforts must be implemented alongside the repair work. By understanding and addressing the underlying problems, it is possible to extend the life of concrete structures and avoid the need for more extensive and costly repairs in the future.

v) Evaluate the Extent of Damage

Evaluating the extent of damage to concrete structures is a critical step in the repair process. It involves a systematic approach to assess the severity of the damage and its impact on the structure's serviceability. The evaluation process aims to quantify the damage in terms of dimensions and the extent of the affected area. This includes assessing the rate of deterioration and predicting the likely progression of the damage. The severity of the damage is crucial to understand because it influences the repair strategy.

For a more comprehensive analysis, specialised tools and test methods shall be employed as discussed in Chapter-8.

In summary, the evaluation of concrete damage is a multifaceted process that requires expertise and the use of various techniques to accurately diagnose the condition of the structure and guide the subsequent repair or rehabilitation efforts.

vi) Evaluate the Need to Repair

Assessing concrete damage requires a balanced and judicious approach. Not all defects demand immediate action; the key is to understand the severity and implications of the damage. Surface-level cracks may be superficial, but deeper fissures can threaten structural integrity and necessitate prompt repair. Long-term considerations, such as the repair's durability and cost, are also crucial. A careful evaluation can ensure that the chosen solution endorses the structure's safety and longevity.

vii) Select the Materials for Repair

The selection of materials for concrete repair is critical to the longevity and performance of the repair work. High-quality materials that meet specific application requirements ensure that repairs are durable and effective. Adhering to manufacturers' recommendations and approved methods is crucial for achieving the desired results. A systematic approach to concrete repair, which includes proper mixing, proportioning, and handling, is essential to prevent future failures. By following these guidelines, it can be ensured that the repair work will meet the required workmanship and contribute to the structural integrity. There are various established repair systems, and some of them are outlined in Chapter-10 and choosing the right one depends on the specific conditions and needs of the project.

viii) Select the Repair Method

When it comes to repairing old concrete structures, the selection of the repair method is very crucial. It's not just about patching up the damage; it's about making informed decisions that consider the long-term implications and performance of the repair materials. A thorough evaluation of the damage includes considering the type of structure, the characteristics of the materials, and the environmental conditions. The compatibility of the new repair materials with the old concrete is also a significant factor, as it can prevent future problems like differential expansion or adverse chemical reactions.

In essence, the right repair method is one that is chosen based on a comprehensive understanding of all these factors, leading to a repair that is not only durable but also cost-effective. This strategic approach to concrete repair ensures the safety and longevity of the structure, safeguarding it against future damage and degradation. Also, the repair strategy shall be based on the availability of repair material, machinery, and the duration involved in the execution of the work.

11.2.5 Prepare the Old Concrete for Repair

i) Removal of Damaged Concrete

For concrete repair, the removal of deteriorated or damaged concrete is a critical step to ensure a successful and long-lasting repair. It's a common misconception that time or money can be saved by avoiding this process. However, this is a false economy, as the longevity and effectiveness of the repair are compromised when the damaged concrete is not thoroughly removed.

The preferred method for removing concrete in marine structures is high-pressure hydro-blasting or hydro-demolition, with pressures ranging from 8,000 to 15,000 psi. This technique is highly effective because it selectively removes only the unsound concrete, preserving the integrity of the sound concrete that remains. Additionally, it avoids creating micro-fractures on the surface of the old concrete, which can be detrimental to the bonding process.

In contrast, impact removal techniques like bush-hammering, scrabbling, or jackhammering can be less desirable. These methods tend to induce micro-fractures within the concrete surface, which can significantly weaken the bond between the new repair material and the existing concrete. To address this issue, it's often necessary to perform additional surface preparation, such as hydro-blasting, shot blasting, or sandblasting (either wet or dry), to remove the micro-fractured layer and provide a sound surface for the repair material to adhere to.

For concrete removal, the traditional impact methods, are as jackhammering and bush-hammering, are time-tested for their speed and cost-effectiveness. However, they come with the downside of potentially creating micro-fractures on the surface, which can add to the overall cost due to the need for additional surface treatment. To minimize damage, it's recommended to use jackhammers no heavier than 60 pounds and to opt for pointed bits that break concrete more cleanly. For shallow surface issues, shot blasting or sandblasting are preferred methods, with shot blasting being particularly efficient due to its integrated vacuum pickup system. These techniques, when chosen and applied correctly, can ensure the integrity of the remaining structure and the effectiveness of the concrete removal process.

In summary, for a durable and effective concrete repair, it's essential to remove all compromised concrete thoroughly and to choose a removal technique that maintains the quality of the remaining concrete and provides a solid foundation for the new material to bond with.

ii) Preparation of Reinforcing Steel

The preparation of reinforcing steel during concrete repair is a critical process to ensure the longevity and structural integrity of the repair. The steel must be cleaned to remove any debris, such as scale, rust, corrosion, and concrete, that could impair the bond between the new repair material and the existing steel. This cleaning can be achieved through methods like wire brushing, high-pressure water, or sandblasting. It's important to note that the steel does not need to be cleaned to a white metal condition; the goal is to remove loose material that could compromise the bond between the concrete and steel.

In cases where corrosion has significantly reduced the diameter of the steel bars, it is necessary to follow the guidelines set out in ACI 318 or other equivalent Standard Code of Practice. According to this, if the cross-sectional area of the steel has diminished to less than 75% of its original size, the compromised bars should be removed and replaced. Additionally, when steel is exposed beyond one-third of its perimeter, it should be cleared to maintain a minimum of 1-inch clearance from the concrete to allow for proper encapsulation by the repair material.

These procedures are essential for repairing issues such as delamination at the top mat of reinforcing steel in concrete members. Proper preparation ensures that the repair material adheres effectively to the reinforcing steel, which is crucial for the restored section to perform as intended and withstand the loads and stresses alongside the original structure.

iii) Finalisation of Methods and Materials for Rehabilitation Plan

To combat these issues, various rehabilitation techniques have been developed. Traditional methods like patch repairs, section enlargement, and concrete jacketing have been employed to restore structural integrity. However, these solutions often provide only a temporary respite from the relentless advancement of corrosion. In response to the limitations of conventional methods, there has been a shift towards the adoption of non-corrosive strengthening systems. Among these, fibre reinforced polymers (FRPs) have emerged as a promising alternative. FRPs offer several advantages over traditional materials, including high strength-to-weight ratios, resistance to corrosion, and ease of installation. Their application in marine environments is particularly advantageous as they are impervious to the corrosive elements that typically assail RC structures.

The implementation of FRPs can significantly extend the service life of RC structures while simultaneously reducing maintenance costs. This innovative approach not only restores the lost strength and stiffness but also enhances the overall durability of the structure. As the industry continues to evolve, the integration of such advanced materials and techniques is pivotal in ensuring the longevity and reliability of marine infrastructure. The proactive adoption of FRPs and similar technologies represents a forward-thinking strategy in structural engineering, one that prioritises long-term performance and sustainability in the face of environmental adversities.

11.3 Repair and Strengthening Methods

11.3.1 Strengthening by Installation of FRP

Fibre-reinforced polymer composites (FRPCs) have revolutionised concrete strengthening by offering superior mechanical properties and environmental resilience. Their high specific strength, stiffness, and fatigue resistance, combined with low density and thermal expansion, make them ideal for reinforcing structural elements.

FRPC sheets, when bonded to concrete beams, girders, or wrapped around columns, significantly enhance strength, ductility, and energy absorption. However, long-term durability under environmental stressors—such as freeze-thaw cycles, sulphate attack, and salt erosion—remains a concern, as these conditions can degrade the bond and reduce structural performance. Debonding, especially in high-stress zones, is a common failure mode, often leading to brittle failure, and is exacerbated by temperature and moisture fluctuations.

Figure 11.1: Photo of CFC (Carbon Fibre Cloth)

11.3.1.1 Features of CFC

Carbon Fibre Cloth (CFC) is a material commonly used for reinforcing concrete. It consists of carbon fibres that are impregnated with resin, which is then bonded and hardened onto the surface of concrete structures. The key features of CFC include its high strength and stiffness, which are comparable to high-strength steel, and its light weight, which makes it an excellent material for strengthening and repairing concrete without adding significant weight. CFC is also known for its crack resistance, shock resistance, and non-corrosiveness, making it a durable choice for various engineering applications. Additionally, CFC has superior creep resistance, high thermal conductivity, and chemical stability, which contribute to its performance in dynamic environmental conditions. Thus, the merits of the CFC reinforcement method are listed below;

- Excellent material properties.
- Lightweight
- High strength, high elasticity (Tensile strength: 3,400N/mm^2, 10 times more compared with iron)
- High durability against ultraviolet rays
- Excellent workability

Carbon Fibre Composite (CFC) is a material known for its high strength-to-weight ratio, which is significantly higher than that of traditional building materials like steel. With a density of approximately 1.6 g/cm^3, CFC is much lighter than steel, which has a density of around 8 g/cm^3. This makes CFC an excellent choice for reinforcing structures without adding much weight.

The tensile strength of high-strength CFC can exceed 3,400 N/mm², which is more than ten times the tensile strength of steel. This property allows for maintaining the structural integrity of concrete structures while minimizing the increase in size and weight. The thickness of CFC used for strengthening is typically less than 0.35mm, which is negligible compared to the dimensions of most structural elements.

CFC can be designed to be unidirectional or bidirectional. Unidirectional CFC is optimized to provide strength in one direction and is used when reinforcement is needed in a specific orientation. Bidirectional CFC, on the other hand, offers reinforcement in two directions, making it versatile for various applications where multidirectional strength is required. The choice between unidirectional and bidirectional CFC will depend on the specific requirements of the structure being reinforced.

i) Outline of Repair

This repair method strengthens concrete structures using Continuous Fibre Composite (CFC). CFC is lightweight and can be applied manually without heavy equipment, and it can be easily cut with scissors to fit the concrete surface. However, it is essential to apply the specified amount of epoxy resin per unit area (Table 11-1). Using too little resin can weaken the reinforcement and reduce its effectiveness. Proper application is key to ensuring durable and long-lasting CFC reinforcement.

ii) Work procedure

The procedure for deck reinforcement typically involves a series of steps to ensure structural integrity and safety. Initially, any necessary repair work must be completed to address existing issues with the deck.

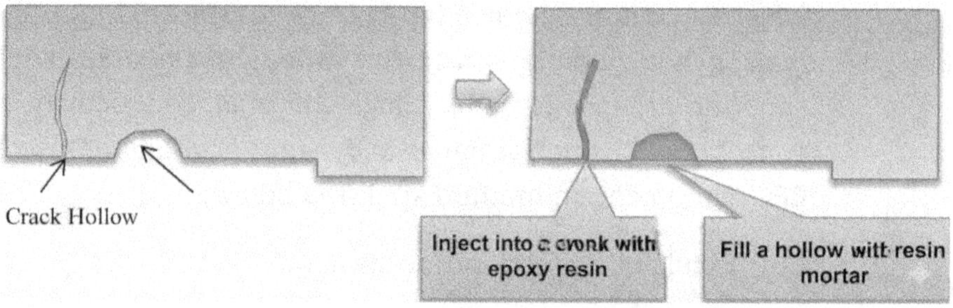

Figure 11.2: Repair of defects in concrete members before application of CFC

Accurate measurements of the areas requiring repair and reinforcement are crucial and should be incorporated into the work plan. This plan should be finalised before any on-site work begins to ensure a smooth workflow. Additionally, for specific design varieties and loading characteristics, it's advisable to refer to the specifications provided by manufacturers. This workflow of CFC is shown in Table 11-2.

Table 11-2: The standard usage weight of epoxy resin (per unit area)

Type of Resin	Areal weight of CFC (g/m^2)	Standard usage weight per unit area (Kg/m^2)		
		Under applying	Over applying	Total
Primer	-	-	-	0.2
Smoothing Agent	-	-		1.00
Impregnating resin	400	0.50～0.60	0.50～0.60	1.00
CFC	-	1	-	-

Note: The Above details are indicative based on projects executed in the past. It depends on the type of CFC and site requirements.

iii) Advance reinforcement work

Deck inspection is the first key step in maintenance and repair. The deck is visually checked for cracks, which must be repaired if wider than 0.3 mm. A hammer test is also carried out to detect any loose or detached concrete, which is then removed. After inspection, the required reinforcement area is measured and recorded to plan the repairs. This process ensures the deck remains safe and structurally sound.

iv) Sectional Repair

In the context of concrete repair, it is considered crucial that any hollows, cracks, or missing parts are addressed to ensure proper bonding and structural integrity. Hollows or gaps in the concrete are effectively filled and repaired using resin mortar. For cracks wider than 0.3 mm, epoxy resin is recommended, as it is known to provide a strong and durable bond capable of withstanding internal stresses within the concrete structure. Regular maintenance and timely repair of such defects are recognised as essential measures for significantly extending the service life of concrete structures.

v) Surface Preparation

Surface preparation is essential for proper bonding of carbon fibre to concrete. Uneven surfaces can lead to poor adhesion and failure. Small level differences (under 1 mm) are smoothed using a disk grinder, while larger uneven areas are corrected with resin mortar. The surface is then thoroughly cleaned to remove dust, oil, or loose mortar using a grinder or wire brush, ensuring a clean and sound base for carbon fibre application.

(a) To flatten by disk grinder (b) To remove dust by polishing, scrubbing

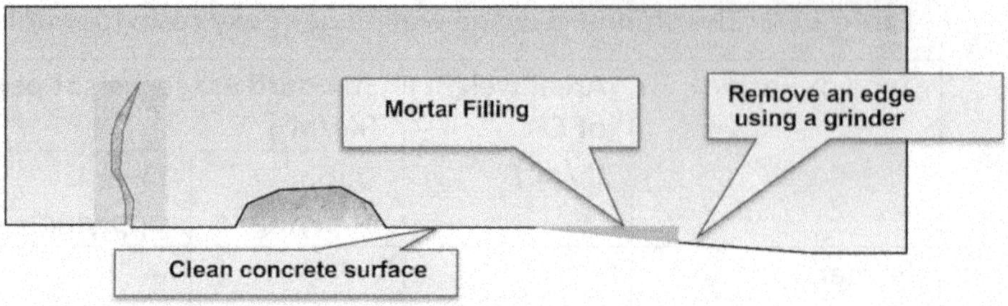

Figure 11.3: Surface Preparation for Application of CFC

vi) Preparing reinforcement work

Reinforcement work begins by measuring and marking the required areas on site. Based on these measurements and inspection results, a detailed CFC reinforcement plan is prepared. The layout is planned to ensure timely placement within the resin's 30-minute working time, with proper splicing to maintain strength and continuity, ensuring overall structural safety and durability.

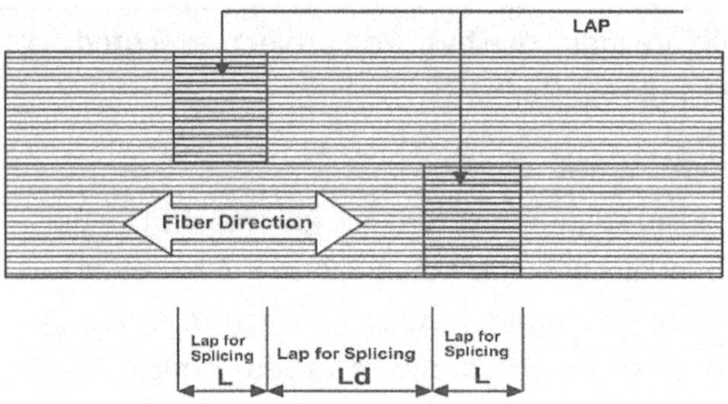

Figure 11.4: Splicing Arrangement for CFC

Table 11-3: Standard Splice length for CFC

Purpose	Splice length (L)	Detached (Ld)
Strengthening of bending, shearing	≥ 200mm	≥ 300mm

In construction applications, CFC may be installed by butting the pieces together in the longitudinal direction without gaps or by wrapping them around each other, with a typical wrapping allowance of 0 to 20 mm. This method is employed to ensure a tight fit and material continuity, which are essential for maintaining the structural integrity and performance of the composite. Proper execution of this technique is required to achieve the intended strength and durability of the construction.

vii) Surface cleaning

Surface cleaning is considered a crucial step in the preparation and maintenance of materials such as concrete. Polishing sheets or disk grinders are commonly used to remove dust, dirt, and contaminants. This process also aids in smoothing surface irregularities for an improved finish. Appropriate abrasive grades must be selected, starting with coarse grit and progressing to finer grits to achieve the desired polish. To ensure uniformity and avoid overgrinding, the grinder should be kept flat and moved in circular motions.

viii) Applying primer

The preparation of primer involves the careful mixing of a 'Main Material' with a 'Hardener' in specified proportions to achieve a uniform texture. Once mixed, the primer begins to harden, making it essential that it is thoroughly mixed and applied within 30 minutes, particularly at temperatures around 30°C, to ensure optimal performance. A uniform coat of primer is then applied to the concrete surface as a preparatory layer to enhance the adhesion of CFC reinforcement. An application roller or brush is used to ensure even distribution. This process is guided by specifications regarding the quantity of primer, application method, and required drying time before subsequent layers are applied.

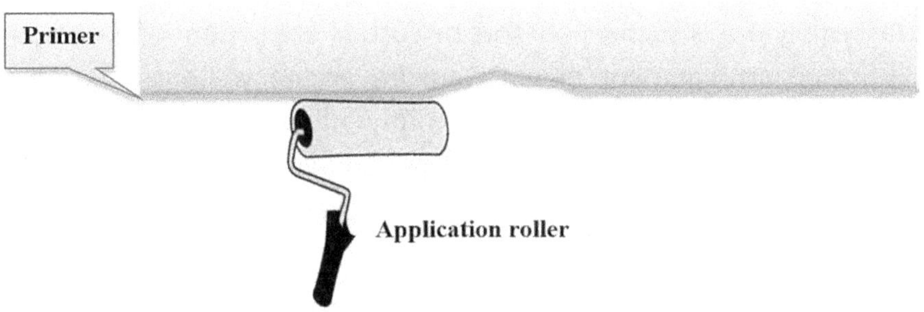

Figure 11.5: Application of Primer on Prepared Concrete Surface

ix) Confirmation of primer surface flatness

The flatness of a primer surface shall be confirmed before applying a smoothing agent. If unevenness is detected, it should be addressed by removing high spots or filling in low areas to achieve the desired level of flatness. This ensures a proper foundation for the smoothing agent and contributes to the quality of the final product.

x) Applying Smoothing Agent

The application of a smoothing agent is considered a critical step in ensuring proper bonding of carbon fibre to concrete surfaces. The agent must be applied evenly using a spatula or trowel to achieve a surface level difference of less than 1 mm. This careful process is intended to prevent poor bonding and potential failure of the carbon fibre due to uneven surfaces or sharp edges. Prior to application, the hardening of the primer must be confirmed. After application, the smoothing agent is required to harden for at least 24 hours, particularly at a temperature of 30°C, to ensure a stable foundation for subsequent carbon fibre installation..

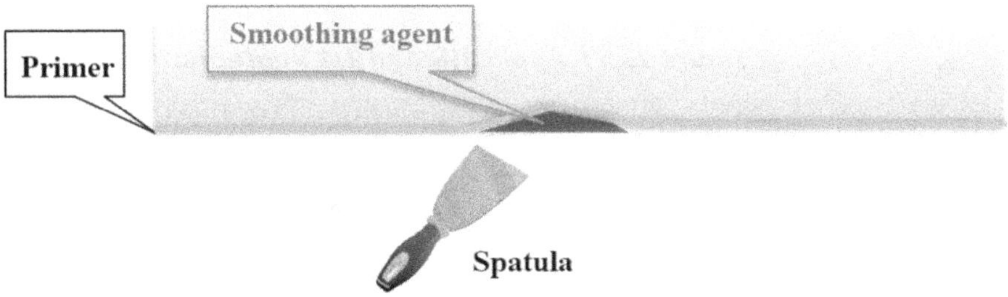

Figure 11.6: Application of Smoothening Agent for Finishing of Concrete Surface

xi) Confirmation of smoothing agent surface flatness

Before applying the impregnating resin, it is necessary to confirm the smoothing agent's surface flatness. If the unevenness is found, it is necessary to remove it by using a polishing sheet or disk grinder as much as possible.

xii) Cutting CFC

Cutting of Carbon Fibre Composite (CFC) materials must be carried out with precision to preserve fibre integrity and structural performance. The cutting tool is required to be aligned at a right angle to the fibre direction to minimise damage and maintain strength. The cut length should include the necessary splice allowance to ensure proper fit during assembly. Tools such as scissors or cutters are commonly used, and cutting is guided by a detailed arrangement plan to ensure accuracy. Consideration must be given to fibre orientation angles, as they significantly influence the cutting process and surface quality. Optimising the feed direction is recommended to reduce cutting force and minimise material damage. Proper planning and execution are essential for achieving high-quality results in CFC machining.

xiii) Applying 1st Impregnating Resin

Impregnating resin is a two-compound material, such as 'Main material' and 'Hardener'. Before application, it shall be mixed and prepared in the specified proportion. The application of impregnating resin is a critical process in construction, where it's used to enhance the strength and durability of materials. It's typically applied with an application roller to ensure even distribution. Insufficient application can result in a lack of strength, making it crucial to apply the resin in adequate amounts. After the initial application, the resin requires a curing period of less than 24 hours at 30 degrees Celsius before proceeding with the next steps. This curing time is essential for the resin to harden and provide the necessary reinforcement to the material.

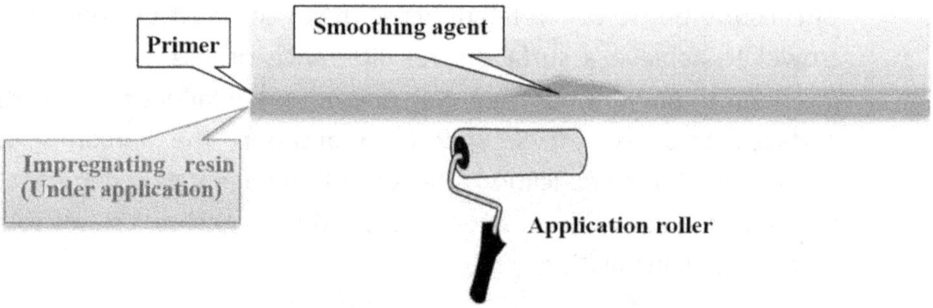

Figure 11.7: Application of 1st Impregnating Resin

xiv) Setting CFC

The setting of Carbon Fibre Composite (CFC) involves several critical steps to ensure proper installation. A detailed arrangement plan, including splice lengths, must be prepared in advance. During installation, CFC is placed along pre-marked lines while impregnating resin is applied. A paddle roller is used to remove trapped air, which must be completed within 24 hours at 30°C to avoid compromising bond quality. The roller should be moved along the fibre direction to ensure full saturation. Adherence to these steps is essential for achieving structural integrity and durability.

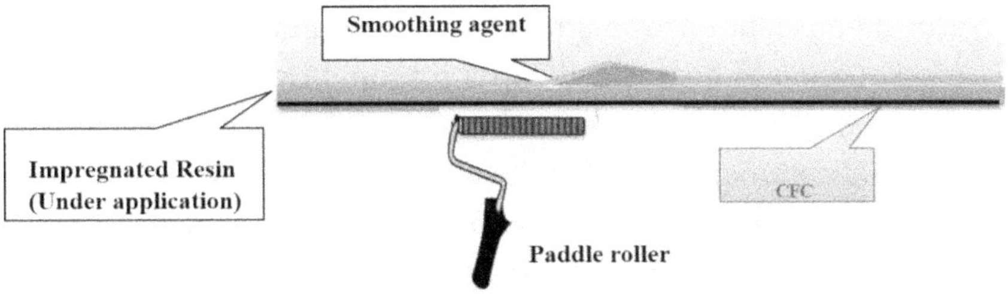

Figure 11.8: Setting out of CFC and pushing out air bubbles by paddle roller

The application of impregnating resin on Carbon Fibre Composite (CFC) is carried out in two stages. An initial coat is applied after the CFC is placed, followed by a second coat 20 minutes later using an application roller to ensure even distribution without displacing the material. The usage time of the second resin must be monitored. After application, a curing period of over 24 hours at 30°C is required to ensure proper hardening and performance of the CFC.

11.3.2 The Usage Of CFRP in Strengthening Beams For Shear

11.3.2.1 Shear In Concrete Beams

Some reinforced concrete beams have shear weaknesses due to various reasons and need strengthening. Reasons such as insufficient shear reinforcement, reduction in cross-section of reinforcement due to corrosion, increase in service loads, design and or construction defects. Normally, shear reinforcement of the beam is provided by stirrups, and in some cases, inclined bent-up bars are also used along with stirrups to strengthen the shear resistance of the beam in support locations (Figure.11.7).

Since shear failure is sudden, brittle and without any advanced warning, beams must be designed in a way that the shear strength of the member exceeds the flexural strength. Shear failure occurs when the beam has shear resistance lower than flexural strength, and the shear force exceeds the shear capacity of the different materials of the beam. Once a diagonal shear crack appears, the worst thing is that the crack propagates throughout the compression zone and makes a sudden, brittle and complete failure. Therefore, in this article, the beam's shear strengthening with CFRP to stop diagonal cracks from developing or at least to limit their widths as a result of increasing the shear strength of the beam has been reviewed.

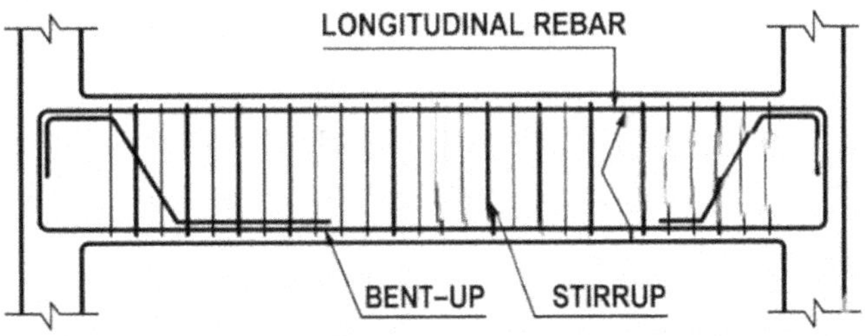

Figure 11.9: Shear reinforcement of a Concrete Beam

11.3.2.2 Shear Strengthening of Concrete Beam

For shear strengthening of concrete structures, we should be familiar with shear behaviour and shear failure of the structure. Figure 11.8 is a concrete beam and its possible shear crack. Concrete beams develop a diagonal crack from the loading point to the bottom of the fixed end, leading to the shear failure of the whole structure. The location of the crack is an indication of the failure mode of the structural element. Shear failure of the beam is brittle and without any forewarning, so beams should be designed in a way that bending failure with large deformations occurs before shear failure.

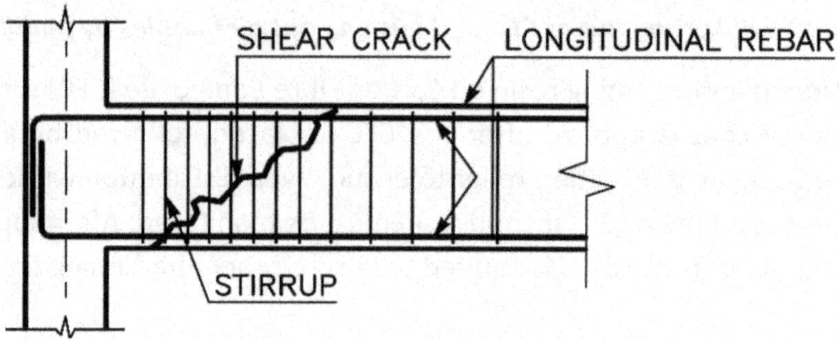

Figure 11.10: The Shear Crack in a Concrete Beam

However, according to various investigations on this issue, it has been proven that over-strengthening in shear resulted in bending failure in some cases, but after strengthening the beams for bending, shear failure was achieved. The experimental results, which evaluated the contribution of CFRP towards the shear strength of continuous reinforced concrete (RC) beams, have shown that the shear strength of the beams was significantly increased by using CFRP and placing the CFRP strips at 45° to the axis of the beam is beneficial (I. A. Bukhari, 2009).

11.3.3 Patch Repairing

11.3.3.1 Use of Epoxy-Bonded Epoxy Mortar

Epoxy-bonded epoxy mortar is a specialized material for concrete repairs, best suited for situations where;

i) the repair depth is less than 1-1/2 inches and the temperature remains relatively stable.

ii) It's not ideal for environments with frequent or significant temperature changes, as it has a different thermal expansion coefficient than conventional concrete, which can lead to failures at the bond line.

iii) Its use is also discouraged in areas affected by corroding steel reinforcement, due to the creation of different electrical potential zones that can be problematic.

iv) Ideal applications include indoor settings, tunnel linings, the underside of wharf decks, trenches, and submerged structures, where temperature conditions are controlled. However, it's not recommended for areas that experience daily temperature shifts, especially those in direct sunlight.

v) When used correctly, epoxy mortar has a proven track record for durable repairs in consistent conditions, as evidenced by long-term inspections.

i) **Materials**

Epoxy resins are essential components in the preparation of epoxy mortar for concrete repair. They are supplied as a two-component system and completely solid, conforming to ASTM C-881 specifications for type III, grade 2, class B or C. Additionally, the sand mixed into the epoxy mortar needs to be clean, dry, and well-graded, ensuring that only sound particles are included in the mix for optimal performance and durability of the repair work.

ii) **Mixing**

Epoxy mortar is prepared by accurately mixing epoxy resin and hardener, followed by the gradual addition of sand. The components must be proportioned by weight, typically at a ratio of 5.5–6 parts sand to 1 part epoxy (ASTM C-881), or 4–4.5 parts by volume. Mixing must be done in small batches using a slow-speed mechanical mixer, and the mixture should be used within 30 minutes to avoid hardening. The pot life varies with temperature and no thinners should be added, as they compromise strength. Proper mixing and timely application are essential for achieving a durable and workable mortar. Manufacturer's instructions must be followed for the best results.

iii) **Application**

When applying a resin bond coat for epoxy mortar, it's crucial to use the same type of epoxy resin as in the mortar mix. This coat should be applied directly to the prepared concrete surface just before the epoxy mortar is placed. The freshly mixed bond coat resin needs to be spread evenly over the dry, existing concrete. Coverage rates should not exceed 2 square meters per Liter, but this can vary based on the surface roughness. In fact, if the surface is particularly rough, the actual coverage may be significantly less than this maximum guideline. It's all about ensuring a strong, durable bond for the epoxy mortar to adhere to.

iv) **Curing**

The curing of epoxy mortar is critical to ensure durability and performance. After application, it must be cured at or above the temperature specified by its classification until hardened. Post-curing, if required, is conducted by maintaining the epoxy mortar and substrate at 33°C to 43°C for a minimum of 4 hours. In cold conditions, approved heating sources such as propane-fired heaters or infrared lamps must be used to achieve the necessary surface temperature. Moisture must be strictly avoided during post-curing to prevent compromising the mortar's integrity. Adherence to these procedures ensures the intended strength and longevity of the repair.

11.3.4 Cross-Section Restoration Methods

The restoration of damaged or deteriorated reinforced concrete (RC) sections is a critical process aimed at reinstating the structural integrity, enhancing durability, and extending the service life of the structure. This section outlines the standard and widely adopted techniques for restoring cross-sections in RC elements such as decks, beams, and slabs. Each step shall be suitably verified or selected based on the nature and extent of deterioration, ensuring effective and long-lasting rehabilitation.

i) **Concrete Removal and Reinforcement Exposure**

As a first step in cross-section restoration, all deteriorated concrete—especially around corroded reinforcement—must be carefully removed. The removal depth should extend at least 50 mm behind the reinforcing steel bars, or further if required, to ensure complete exposure of the steel for cleaning and corrosion treatment.

a. This process facilitates proper inspection and surface preparation of the reinforcement.

b. Mechanical tools or hydro-demolition may be used depending on site conditions and the extent of deterioration.

c. Care must be taken to avoid damaging adjacent sound concrete or reinforcement.

ii) **Reinforcement Augmentation**

If inspection identifies a cross-sectional loss exceeding 10% in a reinforcing steel bar, it must be reinforced by adding a supplementary bar.

- The additional bar should be securely fixed using lapping (with adequate development length) or welding, depending on design requirements and accessibility.
- The augmentation ensures that the structural capacity of the member is restored to its original or required level.
- All reinforcement must be properly anchored and aligned to maintain continuity and load transfer.

iii) **Cover Depth Requirements**

To protect the reinforcement from future corrosion and environmental exposure, the repaired section must provide a minimum concrete cover of 75 mm.

a. This is typically achieved by enlarging the cross-section at the repair location.

b. The increased cover enhances durability, especially in aggressive environments such as marine or industrial settings.

c. The repair material should be compatible with the existing concrete in terms of thermal expansion, shrinkage, and bonding characteristics.

iv) **Selection of Restoration Technique**

The choice of restoration method should be based on efficiency, practicality, and site-specific conditions, including:

1. Extent of damage
2. Volume of repair

3. Accessibility and working environment

Recommended techniques include:

a. **Micro-concrete or mortar spraying** for **slab repairs**, offering good flowability and surface finish.

b. **Mortar grouting** for **beam repairs**, ensuring deep penetration and bonding in confined spaces.

Each method should be selected to optimise performance, minimise downtime, and ensure long-term durability.

v) Surface Protection

To further enhance the durability of the restored RC deck, a protective surface coating must be applied to the entire underside of the structure.

- The coating acts as a barrier against moisture, chloride ingress, and other aggressive agents.
- Suitable materials include epoxy-based coatings, polymer-modified cementitious coatings, or silane/siloxane treatments.
- Surface preparation prior to coating is critical to ensure adhesion and effectiveness.

11.3.5 Replacement of Obsolete Deck Beam Slabs

If the existing deck beam slabs are no longer suitable, they can be replaced with new, more efficient structural configurations that are designed to handle increased loads.

When it comes to replacing obsolete deck beam slabs, the focus is on selecting new structural configurations that can handle increased loads efficiently. Modern construction practices often recommend the use of high-strength materials or composite slabs and beams, which combine the strength of steel with the durability of concrete. This method is not only structurally efficient but also speeds up the construction process, offering an economical solution for a wide range of structures.

Replacement of obsolete deck beam slabs is a structural strengthening strategy that involves removing deteriorated or underperforming slabs and installing new, more efficient systems capable of handling modern load demands. This method is particularly relevant when existing slabs can no longer meet safety or serviceability requirements. Modern replacements often utilise high-strength materials or composite construction, where concrete slabs are cast over steel decking supported by steel beams. These systems are designed for composite action, ensuring the concrete and steel act together under load. This integration significantly enhances load-bearing capacity, stiffness, and structural efficiency.

It must be noted that replacement of a structural member becomes essential when it can no longer safely perform its intended function due to one or more of the following conditions:

i) **Severe Structural Damage**

When a structural member exhibits extensive physical deterioration—such as deep cracking, spalling, delamination, or crushing—its integrity is compromised. Significant loss of cross-sectional area due to corrosion or erosion, and exposed reinforcement with advanced rusting, reduces its ability to safely carry loads. In such cases, repair may not restore full functionality, making complete replacement essential to ensure safety and performance.

ii) **Irreversible Material Degradation**

Exposure to aggressive environmental agents like chlorides, sulphates, or carbonation can cause permanent chemical damage to concrete and steel. Reactions such as alkali-silica expansion lead to internal fracturing and scaling. Fire damage can alter the microstructure of concrete and reduce steel strength. When these effects are beyond repair, replacement is necessary to restore structural reliability.

iii) **Inadequate Load-Carrying Capacity**

If a structural element cannot support increased loads due to changes in usage—such as heavier traffic or machinery—it may fail under service conditions. Outdated design standards or insufficient reinforcement detailing further limit its capacity. In fatigue-prone environments, repeated loading can cause progressive weakening. Replacement with a redesigned member ensures compliance with current load requirements and safety codes.

iv) **Loss of Structural Continuity or Stability**

Structural elements that have already experienced excessive deflection, shift, tilt, or lost alignment due to foundation settlement, seismic activity, or other dynamic forces may no longer function as intended. Furthermore, loss of anchorage or support at critical connections can lead to instability or progressive collapse. In such cases, replacement is the only viable solution to restore structural integrity and prevent further failure.

v) **Economic Non-Viability of Repair**

When the cost of repair approaches or exceeds the cost of replacement, and the repair offers only temporary or limited improvement, replacement becomes the more practical option. Additionally, if repair techniques are technically infeasible or do not meet long-term performance expectations, replacing the member ensures better durability, efficiency, and compliance with modern standards.

vi) **Functional or Geometric Obsolescence**

Structural elements may require replacement when they no longer meet the evolving functional demands or spatial requirements of a facility. This can arise from the need for increased vertical or horizontal clearance, reconfiguration of layouts, or integration with modern systems and technologies. In such cases, the existing member becomes incompatible with current usage or design standards. Replacement enables the adoption of advanced structural solutions—such as composite systems or high-performance materials—that offer superior strength, reduced weight, and faster installation. These modern alternatives not only improve structural efficiency but also enhance adaptability, ensuring the infrastructure aligns with contemporary operational and architectural needs.

During replacement, additional supports must be provided to ensure that cutting and installation do not compromise the local or global structural integrity.

vii) Concrete Removal and Bar Cleaning

Concrete removal and bar cleaning are critical processes in the maintenance and repair of structures. They involve the careful extraction of damaged or contaminated concrete to allow for effective patching and rehabilitation. The depth, area, and specific location of the concrete to be removed are essential considerations in these operations. Techniques vary depending on the extent of damage and the structure's requirements, ranging from manual chipping to more advanced methods like hydro-demolition. Once the concrete is removed, cleaning the exposed reinforcement bars is crucial to ensure a good bond with new concrete during the repair. This step often involves rust removal and surface preparation to restore the structural integrity of the reinforcement bars. Proper execution of these processes is vital for the longevity and safety of the structure undergoing repair.

viii) Method Used to Identify the Concrete to be Removed

The methods used to identify concrete for removal are crucial in ensuring the integrity and safety of a structure during renovation or repair. Visual inspection and sounding are the more traditional approaches, relying on observable damage and changes in sound to detect deterioration. These methods are often quicker and less invasive, allowing for immediate identification of problem areas. On the other hand, concrete core sampling and half-cell potential measurement provide a more in-depth analysis. They can detect issues that are not visible to the naked eye, such as chloride contamination, which can lead to corrosion of steel reinforcement within the concrete. The removal process for concrete identified by these latter methods is typically more complex, as it requires not only the removal of visibly damaged concrete but also areas that may appear sound yet are at risk of future deterioration. This ensures a more comprehensive approach to maintaining the structure's longevity and stability.

Visual inspection and sounding are essential non-destructive techniques for assessing the condition of concrete structures. They help in identifying surface-level deterioration such as cracking, delamination, and spalling. These methods are particularly useful for pinpointing small and irregularly shaped areas where the concrete has deteriorated. Although the localised nature of the damage may limit the methods available for concrete removal and reinforcement cleaning, the presence of cracks and fracture planes can facilitate the removal process. It's important to note that while visual inspection provides valuable information about the surface condition, it may not reveal internal defects. Therefore, it is often supplemented with other non-destructive testing methods or partial destructive testing, such as core drilling, to gain a comprehensive understanding of the concrete's condition.

ix) Location of the Concrete to be Removed

The process of concrete removal is highly dependent on the location and orientation of the concrete to be removed. On structure decks, the horizontal layout allows for the use of mechanised methods, which can enhance productivity and efficiency. The flat surface facilitates not only the removal process but also the subsequent clean-up and containment of debris. In contrast, concrete removal from vertical or overhead structural elements presents a set of challenges. The limited surface area, coupled with the difficulty in reaching these spaces, often necessitates manual methods of removal. This can significantly reduce the pace of work and increase costs due to the slower, more labour-intensive process. Additionally, the safety risks are heightened, and the need for scaffolding or other access solutions can complicate the task further. The variability in the size and location of the areas requiring concrete removal means that each project must be assessed individually, with methods tailored to the specific conditions of the structure component in question. Planning and executing concrete removal operations require careful consideration of these factors to ensure safety, efficiency, and cost-effectiveness.

x) Depth of Concrete Removal Required

The depth of concrete removal is a critical factor in the repair and rehabilitation of concrete structures. It influences the choice of removal method, such as impacting, blasting, cutting, milling, presplitting, or abrading, and significantly affects the project's cost. For instance, portable crushers are suitable for shearing plane depths of 300 mm or less. The process involves not only the elimination of unsound concrete but may also require the removal of sound concrete to achieve the necessary repair geometry or to address reinforcement issues. Determining the precise depth of removal is essential for ensuring the compatibility of the existing concrete with the repair materials and for achieving the desired substrate conditions, such as cleanliness, roughness, and moisture content. A thorough assessment of the concrete's condition, including factors like spalling, delamination, and cracking, is crucial for establishing the appropriate depth of removal and selecting the most effective and cost-efficient removal technique.

xi) Surface Removal

Surface removal is a critical process, particularly when dealing with the degradation of concrete or the presence of salt deposits. The goal is to eliminate any contaminants that could prevent a secure bond between the old and new materials. Techniques for surface removal can vary, from mechanical methods like sandblasting or water jetting to chemical methods using acids or other cleaning agents. The chosen method must be thorough enough to clean without causing additional damage to the underlying material.

xii) **Cover Concrete Removal**

Cover concrete removal is a process that involves the extraction of the concrete layer above the reinforcement without disturbing the underlying steel. This procedure is crucial for maintaining the integrity of the structure while addressing any potential issues with the concrete itself, such as cracking or spalling. The cover provides essential protection for the steel reinforcement from environmental factors and contributes to the overall durability of the structure. It's important to ensure that the removal process is carried out carefully to prevent any damage to the reinforcement, which could compromise the structural integrity.

xiii) Matrix Concrete Removal

Matrix concrete removal is a critical process in the maintenance and repair of reinforced concrete structures. It involves the extraction of concrete that surrounds and interlocks with the steel reinforcement, often located in challenging, confined spaces. Techniques for effective removal include blasting, crushing, cutting, impacting, milling, and presplitting, each suitable for different scenarios and concrete conditions. For instance, deteriorated concrete can be more easily removed due to existing fracture planes, while contaminated concrete in tight zones requires more precise methods. The goal is to reach sound concrete, ensuring a clean surface for the flow of replacement materials and the successful adhesion of new concrete. Proper assessment, tool selection, and disposal are essential steps in this process. The depth of removal typically extends slightly below the steel reinforcement to guarantee complete void filling during the repair.

xiv) Core Concrete Removal.

Core concrete removal is a specialised task that requires careful consideration of the structural element's integrity and the surrounding reinforced zones. Techniques such as mechanical bursting, where hydraulic pressure machines split the concrete, or chemical bursting, involving an expansive slurry to split the concrete, are often employed. Once split, the concrete can be removed manually or with a crane. For precise cuts, core drilling machines with diamond bits are used, which can be wet or dry, with wet drilling being more common due to better bit protection.

The process must avoid damaging the reinforcing steel and consider the volume of material affected by chloride contamination. Safety measures and proper equipment are paramount to ensure efficient and effective removal without compromising the structure's stability or worker safety. For detailed guidance on the process, professionals often refer to tutorials and industry standards that outline step-by-step procedures and necessary precautions.

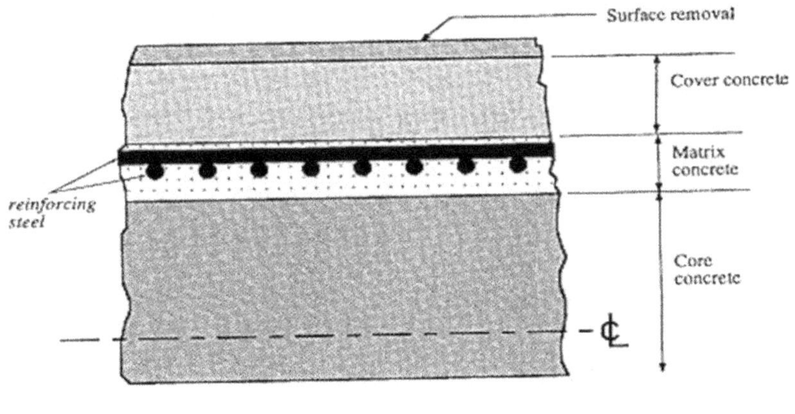

Figure 11.11: Depth classification for concrete removal

xv) Bar Cleaning

Bar cleaning is a crucial process in construction and maintenance, ensuring that the reinforcing steel bars are free of rust and chlorides. This is essential for the integrity of the structure, as it allows for a strong bond between the steel bars and the concrete or repair material. The goal is to remove only as much material as necessary to maintain the bar's strength while providing a clean surface for effective adhesion. Proper bar cleaning contributes significantly to the longevity and safety of the constructed structure.

xvi) Quality Constraints

Concrete removal and bar cleaning are critical tasks in the renovation or rehabilitation of existing reinforced concrete structures. These tasks are distinct from demolition because they require very careful and controlled execution to maintain the structural integrity and ensure that the cleaned areas are suitable for new concrete to bond effectively. These constraints are essential for the post-repair safety and longevity of the structure.

xvii) Area Demarcation

In concrete repair, area demarcation is a vital process that involves marking out the exact sections of concrete and steel reinforcement needing removal due to deterioration. This is an important step to preserve the integrity of the structure and to avoid the extra costs in removing more than necessary. The aim is to remove only the compromised materials, clean the area, and create a rough surface that ensures a strong bond with the new repair materials, all while keeping the structure sound and avoiding unnecessary expenses. It's a delicate balance between removing what's necessary for a robust repair and retaining as much of the original structure as possible. Methods like saw-cutting to outline the repair boundaries may be employed as shown in the figure below.

xviii) Controlled Cutting and Breaking

Controlled cutting and breaking of damaged concrete are essential in the demolition and renovation work of structures. The goal is to remove damaged sections without compromising the structural integrity of the remaining concrete and steel. Specialised equipment and methods are employed to ensure that the structure is not subjected to high impact forces, which may cause additional cracking or damage. Additionally, methods such as diamond wire saws for pre-cutting and hydraulic breakers for precise breaking are recommended to manage the demolition process with minimal damage.

xix) Bond Quality

After cutting, controlled breaking, and removal of deteriorated concrete from the structure, preparing for the addition of new material is also important. The new material must adhere well to the existing concrete and steel, which requires the surfaces to be clean and properly textured

11.3.6 Methods for Performing the Tasks

11.3.6.1 Methods for Surface Removal

It is necessary to remove surface contaminants such as oil, chemicals, and rust from the work area in order to provide a sound, long-lasting bond between the existing structure and the new materials used to repair or rehabilitate the structure. The objective is to clean rather than to remove material. The following four methods are frequently used:

i) Scrabbling

Scrabbling is a process used to remove layers of concrete from a surface. Scrabbler is a tool equipped with pneumatically driven bits that pound the concrete surface, and the depth of removal can range from 1 mm to 6 mm, allowing for precise adjustments to the surface level. Scrabblers come in various sizes, with larger, self-propelled models designed for big, flat areas, and smaller, handheld versions for tight spaces or uneven surfaces. To manage the resulting debris, vacuum collection systems are often employed, ensuring a cleaner work environment and easier disposal of material.

ii) Planing

A plane or diamond grinder removes concrete by abrasion. The planing machine is equipped with diamond-tipped blades to remove layers of concrete. This method is effective for levelling out uneven surfaces, removing coatings, or preparing the concrete for further treatment. The blades are mounted closely together and rotate to shave off concrete, with water used to cool the blades and minimize dust. This technique can remove up to 12 mm of concrete in a single pass, making it a powerful tool for concrete surface preparation.

iii) Sandblasting

Sandblasting is a powerful technique used to clean and texture surfaces by blasting them with sand particles. This process utilises compressed air to accelerate sand particles towards surfaces like concrete or steel, effectively removing rust, paint, or other coatings, and creating a roughened texture that's ideal for refinishing. The equipment used in sandblasting can range from small, portable devices for detailed work on irregular surfaces to large, industrial machines for bigger projects. Additionally, modern sandblasting setups often include vacuum systems to reclaim the sand and keep the work environment clean from debris. It's a versatile and efficient method widely used in various industries for surface preparation and cleaning.

iv) Shot Blasting

Shot blasting is done using a machine that propels steel shots at high speeds onto the concrete, which can strip away layers up to 12 mm deep. The process not only cleans the surface but can also create a rough texture, which is useful for various applications. The size of the steel shots can be varied to control the roughness of the final surface. Although these machines are versatile, their design, which includes a collection chamber for the steel shots, means they are primarily used on flat, horizontal surfaces. Additionally, they are equipped with a vacuum system to collect the debris and steel shots, allowing for the recycling of the shots, making the process more sustainable.

v) Methods for Removing Cover Concrete

The removal of the cover of concrete over a relatively wide area is frequently necessary in structural rehabilitation projects. The work involves removal to a depth less than the cover depth of the steel, and thus, no work between, around, or under the reinforcing mat is included in the task.

Scrabbling, planning, sandblasting, and shot blasting all can be used in repeated passes to achieve the required depth. This is an inefficient use of these methods, and the only really effective way of doing the work when large areas are involved is by using a concrete milling machine. A milling machine removes concrete by the impact of numerous tungsten-tipped teeth mounted on a rotating drum or mandrel, as shown in Figure 11.10.

Milling machines can achieve very high levels of production, but they can only be used on horizontal surfaces such as structure decks. They are also only capable of removing concrete above the reinforcing steel and severe damage can occur when milling teeth cut or snag the reinforcement. Despite these limitations, the use of milling for the removal of cover concrete is an emerging technology suited to the large volume of high-production work.

Hydro-demolition can also be used as a high-production, equipment-intensive method for the removal of cover concrete. It is, however, more frequently used to remove cover and matrix concrete simultaneously and is discussed in successive sections.

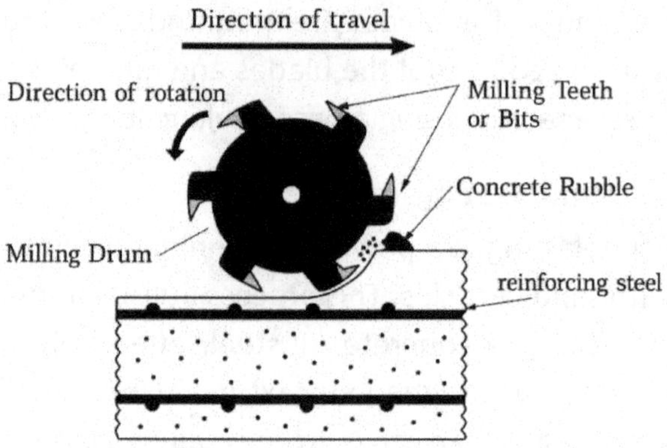

Figure 11.12: Action of a milling machine on a structure deck

Hand-held pneumatic breakers are frequently used to remove cover concrete when the areas involved are inaccessible to a milling machine or too small to be milled. As with hydro-demolition, hand-held breakers are primarily used for matrix concrete removal and are discussed. Each method has its specific applications and shall be chosen based on the project's requirements, the area's accessibility, and the depth of concrete removal needed. The selection of the appropriate technique is crucial to ensure the efficiency and safety of the rehabilitation process.

vi) Methods for Removing Matrix Concrete

It is necessary to remove matrix concrete when contamination, spalling, and delamination have progressed into the concrete layer that surrounds and encases the reinforcing mat. The work shall be performed between, around, and under the steel without damaging the steel, cracking the substrate concrete, or destroying the bond between steel and concrete in areas where the concrete is not to be removed.

In such cases, removing matrix concrete is a delicate process that requires careful consideration to avoid damaging the reinforcing steel, cracking the substrate concrete, or compromising the bond between the steel and concrete. The two primary methods for this task are pneumatic breakers and hydro-demolition. Pneumatic breakers are powerful tools that use compressed air to break the concrete, while hydro-demolition uses high-pressure water jets to remove concrete without generating dust. Both methods can be employed either to remove both cover and matrix concrete in one go or to target only the matrix concrete after the cover concrete has been removed by another method, like milling. It's essential to choose the right method based on the specific conditions of the concrete and the desired outcome of the removal process.

Pneumatic Breakers.–Pneumatic breakers, often called jack-hammers, are essential tools in structural rehabilitation work for removing concrete. They are handheld devices that use compressed air to deliver high-frequency blows, breaking the concrete in a controlled area. The effectiveness of these breakers is influenced by their weight and the operator's skill. Weights range from 8 to over 120 pounds, with heavier breakers being more productive yet more destructive. To prevent damage and maintain the integrity of the structure, there's often a weight limit of less than 45 pounds for breakers used in selective concrete removal. Additionally, the angle of attack is typically restricted to 45 degrees. The operator's proficiency is crucial for both the amount and quality of the removal work, making it a significant factor in evaluating the method's success. Overall, pneumatic breakers are pivotal in concrete removal across various structural rehabilitation projects.

Hydro-demolition: **Hydro-demolition** is a high-output but capital-intensive method of concrete removal that uses very high-pressure water jets (12,000–35,000 psi) to break down the cement matrix and release the aggregate. It is commonly used on decks and can be adjusted to remove concrete to various depths based on material condition. While it can be applied to horizontal, inclined, and overhead surfaces, costs increase due to the need for specialised equipment and debris control. Despite the high initial investment, hydro-demolition is an effective and increasingly adopted technique that delivers excellent results under suitable conditions.

vii) Core Concrete Removal

Core concrete removal necessitates the removal of concrete at the core of the structural element in a manner that respects the three quality constraints of selectivity, residual damage and bond quality. Traditional methods like pneumatic breakers and hydro-demolition are limited by their productivity rates due to the difficulty in accessing the material. Innovative techniques for concrete removal focus on precision and minimal residual damage, keeping the surrounding mats of reinforcing steel intact. Hydraulic breakers, saws, and crushers can dismantle concrete structures effectively, while concrete core drilling offers accuracy for creating openings without compromising structural integrity. For larger projects, heavy equipment like jackhammers or mini-excavators might be necessary. It's crucial to plan carefully, prioritise safety, and choose the right tools for the job. Considering professional services may also be beneficial for complex or large-scale removals.

viii) **Concrete Cutting - Sawing**

Concrete sawing is a practical and efficient method for cutting through concrete and steel, utilising blades coated with industrial diamonds for maximum durability. These saws come in a wide range of sizes and can be either handheld or mounted, catering to various project needs. The use of water during the cutting process is crucial as it cools the blade and minimises friction, which helps in extending the blade's life and ensuring a cleaner cut. This technique is not only cost-effective but also highly adaptable, making it suitable for tasks such as creating boundaries for demolition areas, removing specific sections of concrete, and installing joints in freshly poured concrete.

a. Bar Cleaning

Bar cleaning is a critical process in construction that involves the thorough removal of rust, chlorides, and other contaminants from exposed reinforcing steel. This step is essential after the removal of matrix concrete to ensure that all chloride-contaminated rust and cement paste are eliminated. This prevents further corrosion within the repaired concrete structure. The three commonly employed methods for this task include wire brushing, either manually or with machines, which is effective for minor rust stages; high-pressure water jetting, suitable for removing thin layers of rust; and sandblasting, which is highly effective for more significant corrosion.

- Wire brushing can be done manually or with machines, depending on whether the rebar is tied to the slab or not. For rebars that are not tied, machine brushing is advisable for efficiency. However, for those tied to the slab or in hard-to-reach areas, manual brushing is necessary. It's important to remove all rust to prevent further corrosion and to maintain the bond strength between the concrete and the reinforcement bar.
- High-pressure water jets can also be used for rust that is not significant, ensuring the entire surface is cleaned.
- The abrasive impact of sand particles under compressed air is an effective method for cleaning exposed steel because it is able to remove cement deposits and rust to leave a bare metal finish.

11.3.7 Strengthening by Section Enlargement

Section enlargement is an effective technique to increase the stiffness and rigidity of a structure, particularly when it comes to retrofitting and strengthening works. This method involves RCC jacketing, the addition of a new reinforced concrete layer, which can be applied either to the bottom or the top of the structural element, depending on the specific requirements. The process typically includes several key steps:

- Removal of deteriorated concrete and thoroughly cleaning surfaces for optimal bonding
- treating exposed reinforcement to prevent further corrosion, adding or replacing reinforcement as needed before the new concrete layer is applied.

While this method significantly increases the load-carrying capacity and stiffness of a structure, it also requires careful consideration of the increased weight of the member, as well as the logistics of constructing new formwork. Despite these challenges, section enlargement remains a valuable tool to extend the life and functionality of concrete structures.

11.3.7.1 Reinforced Concrete Jackets

Concrete jacketing is one of the widely used techniques for the strengthening of RC members. It is constructed either with cast-in-place concrete or, more often, with shotcrete. The method involves the addition of a layer of reinforced concrete in the form of the jacket using longitudinal steel reinforcement and transverse steel ties outside the perimeter of the existing member (Figure 11.11)

With reinforced concrete jacketing, a considerable increase in the strength in terms of flexural, shear, and axial load-carrying capacity of the member can be achieved. There is a significant increase in the ductility and the deformation capacity of the member through the confinement and the anti-buckling action of the new stirrups. The bearing capacity and the flexural strength of the member are enhanced, due to additional longitudinal reinforcement, whilst the improvement in the shear strength and the ductility are achieved through the additional transverse reinforcement.

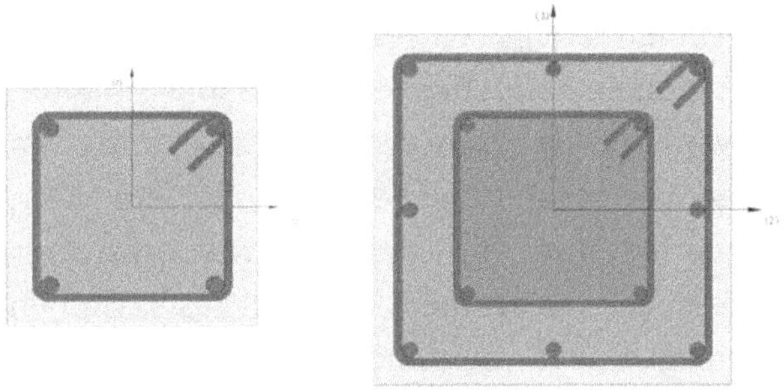

Figure 11.13: Schematic Details of Concrete Jacketing

The thickness of the jacket shall generally be more than 10 cm, in order to allow the casting of the concrete without voids and gaps. On the contrary, shotcrete allows for jackets of thickness as low as 5cm, although typically the jackets are 7.5 cm thick or more. In order to allow for a cover of adequate thickness, the positioning of the longitudinal and transverse reinforcements, and some space between the new rebars and the existing member.

To add a new concrete section or jacket around an old one, preparation of the surface on the existing member is very important to ensure a clean and sound concrete substrate to achieve good bonding conditions with the jacket. The connection between the new and the existing concrete is enhanced with the roughening of the surface and the introduction of steel dowels.

The new vertical steel bars and stirrups of the jacket are installed according to the designed dimensions and diameters, paying particular attention to the good closing of the hoops. If bending to the hoops at 135 ° angles is not possible due to the presence of the existing member and the small thickness of the jacket, very often welding is applied.

The alternative of RCC jacketing is FRP wrapping to save the time and cost, however, it should be noted that RC jackets feature some significant advantages over FRP wraps, providing enhanced strength and stiffness in both bending and shear, contrary to the latter that in its usual application (FRP wraps with the fibres perpendicular to the axis of the member) increases only the shear strength, the deformability and the confinement, but not the bending capacity.

Advantages and Disadvantages

With reinforced concrete jackets, a considerable increase in the strength, both flexural and shear, and the vertical bearing capacity of the member can be achieved. Furthermore, there is a significant **increase** in the ductility and the deformation capacity of the member through the confinement and the anti-buckling action of the new stirrups. The bearing capacity and the flexural strength of the member are enhanced, due to additional longitudinal reinforcement, whilst the improvement in the shear strength and the ductility are achieved through the additional transverse reinforcement. Furthermore, because of the increase in the section's dimensions, an increase in the member stiffness is also achieved.

11.3.7.2 Ferrocement Lining

Ferrocement lining as a section enlargement method for strengthening concrete structures is a precise and effective technique that enhances both structural capacity and durability. It involves applying a thin layer (typically 10–25 mm) of cement mortar reinforced with closely spaced steel meshes over existing concrete elements such as beams, columns, or slabs. This method increases the cross-sectional area, thereby improving load-bearing capacity and stiffness. The ferrocement layer acts as a confining jacket, enhancing compressive strength, providing tensile resistance, and improving crack control. Its excellent bond with the substrate and adaptability to complex shapes make it ideal for retrofitting, seismic upgrades, and repairs. The process includes surface preparation, mesh anchoring, and mortar application, ensuring full infiltration of the mesh for composite action. Materials used conform to ACI 549.1R-93 standards, including fresh Portland cement, fine washed sand, clean water, various steel meshes, and admixtures like superplasticisers and waterproofing compounds. Design considerations include mesh spacing (≤12 mm), bonding quality, and load-sharing mechanisms. Ferrocement offers a lightweight, cost-effective, and durable solution for structural enhancement.

11.3.7.3 Strengthening by Placing Additional Overlay of RCC

Placing an additional layer of reinforced concrete is another method to strengthen deficient deck slabs, enhancing their ability to withstand higher loads.

It is a well-established method for enhancing the load-bearing capacity of deficient deck slabs. This technique involves casting a new layer of concrete, reinforced with steel bars or mesh, directly over the prepared surface of the existing slab. The added reinforcement improves tensile strength, while the overlay increases compressive capacity and helps distribute dynamic loads more evenly across a larger surface area. This results in improved structural performance, enabling the slab to support heavier traffic loads than originally designed. Additionally, the overlay enhances durability and extends the service life of the structure by providing a protective layer against environmental and mechanical wear.

An additional reinforced concrete overlay can be applied to deficient floor slabs to enhance structural performance. This method improves tensile and compressive strength, enabling better load distribution and increased service capacity.

For placement of concrete overlay, the fabricated reinforcement shall be accurately placed, and upon completion of inspection, the concrete overlay shall be cast and uniformly distributed to achieve the required thickness. Adequate curing shall commence after finishing to ensure proper hydration and strength development.

11.4 Patch Repair (concrete)

At first, the defective concrete must be removed until sound concrete is found. If an exposed reinforcement bar exists, it must be cleaned and protected by painting with appropriate paint to avoid rusting. Next, the affected areas must be thoroughly cleaned and applied with a bonding agent. Finally, patching should be carried out using resin mortar.

For extensive repair, the use of formwork is required, and the mortar is pumped into the formwork.

Maintenance strategy

- Structural cracks not exceeding 5mm in width and minimal spalling of concrete - Epoxy injection and mortar application. Application of the FRP system.
- Corroded rebars and spalled concrete. Temporary concrete plaster with marine organisms. Corrosion treatment for rebars. Removal of plastered concrete and cleaning of the surface. Application of the FRP system.
- A portion of severely corroded rebars and spalled concrete. - Subject for rebars replacement and re-concreting of the damaged section. Application of the FRP system.

11.4.1 Methods of Crack Repair

The appropriate crack repair method should be selected based on the nature and cause of the cracking in the structure. If the cracks are primarily due to **drying shrinkage**, they typically stabilise over time and may not require immediate intervention. However, if the cracks result from ongoing foundation settlement, any repair will be ineffective unless the underlying settlement issue is first resolved.

Once the cause is identified, suitable repair techniques can be applied. The following methods are commonly used to address structural cracks, depending on their type, severity, and the environmental conditions surrounding the structure.

11.4.2 Common Methods for Crack Repair

For lightly-damaged members where cracks are observed, notches are cut along the line of cracks, then holes are drilled 150mm apart to insert the injection ports. Apply epoxy mortar along the notch and let it cure. Pump/inject the epoxy grout through the injection ports starting from the lowest port, going upward to the highest port.

For highly-damaged members where rebars are exposed and corroded, evaluate whether rebars should be replaced, or otherwise, clean and apply rust converter prior to application of epoxy mortar to their original geometry.

11.4.2.1 Repairs For Narrow Cracks (Epoxy Injection)

Epoxy injection is a precise method used to repair narrow cracks in concrete structures, with the capability to bond cracks as fine as 0.05 mm. The process involves setting up entry and venting ports along the crack, sealing the crack's exposed surfaces, and injecting epoxy under pressure. It's a proven technique for repairing cracks in various concrete structures.

- To prevent recurrence of cracks, it's crucial to address the underlying cause of the cracking before applying epoxy injection. If the root cause cannot be resolved, simple grouting work may not give a long-lasting solution.
- The ambient temperature can influence the application of epoxy injection, as certain epoxies are not effective in specific temperature ranges.
- Care shall be taken for actively leaking cracks or those that cannot be dried; moisture-tolerant epoxies must be used, as standard epoxies may not provide a reliable repair.

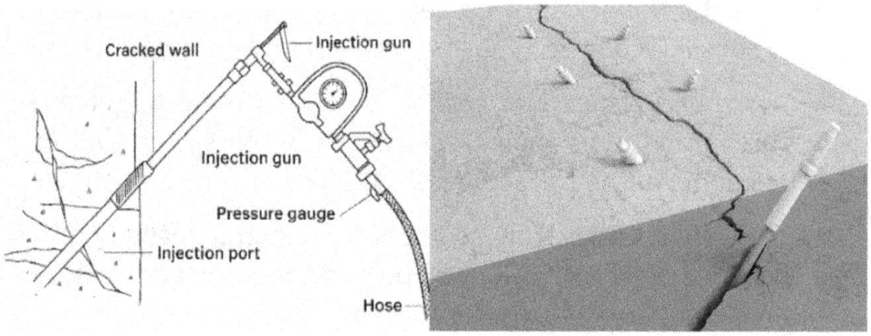

Figure 11.14: Procedure for Epoxy Injection

- **Clean the Cracks**

  Remove contaminants (oil, dirt, etc.) using vacuum, water, or cleaning solutions. Dry the crack thoroughly.
- **Seal the Surface**

  Apply a sealant (epoxy, polyester, or cementitious) to prevent epoxy leakage during injection. Use strippable sealers if appearance matters.
- **Install Entry & Vent Ports**

  Three methods:
  - **Drilled fittings**: Inserted into holes near the crack.
  - **Flush fittings**: Bonded directly over the crack.
  - **Seal interruption**: Leave gaps in the seal for injection access.
- **Mix the Epoxy**

  Mix components either in batches or continuously. Use only as much as needed before it begins to gel.

- **Inject the Epoxy**

  Use pumps or caulking guns. Inject from the lowest point upward (for vertical cracks) or end-to-end (for horizontal cracks). Avoid excessive pressure to prevent further cracking.

- **Remove Surface Seal**

  After curing, remove the seal by grinding or other suitable methods.

Limitations:

- Not suitable for actively leaking cracks unless moisture-tolerant epoxy is used.
- Requires skilled application and may be affected by ambient temperature.
- Flexible adhesives don't allow significant movement due to the concrete's restraint.

When the root cause cannot be eliminated, following options shall be explored:

11.4.2.2 Routing and Sealing: A Non-Structural Crack Repair Method

Routing and sealing is a preventive maintenance technique used to address non-structural cracks in concrete. These are cracks that do not affect the load-bearing capacity of the structure. The primary goal is to block moisture ingress, which can lead to corrosion of reinforcement, surface staining, and deterioration of concrete over time. This method is especially useful in environments where waterproofing and durability are critical.

Detailed Process

1. Routing (Crack Enlargement):

- The crack is mechanically widened using tools like a saw, grinder, or router.
- The groove is shaped into a V or U profile, which:
- Enhances sealant adhesion.
  - Allows for better sealant penetration.
  - Facilitates movement accommodation in the crack.

The dimensions of the groove are determined based on a) Crack width and depth, b) Type of sealant used and c) Expected movement or environmental exposure.

2. Cleaning:

i) The groove is cleaned using compressed air, vacuum, or brushes to remove:
- Dust
- Loose particles
- Moisture

This step is crucial for ensuring a strong bond between the sealant and concrete.

3. Sealing:

i) A flexible, durable sealant is applied to fill the routed groove.

ii) Common sealant materials include:
- Polyurethane – good elasticity and adhesion.

- Silicone – excellent UV resistance.
- Polysulfide – high chemical resistance.

iii) The sealant is designed to:

- Accommodate minor movements due to thermal expansion/contraction.
- Maintain watertightness over time.
- Resist environmental degradation.

Applications:

i) Horizontal Surfaces:

- Deck slabs
- Pavements
- Floors

ii) Vertical and Curved Surfaces:

- Retaining walls
- Columns and piers

Crack Types:

- Fine Pattern Cracks:
  - Often caused by shrinkage or thermal effects.
  - Typically shallow and widespread.
- Larger Isolated Cracks:
  - May result from minor settlement or drying shrinkage.
  - Still non-structural but requires targeted sealing.

Advantages:

- Cost-effective compared to structural repair methods.
- Quick application with minimal disruption to service.
- Improves durability by preventing moisture-related damage.
- Versatile – applicable to various surface orientations and crack types.

Limitations:

- Not suitable for active or structural cracks that require load transfer or reinforcement.
- Sealant performance is highly dependent on proper surface preparation and environmental conditions during application.
- May require periodic inspection and maintenance to ensure long-term effectiveness.

11.4.2.3 Gravity Filling

Low viscosity monomers and resins can be used to seal cracks with surface widths of 0.001 to 0.08 inches by gravity filling. The lower the viscosity then the finer the cracks that can be filled. The typical procedure is to clean the surface by air blasting and/or water blasting. Wet surfaces should be permitted to dry for several days to obtain the best crack filling. Water blasting followed by a drying time may be effective in cleaning and preparing these cracks. Cores taken from cracks can be used to evaluate the effectiveness of the crack filling. The depth of penetration of the sealant can be measured. Shear tests can be performed with the load applied in a direction parallel to the repaired cracks. For some polymers, the failure crack will occur outside the repaired crack.

11.4.2.4 Grouting

The wide cracks, particularly in thick concrete walls, may be repaired by filling with Portland cement grout. This method is effective in stopping water leaks, but it will not structurally bond cracked sections. The procedure consists of cleaning the concrete along the crack, installing built-up seats at intervals astride the crack, sealing the crack between the seats with a cement paint, sealant, or grout, flushing the crack to clean it and test the seal, and then grouting the whole area. Grout mixtures may contain cement and water or cement plus sand and water, depending on the width of the crack.

However, the water-cement ratio should be kept as low as possible to maximize the strength and minimize the shrinkage. Water reducers or other admixtures may be used to improve the properties of the grout. For small volumes, a manual injection gun may be used; for larger volumes, a pump should be used. After the crack is filled, the pressure should be maintained for several minutes to ensure good penetration.

11.4.2.5 Overlay And Surface Treatments

Fine surface cracks in structural slabs and pavements may be repaired using either a bonded overlay or surface treatment if there will not be further significant movement across the cracks. Unbounded overlays may be used to cover, but not necessarily repair, a slab. Overlays and surface treatments can be appropriate for cracks caused by one-time occurrences that do not completely penetrate the slab.

Low solids and low viscosity resin-based systems have been used to seal the concrete surfaces, including the treatment of very fine cracks. They are very much suited for surfaces not subject to significant wear. Bridge decks and parking structure slabs, as well as other interior slabs, may be coated effectively after cracks are treated by injecting with epoxy or by routing and sealing. Materials such as urethanes, epoxies, polyesters, and acrylics have been applied in thicknesses of 1mm to 50mm, depending on the material and purpose of the treatment. Skid-resistant aggregates are often mixed into the material or broadcast onto the surface to improve traction.

Dormant cracks in slabs can be repaired using an overlay such as polymer-modified Portland-cement mortar or concrete, or silica-fume concrete. Slabs with active (working) cracks may also receive an overlay, provided that joints in the overlay are aligned directly above the existing cracks. In highway bridge applications, overlay thicknesses as low as 30 mm have performed successfully. Suitable polymers include styrene-butadiene or acrylic latexes, and the polymer resin solids should constitute at least 15% by weight of the mix.

Stability Assessment for Structural Upgrade of Quay Wall

Chapter Summary

The assessment of structural stability and the definition of engineering requirements for strengthening the existing jetty/quay wall have become critical due to increased operational loads, modified dredge levels, and ageing structural components. Dredging-related reductions in passive soil resistance, greater free-standing pile length, higher lateral earth pressures, and the potential for both global and local failure modes demand a systematic evaluation and appropriate reinforcement. The study addresses key stability checks—including global failures such as deep-seated rotational slip and sliding/overturning, and local failures such as pile bending, shear, buckling, inadequate passive resistance, and anchor capacity—to ensure structural reliability under enhanced loading. Proposed strengthening measures include additional portal frames, batter piles, RCC overlays or voided slabs, RCC jacketing of beams and piles, relieving platforms, and tiebacks/ground anchors. Overall, the upgradation strategy prioritizes structural integrity, safety under revised operating conditions, and extended service life while minimizing disruption to port operations.

Learning Objectives

- Understand the need for stability assessment of existing jetty/quay walls under increased operational loads, and modified dredge levels
- Assess the impact of structural configuration and soil conditions on quay wall behaviour,
- Understand the engineering principles and applicability of various strengthening solutions
- Develop the ability to select appropriate strengthening methods based on site-specific conditions and operational constraints.
- Recognise anchorage principles

12.1 Stability Evaluation Framework

Upgradation or innovation of an existing wharf or quay wall requires a comprehensive stability assessment to ensure that the structure continues to perform safely, reliably, and durably throughout its service life. The assessment establishes the ability of the quay wall to resist all anticipated actions during operation, including environmental and operational loads characteristic of marine conditions. It systematically evaluates potential failure mechanisms covering geotechnical, structural, and functional aspects, with particular attention to risks such as excessive deformation, loss of bearing capacity, and overall instability. Based on the identified deficiencies, suitable strengthening and anchoring measures are developed to match the existing structural configuration and site-specific ground conditions, while maintaining constructability, cost effectiveness, and operational feasibility. The objective is to achieve a stable and well-balanced structural system capable of supporting safe and uninterrupted port operations over the long term.

A prerequisite to defining strengthening or upgrade measures is a thorough understanding of the existing condition of the quay wall. This includes an assessment of age-related deterioration, existing damage, material performance, and the expected structural response following modification. Equally critical is the evaluation of subsoil conditions, as quay wall stability is predominantly governed by soil behaviour. Factors such as historical dredging, seabed morphology changes, settlement, and reduction in passive resistance can significantly influence performance and must be rigorously assessed. This chapter describes how strengthening solutions can be developed to integrate effectively with the existing structure, minimise operational disruption, extend service life, and meet both current and future port operational requirements.

12.1.1 Priority and Severity of Investigations Details

The stability assessment of quay walls for port upgradation is fundamentally dependent on the quality and reliability of input data, particularly geotechnical, geomorphological, structural condition, bathymetric, and metocean investigations. These investigations constitute the core safety-critical inputs for design and therefore require the highest priority. Inaccuracies, omissions, or misinterpretation of such data can result in disproportionately severe consequences, including structural instability, reduced service life, unsafe operating conditions, or major operational disruptions.

To ensure safe design, cost efficiency, regulatory compliance, and long-term asset sustainability, investigations must be prioritised through a structured framework linked to the severity of potential consequences.

Priority defines the importance and sequencing of investigations during planning, design, and execution, with high-priority studies required at early project stages to establish reliable design parameters and reduce technical uncertainty.

Severity represents the magnitude of adverse impact on structural safety, serviceability, durability, port operations, or overall project viability if an investigation is omitted, inadequately performed, or incorrectly interpreted. In port engineering practice, investigations associated with high-severity risks inherently demand high priority, as failures arising from deficient inputs may be irreversible and lead to serious structural or operational consequences.

12.1.2 Necessity of Precise Investigation Stability Assessment

In continuity with the stability assessment and risk-based prioritisation framework described above, the accuracy and reliability of investigation results are fundamental to any quay wall stability evaluation. All design verifications and safety-critical decisions are directly dependent on the quality of input data. Deficiencies arising from assumptions, inconsistencies, or inaccuracies in investigations can lead to erroneous assessment of bearing capacity, settlement behaviour, load resistance, or residual service life, potentially resulting in unsafe designs, premature deterioration, or structural failure.

High-quality investigation results significantly reduce uncertainty, enable realistic modelling of soil–structure interaction and structural behaviour, and support sound engineering judgement. They also ensure that repair, strengthening, or anchoring measures are appropriately targeted, technically effective, and economically justified. Consequently, precise and well-documented investigation reports are essential to achieving safe, durable, and sustainable solutions for quay wall upgradation projects.

Details of Investigation Priority Based on Severity

Priority Level	**Investigation Category**
Priority 1 – Very High Severity	- Geotechnical Investigations - Structural and Condition Surveys - Bathymetric Survey - Metocean Studies
Priority 2 – High to Medium Severity	- Geophysical Investigations - Geomorphological Studies - Mathematical / Numerical Modelling Studies - Topographic Survey
Priority 3 – Medium / Regulatory	- Navigation Simulation and Manoeuvring Studies - Environmental and Marine Ecology Assessments

For effective risk management, all design inputs should be clearly classified as *primary* or *secondary* data. Primary data comprises information critical to structural safety and performance and must be obtained through proper, site-specific, and reliable investigations without compromise. In contrast, secondary data may be adopted where direct information is unavailable due to site constraints or practical limitations. Such data may be derived from credible external sources, historical records, or comparable projects; however, its applicability must be critically evaluated. The influence of secondary data on structural performance must be explicitly assessed, and appropriate conservatism or additional safety factors should be incorporated to address associated uncertainties. This structured approach allows risks to be managed systematically while maintaining overall safety, robustness, and design reliability

12.2 Major Quay Wall Failures Attributed to Geotechnical Deficiency

Experience from past port infrastructure projects demonstrates that a large number of quay wall and wharf failures are directly attributable to geotechnical deficiencies. These deficiencies commonly arise from inadequate site investigations, incomplete or inaccurate soil characterisation, or misinterpretation of ground behaviour. Typical failure modes include excessive settlement, reduction in passive resistance, global instability, and structural collapse, often influenced by dredging operations, scour, or long-term soil deformation.

These failures clearly highlight the critical importance of obtaining accurate, reliable, and site-specific geotechnical data as a primary input to quay wall assessment and design. They further emphasise the need to prioritise safety-critical investigations during the planning and design stages of port upgradation projects. Lessons learned from such cases indicate that shortcomings in geotechnical assessment cannot be adequately compensated through structural design measures alone and remain a dominant risk to long-term structural stability, durability, and safe port operations. Representative examples of such failures are discussed in the following sections.

12.2.1 Failure of the 1916 Gothenburg Harbour Quay-Wall

The quay wall in Gothenburg Harbour, Sweden, was a gravity structure founded on timber piles with granular backfill. The original subsoil consisted of deep soft clay, later reclaimed with granular fill to construct the quay wall. By 1916, the wall had been in service for several decades.

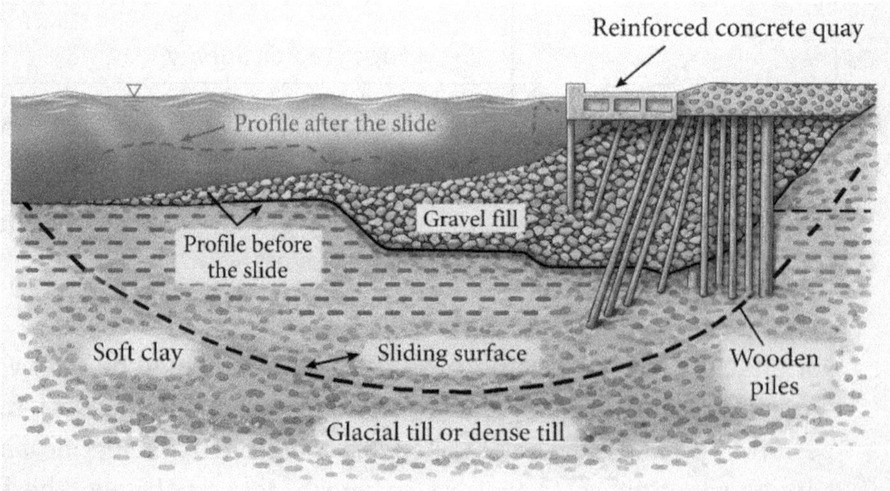

Figure-12.1: Representative Figure of Gothenburg Harbour quay wall failure, 1916.

In 1916, the quay wall suddenly failed by forward rotation and sliding. The observed failure profile indicated a circular slip surface through the soft clay, where shear stresses exceeded the undrained shear strength because of overloading from fill placed too close to the wall.

Key conditions leading to failure:

- Progressive increase in backfill height over time.
- Poor drainage causes excess pore-water pressure buildup.
- Weak, highly deformable soft-clay foundation.
- Limited understanding (at the time) of saturated lateral earth pressures.

The 1916 Gothenburg failure became a classic geotechnical case because it highlighted the importance of **undrained shear strength**, **pore-pressure effects**, **effective stress**, and the role of **creep deformation** in reducing long-term stability. It strongly influenced modern methods for earth pressure evaluation, slope stability, and undrained analysis.

12.2.2 Failure of Anchored Sheet-Pile Quay Wall – Berth WQ-7, Visakhapatnam Port (2004–2005)

Berth WQ-7 at Visakhapatnam Port was built as an anchored sheet-pile quay wall with an RCC relieving platform. The wall was installed on land, and the berthing basin was dredged after near completion. At the waterfront, HZ king piles were driven, and the sheet-pile head was anchored using steel tie-rods at regular spacing, connected to an AZ-type sheet-pile anchor wall placed behind the main wall in the passive zone.

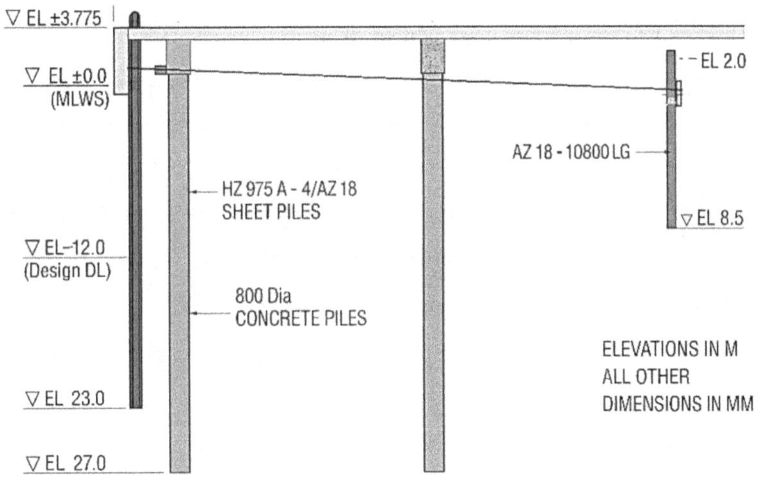

Figure-12.2: Anchored Quay Wall – Berth WQ-7, Visakhapatnam Port.

Failure Timeline and Initial Observations

Soon after completion, a section of the quay wall showed unexpected forward rotation at the base toward the waterfront. Post-construction study and evaluation suggested that the failure was geotechnical in origin. The governing mechanism was a basal heave of the soft seabed clay, triggered by inadequate passive resistance at the sheet-pile embedment depth. Local and global stability checks had not been verified at all critical points. The failure initiated at a localised kink in the alignment, which caused excessive tensile force in adjacent tie-rods. This led to progressive loss of anchor support and forward rotation of a larger portion of the wall.

12.2.3 Mud Slide Failure at 7th General Cargo Berth, Kandla Port (1990)

A major deep-seated mud-slide (slope failure) occurred in 1990 during the construction of the 7th General Cargo Berth at Kandla Port, India. The berth comprised a 65 m-wide RCC deck supported on 1.0 m diameter bored piles at 5 m × 5 m spacing, with a 190 m × 250 m back-stacking area designed for 200 kN/m² loading.

The subsurface consisted of soft, highly compressible marine clay extending to significant depth, underlain by stronger strata. Ground improvement involved sand drains and preloading to accelerate consolidation of the soft clay. The clay was characterised by:

a) Low undrained shear strength,

b) High compressibility and high water content, and

c) Very slow drainage.

These properties made the site highly susceptible to excess pore-pressure buildup during preloading. As the clay was soft and normally consolidated, substantial lateral movements were possible unless instrumentation provided continuous monitoring. However, many installed devices—piezometers, settlement plates, and lateral movement gauges—were non-functional or failed during the preload stage, eliminating critical early-warning capability.

Mechanism of Failure

With roughly 50% of the piles installed, the site exhibited large lateral movement of the clay mass, typical of a rotational mud-slide in soft, saturated soils. Root-cause assessment showed that the preload fill induced a rapid rise in pore-water pressure, reducing shear strength and initiating a deep-seated circular slip surface beneath the pile group.

Unable to sustain the increasing preload, the entire soil mass moved seaward, dragging the installed piles—consistent with undrained slope failure under surcharge in soft clay.

The failure caused significant structural displacement:

- Eight piles had head displacements up to 2.5 m, leading to breakage.
- ~80 piles displaced more than 30 cm, exceeding allowable limits and compromising capacity.
- Many piles suffered cracking, bending, or lost load-transfer function.
- The back area showed large lateral ground movement and differential settlement.

These observations confirm a deep circular slip, mobilising a large soil mass rather than a shallow surface slide. The stage-wise preload placement without effective pore-pressure monitoring significantly increased the risk of undrained failure.

12.2.4 Kobe Quay Wall Failure (1995 Great Hanshin Earthquake)

Background

The quay walls at Kobe Port were caisson-type gravity structures founded on sand fill and loose reclaimed soils, much of which consisted of loose to medium-dense cohesionless material highly prone to liquefaction. The broader waterfront was developed on reclaimed land, making the port especially vulnerable under strong seismic-induced loading.

Seismic Loading and Liquefaction Effects

The 1995 Great Hanshin (Kobe) Earthquake imposed strong horizontal ground accelerations (~0.54g) and vertical accelerations (~0.45g), triggering widespread liquefaction of the reclaimed fill beneath and behind the quay walls. This liquefaction caused excess pore-water pressure, a drastic drop in effective stress, and extensive lateral spreading toward the waterfront. Field observations confirmed sand boils, ground fissures, seaward ground flow, and deformation of waterfront facilities, all hallmarks of liquefaction-induced instability.

Structural Failure Mechanisms

Liquefaction-induced loss of soil strength caused the caisson quay walls to lose bearing capacity, resulting in major deformation patterns:

- Seaward displacement up to ~5 m
- Tilting of approximately 4° toward the sea
- Vertical settlement of 1–2 m behind the quay wall due to ground softening and compaction
- Formation of graben-like depressed zones behind the quay wall and cracking/separation of the deck structures
- **Collapse of bridge ramps** where backfill liquefied and the quay wall moved outward
- **Submergence of older piers** due to excessive settlement from liquefaction-induced ground weakening
- **Buckling of crane rails** and **toppling of cranes** caused by differential ground movement and rapid lateral spreading.

Design Lessons Learned from Kobe Quay-Wall Failure

The 1995 Kobe earthquake provided one of the clearest real-world demonstrations of how liquefaction, lateral spreading, and inadequate seismic detailing can critically compromise quay-wall performance. The fundamental design lessons learned are:

- The Kobe incident showed that **loose to medium-dense reclaimed sand** can liquefy completely under strong seismic activity, eliminating lateral and vertical soil support. Quay walls must therefore be designed for **post-liquefaction conditions**, not just traditional seismic earth pressures.
- Gravity and Sheet-Pile Quay Walls Must Be Checked for Stability with Zero Passive Resistance. After liquefaction, the quay-wall toe soil effectively lost shear strength.

Walls must be designed to remain stable when toe support vanishes, backfill liquefies, and Passive pressure drops to near zero.

- Backfill Quality and Treatment Are Critical to Seismic Performance. Kobe earthquake revealed major settlements (1–2 m) and formation of depression behind walls due to liquefaction and reconsolidation of untreated backfill. It emphasises the use of **engineered, densified, or improved backfill** and improving the loose fill foundation strata employing **deep soil mixing, stone columns,** etc., as per site requirements.

12.2.5 Barcelona Harbour Quay Wall Failure (2007)

The 2007 failure at the Prat Quay (first alignment) in Barcelona Harbour represents a significant modern case of gravity caisson quay-wall instability driven by geotechnical weaknesses and improper construction sequencing. The site was underlain by soft seabed muds of varying thickness, typically exceeding 5 m, underlain by layers of sand, gravel, and stiffer clay. These soft muds exhibited low undrained shear strength, high compressibility, and high water content, making them highly susceptible to deformation when subjected to rapid construction loading.

The quay wall was constructed as a monolithic reinforced-concrete gravity caisson, designed in accordance with Spanish maritime standards (ROM 0.2/90). Its stability depended on self-weight, seabed contact, and adequate foundation support. The design assumed construction within a sheltered environment. However, due to accelerated project deadlines, caisson installation advanced before completion of the breakwater, leaving the structure exposed to wave action and marine dynamic forces that it was not yet designed to resist. Compounding this, rapid backfilling was carried out from the landward side, imposing sudden surcharge loads. This increased vertical effective stress on the soft mud, generated excess pore-water pressures, and induced large strains that caused short-term strain-softening and loss of undrained shear strength. As a result, the foundation mud layer provided insufficient bearing resistance, leading to differential settlement, rotation or tilting of the caisson, and localised shear failure within the weak strata.

In the unsheltered condition, interaction between the caisson base pressure and the weak underlying mud produced a failure mode characterised by loss of support, onset of bearing failure or rotational slip, significant settlement, and structural cracking and distress along the quay alignment. The incident clearly demonstrates how inadequate consideration of soft marine subsoil, improper construction sequencing, and premature exposure to marine forces collectively undermined the gravity-type quay wall.

Overall, the Barcelona Harbour quay-wall failure stands as a key international case study emphasising the critical importance of rigorous geotechnical investigation, staged construction, controlled backfill placement, and completion of protective breakwater works prior to loading the quay structure. It highlights how rapid construction over weak foundation muds and early exposure to marine forces can make bearing failure and rotational instability unavoidable.

12.2.6 Malaga Harbour Quay Wall Failure (Spain, 2004)

In 2004, a major structural failure occurred during the construction of Container Quay No. 8 at Málaga Harbour, Spain. The incident involved the collapse of a quay wall composed of concrete caissons. The failure took place during construction, prior to commissioning, and was directly linked to geotechnical deficiencies and construction-sequence issues.

Geotechnical investigations identified soft, highly compressible clayey strata beneath the caisson foundations, which were inadequate to support rapid construction loading without proper consolidation or staged loading. The applied caisson loads generated immediate shear stresses that exceeded the undrained shear strength of the soil, resulting in global instability.

During the design stage, key shortcomings included the underestimation of the undrained response of soft marine clay, the absence of measures for pore-pressure dissipation, lack of provision for controlled staged loading, and inadequate modelling of soil–structure interaction for the caisson units. Consequently, insufficient geotechnical characterisation and drainage planning significantly contributed to the failure mechanism.

The collapse resulted from a combination of inadequate consideration of time- and stress-dependent loading effects on soil shear strength under rapid construction, together with insufficient geotechnical design provisions. During construction, backfill was also placed rapidly behind the caisson wall, adding further lateral and vertical loads. This rapid loading induced undrained shear failure in the soft muds,

- increasing horizontal earth pressures behind the caissons, and
- reducing the overall stability envelope of the wall and its foundation.

This sequence of actions created conditions favourable for a bearing-capacity failure coupled with rotational instability. The 2004 Málaga Harbour quay wall collapse was fundamentally a geotechnical instability event triggered by the presence of soft clayey mud foundations, rapid backfill placement, and insufficient consideration of the behaviour of marine soft soils.

In view of the above considerations, the proposed upgradation and strengthening strategy shall generally satisfy the following requirements:

12.3 Design Requirements

Assessment of Existing Structural System: Strengthening methods must align with the materials, geometry, and construction techniques of the existing quay wall. Proposed interventions—such as structural anchors, ground anchors, cut-off systems, or geo-media reinforcement—should be selected only after confirming that they suit the site-specific geotechnical, hydraulic, and geological conditions. This ensures that new additions function effectively without inducing unwanted stresses or incompatibilities.

Minimal Disturbance to Existing Elements: Upgradation works should be planned to cause minimal disruption to the existing quay structure. This includes avoiding unnecessary demolition, limiting intrusive activities, and preserving structural components that are functioning satisfactorily. A low-impact design helps maintain structural integrity and reduces operational downtime.

Environmental and Operational Feasibility: All strengthening options must be environmentally acceptable and feasible within the restricted working spaces typical of port environments. The design should consider nearby utilities, buried services, crane rails, pavements, and other port infrastructure to ensure smooth coordination and prevent conflicts during execution.

Economic Justification: Any proposed modification must demonstrate clear economic benefit, showing that the cost of the upgrade is justified by improvements in performance, reliability, and extension of service life. Efficient designs that maximise reuse of existing structural components and minimise construction time are preferred.

Local and Global Stability of System: The upgraded quay wall system must meet all local (component-level) and global (overall system-level) stability requirements. It must remain safe under operational, environmental, and accidental loading conditions, ensuring long-term reliability and structural safety.

i. **Performance Requirements**

The performance requirements shall be based on project-specific criteria; however, a broad outline is presented below;

Structural Stability: Structural Stability: The upgraded structure shall be designed to safely withstand all applicable permanent, variable, and operational loads—including surcharge effects due to port activities—in accordance with the requirements of relevant design standards such as Eurocode 1 (EN 1991) for loading and Eurocode 7 (EN 1997) for geotechnical design. Structural analysis and design shall comply with material-specific codes, including Eurocode 2 (EN 1992) for concrete structures or Eurocode 3 (EN 1993) for steel structures, as applicable.

Lateral and vertical displacements shall remain within allowable serviceability limits as defined in EN 1990 (Basis of Structural Design), ensuring no adverse impact on structural integrity or operational performance. Total and differential settlements must be controlled in accordance with the geotechnical criteria specified in EN 1997-1, as well as port-specific guidelines such as PIANC recommendations, to prevent structural distortion, functional impairment, or global instability.

Foundation Integrity: The interface between the quay wall structure and its foundation system shall maintain sufficient shear resistance, stiffness, and overall load-transfer capacity throughout the entire design life. The foundation elements must comply with the geotechnical design principles outlined in Eurocode 7 (EN 1997-1), ensuring that limit states related to shear failure, sliding, overturning, and deformation are not exceeded under any combination of permanent, variable, environmental, or accidental loads.

No reduction in allowable bearing capacity or axial/lateral resistance shall occur as a result of dredging operations, local or global scour, tidal or current-induced erosion, or long-term soil degradation processes such as consolidation, creep, or strength loss. Evaluation of hydraulic and scour effects shall follow recognised guidelines such as PIANC recommendations and EN 1997-2 for ground investigation and soil behaviour assessment. Appropriate protective measures must be incorporated where necessary to ensure the long-term stability and performance of the quay wall–foundation system.

12.4 Effects of Additional Deepening on Structural Stability

Dredging, which involves removing soil or sediment from the seabed or water body in front of a quay wall, is often necessary for deepening waterways to accommodate larger vessels. However, this activity can significantly impact the stability of the retaining structure, especially if not carefully planned and analysed. Deepening or dredging in front of a quay wall can change how the soil behaves and can reduce the stability of the structure. Here's how it affects the wharf:

i) More Side Pressure on the Wall

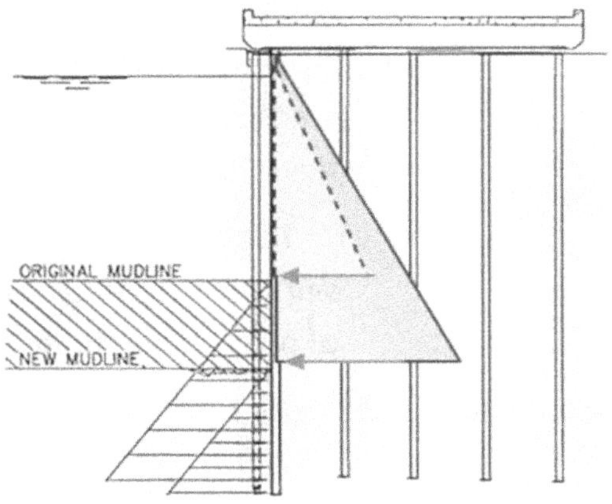

When the depth of retention behind the quay wall becomes higher due to dredging, the sideways (lateral) pressure on the wall increases as shown in Figure-12.3. More soil height means more force pushing outward on the wall, which can affect its stability and must be carefully considered during design

Figure 12.3: Increase in Lateral Earth Pressure on Quay Wall due to additional Dredging and Deepening

ii) Less Soil Support in Front (Lower Passive Resistance)

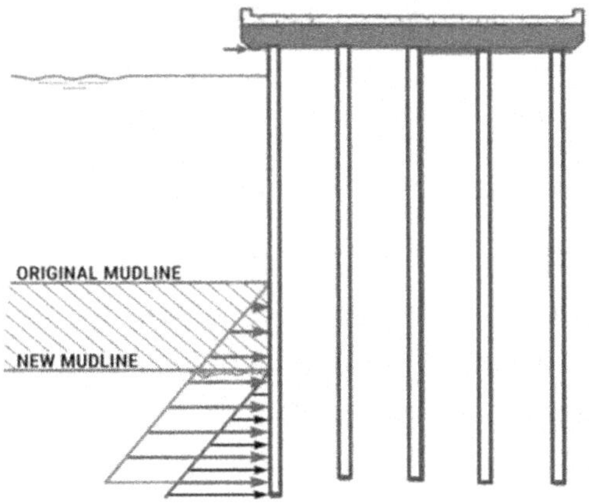

Dredging removes soil in front of the piles or quay wall. With less soil available, the structure loses the passive lateral resistance that normally helps keep it stable and prevents excessive movement. As a result, the wharf becomes more prone to sideways displacement and reduced overall stability. Engineers may need to introduce additional reinforcement or support measures to compensate for this loss.

Figure 12.4 : Decrease in Embedment Depth and Less soil support due to additional Dredging and Deepening at Quay Wall

iii) Reduced Safe Bearing Capacity (SBC)

The soil's capacity to support structural loads may decrease when dredging alters the depth or stress conditions around the foundation. A reduction in soil overburden near pile toes or along the embedded length can lead to a lower safe bearing capacity (SBC). Proper geotechnical investigation is required to assess these changes and confirm that the structure remains adequately supported. Any loss in soil capacity shall be compensated through suitable constructive adjustments or by providing a cut-off arrangement to retain the required soil support

iv) Increase in Free-Standing Length of Piles

When more soil is removed, a greater portion of the pile becomes exposed above the seabed. This leads to three major concerns:

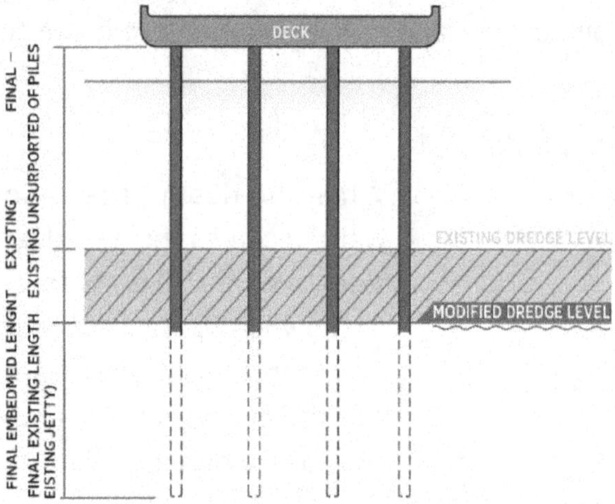

Figure 12.5: Increase in free-standing length due to additional dredging and deepening

i) Buckling Risk

A longer exposed or free-standing length of piles increases their susceptibility to bending under vertical loads, a failure mode known as buckling.

i) Higher Bending Moments

With an increased free-standing length of piles above the seabed.

The piles experience larger bending moments because lateral operational forces—arising from mooring, berthing actions, waves, currents, and soil pressures—act over a greater unsupported length, thereby amplifying the bending effects on the pile.

iii) Loss of Stiffness

Less soil support means the pile becomes more flexible and can sway more easily under lateral loads from mooring and environmental loads like wind, waves, berthing and mooring loads, and loads due to cranes, wind, or earthquakes.

These effects reduce the pile's ability to stay upright and resist sideways movement.

iv) Rotational Failure due to Decrease in Lateral Support

Rotational failure occurs when the reduction of lateral soil support in front of the quay wall compromises its ability to resist overturning. During dredging, the soil at the toe of the wall—which normally provides essential stabilising pressure—is removed.

This loss of passive resistance decreases the opposing forces that counteract the active earth pressures exerted by the retained soil behind the wall. As a result, the wall becomes more susceptible to forward rotation or overturning toward the dredged side. Since the driving forces behind the wall remain unchanged (or may even increase), while the resisting forces in front diminish, the global equilibrium of the quay wall is disturbed. Ensuring stability under these altered conditions requires a detailed geotechnical reassessment, verification of rotational safety factors, and implementation of strengthening or soil-retention measures to mitigate the increased risk of rotational failure.

These factors underline the importance of careful planning and execution when performing dredging operations near quay walls. It's crucial to consider the potential impacts on existing structures and to take appropriate measures to ensure their stability and integrity. This might include reinforcing the piles or using different construction techniques to mitigate the risks.

12.5 Structural Stability Assessment of Quay Wall

Stability analysis is an integral part of structural design, which is a methodical investigation of the stability, strength and rigidity of structures. The primary purpose of a structure is to support the intended loads and transmit it safely to the ground. Therefore, the design process of a wharf or quay wall system involves a set of calculations to identify the response of the structure against the design loads that will be required to resist during its lifetime. In this process,

i) We must verify the geometry of the structure achieving equilibrium under the design conditions, verifying the requirements of all applicable loads, bending moments and shear forces.

ii) We must assess whether the existing sections, material grades, layout configuration, and revised component dimensions are effective and appropriate for the intended purpose, while ensuring overall economy

iii) We shall verify whether the existing structural unit can withstand various load combinations without failure—such as collapse, damage, excessive deflection, or vibration—throughout its intended service life.

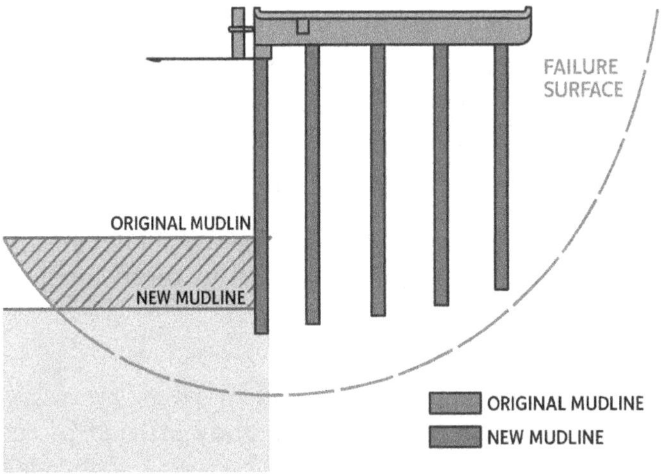

iv) Finally, the installation and execution process for the reconfigured quay wall is finalised based on the overall system requirements

In this stage, we work out how to upgrade the existing structure and understand what changes will happen when we reconfigure it. We look at things like the final layout, what dimensions can realistically be changed, and how much space is available for any rebuilding. Based on these factors, we explore one or more possible design options. We then choose the option that gives the best overall and most cost-effective solution. The main goal is to make sure all the redesigned structural parts meet the new functional needs safely and efficiently.

Figure 12.6: Rotational Instability due to Additional Dredging

12.5.1 Global and Local Stability Verification

Quay walls, commonly constructed as earth-retaining systems, are designed to withstand soil, water, and operational surcharge loads. To ensure safe and reliable performance during construction and throughout their service life, two distinct levels of stability verification are required:

- *Global Stability Verification*
- *Local Stability Verification*

These two categories address fundamentally different failure modes. A quay wall may meet all element-level design requirements yet still experience overall instability if global stability is not confirmed. Therefore, both verifications are essential to achieve a complete assessment of structural and geotechnical safety.

i) Global Stability

Global stability refers to the overall safety of the entire soil–structure system, accounting for large-scale or deep-seated failure mechanisms that may extend far beyond the immediate area of the structure. This assessment verifies whether the quay wall, together with the passive soil zone below the mudline, functions as a stable unit under all critical loading conditions.

Its main Purpose is to ensure that the combined soil–structure system will not undergo catastrophic movement such as rotation, sliding, overturning, or deep subsurface failure. Typical Global Failure Modes are;

ii) Deep-Seated Sliding or Rotational Failure

Deep-seated sliding or rotational failure refers to the movement of a large soil mass surrounding a quay wall structure (configured as a bulkhead, wharf, sheet pile wall, or anchored wall arrangement) along a potential slip surface. This slip surface may be circular (typical in homogeneous soils) or non-circular (common in layered or complex soil profiles). Such failures are critical because they involve global instability of the entire soil–structure system, not just local failure of the structural element.

A deep-seated slip surface may take different paths depending on soil conditions, dredge levels, water levels, and applied loads. The three most common paths are explained below.

The reason behind the slip failure is;

i) The quay wall acts as part of the sliding body instead of resisting it, and the retained soil and the structure rotate or slide as a single block. The whole structure may rotate forward around a deep pivot point.

ii) Due to Insufficient embedment depth, inadequate passive resistance in front of the structure or weak or soft soil strata extending deep behind the wall. Over-consolidated clay transitioning to soft clay due to strain-softening.

i) High surcharge loads or groundwater pressures applied, which are not accounted for during the design process.

iii) The whole structure may rotate forward around a deep pivot point.

iv) Due to excessive dredging or excavation in front of the wall during upgradation, the wall may experience loss of toe support, forward rotation, and excessive lateral displacement.

Global failure usually occurs suddenly, involves a large mass of soil, and results in major structural collapse. Even if the quay wall sections, anchors, and connection components meet all design requirements, the system can still fail if the surrounding soil mass does not remain globally stable. Deep-seated rotational failure behind the wall (often analysed using methods such as Bishop, Janbu, or Morgenstern–Price) as described below.

12.5.2 Analysis Methods

12.5.2.1 Bishop Method (Modified Bishop)

A commonly adopted approach that evaluates stability by performing moment equilibrium along an assumed circular slip surface. The method estimates the Factor of Safety (FS) against rotational failure by comparing the mobilised shear strength to the shear stresses required for equilibrium.

12.5.2.2 Morgenstern–Price, Janbu, and Spencer Methods

These advanced limit-equilibrium methods are used where soil stratigraphy is complex, loads are irregular, or when non-circular slip mechanisms are expected. They satisfy both force and moment equilibrium and provide more robust solutions for layered or heterogeneous soil conditions.

The Purpose of the Check is to confirm that the entire soil–structure system—including the retained mass, foundation zone, and passive resistance zone—remains stable and does not undergo global rotational or translational displacement. This ensures the bulkhead performs reliably under operational and extreme loading conditions.

i) Key Stability Requirement

A global Factor of Safety is defined as:

$$FS = \frac{\text{Resisting Forces}}{\text{Driving Forces}}$$

Typical acceptable values:

- **FS = 1.3 to 1.5** depending on design standards, soil variability, and loading conditions.

This requirement ensures that the resisting shear strength along the slip surface is sufficiently greater than the driving forces generated by earth pressures, surcharge loads, and hydrostatic effects.

ii) **Role of Soil Strength in Deep-Seated Failure**

Shear Strength Determines Stability

Deep-seated failure occurs when the shear strength of the soil is insufficient to resist the driving forces (like earth pressure, water pressure, and surcharge loads, etc.). If the soil's shear strength is low, the factor of safety against sliding or rotational movement decreases, making failure more likely.

iii) **Local Stability**

Local stability assesses the structural integrity of the components of quay wall and wharf and the immediate soil zones surrounding them. It focuses on ensuring that individual elements can resist the design loads without experiencing overstress, deformation, or localized soil failure.

Local failure modes refer to component-level or near-field failures that occur within the structural system or in the immediate surrounding soil. Unlike global failures (which involve large soil masses and deep slip surfaces), local failures affect individual structural elements or the small zones of soil interacting directly with the quay wall. Its main purpose is to verify that each component of the quay wall system is structurally adequate and that the soil directly supporting these components can safely resist applied forces.

The key local failure modes are described below.

12.5.2.3 Bending, Shear, or Local Buckling of the Quay Wall Structure

Quay wall components are subjected to lateral earth pressures and hydrostatic pressures. If these applied loads exceed the structural resistance of the section, the following failure modes may occur:

i. **Bending Failure:** When the applied bending moment exceeds the section's moment capacity, resulting in yielding or formation of a plastic hinge

- Bending Capacity Check : $\boldsymbol{M_{Sd} \leq M_{Rd}}$
- The design bending moment must not exceed the elastic or plastic moment resistance of the section.

ii. **Shear Failure: Occurs when horizontal shear forces exceed the shear capacity of the section, causing shear distortion or web tearing.**

- Shear Capacity Check : $\boldsymbol{V_{Sd} \leq V_{Rd}}$
- The applied shear force must remain within the allowable shear resistance.

iii. **Combined Bending–Shear Interaction**

Required when the shear force exceeds 50% of the shear capacity.

$$\frac{M_{Sd}}{M_{Rd,red}} + \frac{V_{Sd}}{V_{Rd}} \leq 1.0$$

iv. **Local Buckling: Freestanding quay wall components may experience local buckling under compressive stresses, reducing section stiffness and potentially increasing deflections or causing failure.**

Local Buckling /Slenderness Check

$$\lambda = \frac{b}{t} \leq \lambda_{limit}$$

where b= width, t= thickness of structural element.

Buckling must be prevented by ensuring that slenderness limits and compressive stress limits are not exceeded

v. **Deflection Control**

$$\delta \leq \delta_{allowable}$$

Typical limit: H/150 to H/200, where H = retained height.

vi. **Insufficient Passive Resistance at the Toe of the Structure**

The embedded portion of the quay wall elements relies on passive earth resistance to resist horizontal forces. Local failure occurs when the toe:

- is embedded in weak, loose, or soft soils,
- does not penetrate deep enough to mobilize full passive pressure, or
- is subjected to unfavourable hydraulic conditions.

vii. **Passive Earth Pressure Mobilization : $P_p \geq P_{req}$**

Further, compute required embedment using Free-earth support method, Fixed-earth support method, Blum method or Equivalent beam method to ensure:

$$d_{embed} \geq d_{min} + \Delta$$

where Δis a safety allowance (10–20%).

Confirming that available passive resistance exceeds the required reaction. Inadequate passive resistance may cause;

i) Forward displacement or rotation of the wall,

ii) Loss of stability during excavation and dredging

iii) Reduced ability to carry anchor or external tie forces.

Therefore, adequate toe penetration and verification of passive resistance are critical for local stability.

- Local Shear Failure in the Soil Around Anchor Piles or Anchor Blocks

viii. **Anchor piles or deadman systems transfer lateral forces from the quay wall into the surrounding soil. Local failure may occur if the soil:**

- cannot resist the applied tensile force due to lateral loads
- shears along a shallow wedge or plane, or

- experiences progressive yielding.

ix. Shear Resistance Check

In case of Block anchor or Deadman

$$T_{anchor} \leq R_{shear}$$

where:

- T_{anchor}= anchor design load,
- R_{shear}= shear resistance along potential wedge.

x. Failure Wedge Verification

Evaluate geometry using Rankine/Coulomb failure surfaces or Broms' method.

xi. Bearing Resistance of Anchor Block / Pile

$$q_{applied} \leq q_{ult}$$

Ensure block bearing pressure does not exceed soil bearing capacity.

xii. Required Embedment Depth of Anchorage Zone

$$z_{emb} \geq z_{required}$$

Passive resistance may be impaired due to the presence of loose sands, soft clays, or raise in the water table or construction disturbance in the soil. Therefore, Passive Resistance Reduction Factors shall be applied for loose sands, soft clays, or submerged soils and construction disturbance.

Local shear failure generally leads to excessive anchor displacement, or loss of anchor effectiveness and potential forward movement of the wall. The design must therefore consider the shear capacity of the soil mass mobilised around the anchor system, whether it is a wall, pile or block.

xiii. Importance of Global and Local Instability Checks

A quay wall might pass all the individual design checks—like piles, the wall itself, beams, or anchors—but still end up unsafe if its overall (global) stability isn't good enough. On the flip side, the wall might look fine from a global stability point of view, yet still have weak spots locally where certain elements could get overstressed or even fail. So, both global and local stability checks go hand in hand, and you need to satisfy both to be confident that the quay wall will perform safely and reliably as a complete system.

12.6 Options for Upgrading Quay Wall and Wharf

The methodology for the upgradation of the quay structure shall essentially involve a robust design and construction method. While deepening the basin, it shall account for the technique for integrating and strengthening the existing structure, avoiding it from being adversely affected by the construction activities.

If dredging is to be carried out in front of an existing quay wall, a case-by-case assessment shall be performed to determine whether deepening the harbour bottom and reinforcing the waterfront structure is technically feasible and economically viable. Where applicable, the required additional dredging depth shall be determined based on the following factors.

i) **Initial Feasibility Check:** First, an assessment shall be carried out to verify whether the required deepening of the harbour bottom is technically feasible in view of the existing seabed level.

ii) **Structural Suitability Assessment:** Subsequently, an assessment shall be performed to determine whether the design and current condition of the waterfront structure are suitable for operation with a deeper harbour bottom. For this purpose, available design drawings, structural calculations of the quay wall, and results of soil investigations conducted during the original construction phase shall be reviewed. Where necessary, additional soil investigations may be undertaken. Such investigations may identify more favourable soil properties resulting from long-term consolidation and improvement of subsoil characteristics since construction.

iii) **Economic Evaluation:** Finally, the economic viability of the required strengthening measures shall be evaluated by comparing the strengthening costs, in relation to the remaining useful life of the existing waterfront structure, with the costs of constructing a new structure.

The stability and serviceability of the waterfront structure shall subsequently be verified, taking into account the actions arising from the revised load conditions and the increased theoretical harbour bottom level. This verification shall consider the effects of additional earth pressures, hydrostatic forces, and any other load components resulting from the increased dredging depth. Where stability analyses of the existing structure and/or the original design drawings are unavailable or incomplete, a conservative approach may be adopted. In such cases, the additional loads imposed on the wall as a consequence of the increased harbour depth may be compensated by appropriately reducing the imposed operational loads acting on the waterfront structure, in order to maintain adequate safety and performance levels.

12.6.1 Design of Strengthening Measures

A variety of strengthening measures are available to enable quay walls to withstand the increased loads associated with the deepening of the harbour bottom. One effective measure consists of reducing the active earth pressure acting on the quay wall by improving the properties of the backfill material behind the structure. The active earth pressure on a waterfront wall can be reduced, for example, through backfill compaction, installation of soil reinforcement, or the application of grout injection techniques to stabilise and strengthen the retained soil.

12.6.1.1 Soil compaction

Loose, non-cohesive soils may be compacted using vibro-flotation techniques. The spacing of the vibration points (grid spacing) shall be determined based on the in-situ subsoil conditions and the degree of improvement in compaction required. Higher target compaction levels and finer soil grain sizes necessitate closer spacing of the vibration points. As general guidance, an average grid spacing of approximately 1.80 m may be adopted. Where soil replacement is carried out, compaction shall extend throughout the full depth of the newly placed soil, reaching down to the base of the excavation.

Vibro-flotation shall be applied over the entire zone of the active earth pressure sliding wedge in front of the structure and shall extend sufficiently beyond the critical active earth pressure failure plane, measured from the theoretical toe of the quay wall. Where uncertainty exists regarding the location of the failure surface, alternative curved or polygonal failure mechanisms shall also be assessed to confirm that the compaction zone is of adequate extent. During vibro-flotation, temporary liquefaction occurs in the immediate vicinity of the vibrating probe; however, subsequent densification is achieved through the action of soil surcharge. Consequently, the effectiveness of compaction depends significantly on the magnitude of the applied surcharge. The near-surface soil layer, typically about 2 to 3 m thick, can only be adequately compacted if a temporary surcharge, such as placed fill, is provided during the compaction process.

Vibro-flotation may also be employed as a method for the subsequent strengthening of existing waterfront structures. In such cases, the execution procedure shall be carefully selected to ensure that temporary, localised soil liquefaction does not adversely affect the stability of the quay wall. Experience indicates that loose, fine-grained non-cohesive soils, including fine sands, are particularly susceptible to extensive and prolonged liquefaction. Proper compaction of these soils can significantly reduce liquefaction risk, especially in seismically active regions.

12.6.1.2 Strengthening with Stone Columns

Stone columns are formed by installing columns of crushed stone or gravel into predrilled or driven cavities within the backfill area. These columns partially replace the in-situ soil to the installed depth and provide increased stiffness and shear strength compared with the natural ground.

Stone columns improve the shear strength and settlement behaviour of the soil. However, they are generally not suitable for heavily loaded foundations in very soft soils with an undrained shear strength of less than approximately 2 t/m^2. The performance of stone columns depends on the lateral confinement provided by the surrounding soil, which directly influences their stiffness and settlement characteristics.

During installation, a hollow casing or pipe is driven into the ground, displacing the soil laterally and compacting the adjacent subsoil. The cavity is then infilled with granular material, which is compacted in stages to form closely spaced stone or aggregate columns. This process increases the overall stiffness of the ground, enhances bearing capacity, limits post-construction settlements, and improves soil stability under seismic loading.

The effectiveness of stone column improvement depends mainly on:

- the diameter of the stone columns,
- the depth and spacing of the columns, and

- the grading and quality of the granular backfill material.

For offshore or submerged applications, stone columns are typically installed using a vibratory system equipped with a vibro-hammer to drive and subsequently extract the hollow pipe casing.

12.6.1.3 2.14.5 Soil stabilisation by Grout Injection

Where readily permeable, non-cohesive soils such as gravel, gravelly sand, or coarse sand are present within the passive earth pressure zone, soil stabilisation can be achieved through grout injection. In non-cohesive soils with lower permeability, stabilisation by means of high-pressure injection may provide a viable alternative. Chemical stabilisation methods may also be applied, provided that the selected stabilising agent is compatible with the chemical properties of the pore water and is able to cure effectively under site conditions.

In general, chemical stabilisation is associated with high costs and is therefore typically limited to localised applications with specific boundary conditions or stringent construction schedule requirements. It should be noted that stabilisation of the in-situ soil by injection creates a rigid zone that may preclude any future deepening of the harbour basin and/or the installation of a new waterfront structure in front of the existing wall.

A fundamental requirement for all injection-based stabilisation methods is the provision of an adequate temporary surcharge. This surcharge must be applied prior to injection works and may be removed following completion of the stabilisation, as applicable.

12.6.2 Relieving Platform for Quay Wall Stabilisation

The construction of a relieving platform behind a quay wall is an important engineering measure used to enhance the performance and stability of waterfront structures, particularly during port upgrades or expansion works. As ports accommodate deeper berths, higher operational loads, and larger vessels, the structural demands on quay walls increase considerably. A relieving platform functions as a secondary structural system that intercepts, redistributes, and transfers loads away from the quay wall, thereby reducing the stresses acting directly on it.

12.6.2.1 Functional Objectives of a Relieving Platform

A primary role of the relieving platform is to mitigate the lateral earth pressure exerted by the backfill soil. This reduction becomes especially important when the quay wall height is increased or when substantial live load surcharges—such as those from heavy vehicles, handling equipment, or stacked cargo—are applied behind the wall. By limiting the lateral push on the quay wall, the platform helps maintain structural integrity, prevents excessive deflection, and reduces the risk of wall overstressing, particularly in situations where deepening of the quay or intensification of operations places additional demands on the retaining system.

12.6.2.2 Support for Operational Infrastructure

Beyond load reduction, the relieving platform also provides a stable support base for critical operational infrastructure associated with wharf upgrades. It can accommodate equipment such as quay cranes, utility lines, mooring and berthing systems, and other heavy port installations. In doing so, it ensures that operational loads are efficiently transferred to the supporting ground system rather than being imposed directly on the quay wall, which may not have been originally designed to carry such magnitudes of surcharge.

The platform's performance is further enhanced through the incorporation of robust vertical supports such as barrettes or piles, which anchor the structure into deeper, stronger soil strata. These elements increase the overall lateral stiffness of the quay system and provide significant resistance against horizontal forces arising from mooring impacts, berthing loads, wave actions, and potential seismic events. As a result, the combined quay wall–platform system demonstrates improved resilience under both static and dynamic loading conditions.

12.6.2.3 Structural Configuration and Design Considerations

From a design perspective, the relieving platform typically comprises a reinforced concrete deck supported by deep foundations, structurally integrated with the quay wall to ensure coordinated load sharing and controlled deformation. Adequate drainage measures are essential to prevent hydrostatic pressure buildup behind the wall, and the backfill must be appropriately managed to maintain long-term performance.

Overall, a relieving platform represents a multi-functional engineering solution that enhances quay wall safety, operational capacity, and longevity. By reducing lateral earth pressures, providing a stable base for heavy port equipment, and strengthening the system through deep-founded supports, it ensures that the quay wall remains structurally robust even under increased maritime and operational demands. Its implementation reflects a forward-looking engineering response tailored to modern port infrastructure requirements.

12.6.3 Reinforcing Jetty Structures Using Batter Piles

In places where the soil near the jetty is loose or soft for quite a depth, the structure can be made stronger by using batter piles—these are piles installed at an angle into deeper, firmer ground. They're especially useful for handling sideways (lateral) forces caused by things like waves, currents, and ships docking, which can put a lot of lateral loads and stress on the jetty.

Batter piles—inclined piles typically installed at angles ranging from 1H:6V to 1H:3V—are designed to resist significant lateral loads. Unlike vertical piles that primarily carry vertical forces, batter piles are oriented so they can convert horizontal forces into axial compression and shear along their length. This makes them much more effective in resisting sideways movement. Because of this efficiency, structures often require fewer batter piles than vertical ones, and the presence of batter piles helps reduce overall lateral displacement under load.

The inclination angle of batter piles is designed to help them tap into the horizontal part of the axial force, which gives them their strength against lateral loads. Vertical piles, on the other hand, rely more on bending and shear within the pile itself. Batter piles can also handle both pushing (compression) and pulling (tension) forces, depending on how the loads act.

When the angle of the batter is small, bending in the pile also helps resist lateral forces. To understand how these piles behave under such loads, engineers use various analysis methods using soil structure interaction, which treats the pile like a beam resting on elastic soil.

Advantages of the Use of Batter Piles

- **Better Sideways Load Handling:** Their angled setup of batter piles makes them great at resisting lateral forces from waves, wind, and docking impacts. Batter piles help keep the structure more stable with minimal lateral deflection.
- **Fewer Piles Needed:** Because batter piles are much more efficient at carrying lateral loads, fewer piles are typically required to achieve the same level of lateral stability compared to a group of vertical piles. In addition, they transfer the loads at jetty heads into deeper strata, bypassing weaker surface soils and anchoring into stronger layers. This deeper engagement not only increases the overall load-carrying capacity but also enhances the lateral stability of the structural system
- **Flexible Design Options**: The batter angle of a pile can be adjusted in multiple directions, depending on the structural configuration, site geology, and expected load conditions, to achieve optimal performance in all directions—lateral, longitudinal, and vertical.

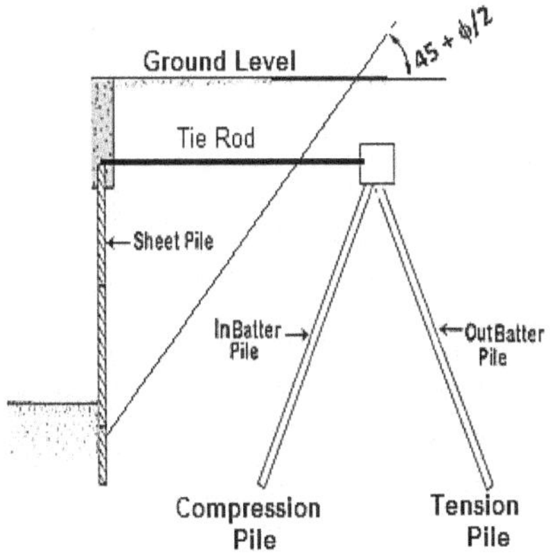

Figure 12.7 (a): A-frame Anchor Pile with Batter Piles

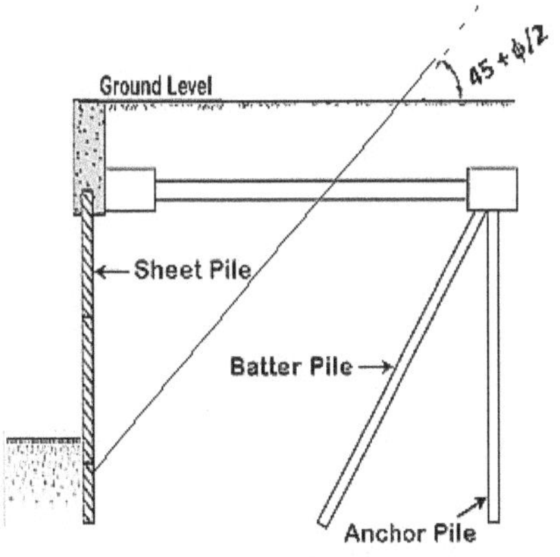

Figure 12.7 (b): A-frame Anchor Pile in combination of Batter Pile & Vertical Pile

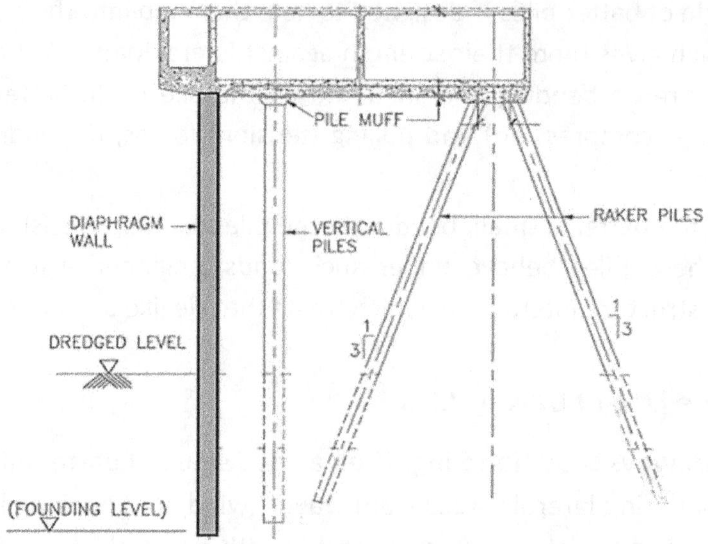

Figure 12.7 (c): Anchor Pile Arrangement in a Quay Wall

12.6.4 Müller Verpress (MV) Pile System

Use of MV Piles for Enhancing Retaining and Mooring Capacity of Wharf Structures

The Müller Verpress (MV) system is a proprietary deep foundation and anchoring technique developed and executed by Soetaert, a specialist geotechnical contractor within the Jan De Nul Group. The system is designed to provide exceptionally high tensile and compressive load capacities, making it particularly suitable for heavy civil, marine, and port infrastructure projects where conventional pile or anchoring solutions are insufficient or impractical.

Conventional raking piles, typically installed at 9°–16°, offer limited lateral capacity due to inclination constraints. Ground anchors can resist tensile forces but depend heavily on favourable soil and geometric conditions.

The Mûller Verpress (MV) system provides an alternative solution through steel tension elements encased in a grout envelope. Installation involves driving an H-shaped steel beam into the ground while continuously injecting cement-based grout. The pile toe incorporates a widened, box-shaped shoe through which grout is delivered via pipes attached to the beam. As the beam is driven, the shoe forms a void around the core that is simultaneously filled with grout. The grout initially lubricates penetration and later hardens to form a high-capacity bond with the soil.

MV piles can be installed to depths of 60–70 m and at inclinations up to 45°, enabling significantly higher tensile and compressive resistance than conventional raking piles or anchors. Full-scale tests have shown MV piles to exceed their design tensile capacity, confirming performance well above typical anchor systems. Their behaviour and proprietary strength characteristics have been validated in major port projects, including deep-sea quay walls in Rotterdam.

MV Pile System Overview

- MV piles are driven steel piles functioning as both structural piles and ground anchors capable of resisting large tensile forces.
- They typically consist of steel H-beams or tubular sections installed using hydraulic impact hammers with continuous grout injection to improve drivability and soil adhesion.
- The grout envelope fully coats the steel core, providing corrosion protection. Corrosion risk is further reduced by the naturally low oxygen environment below ground and below the water table.
- The compact steel section provides excellent strength-to-weight performance, enabling installation to great depths while maintaining high pull-out resistance.
- Grout flow is monitored throughout installation to ensure consistent quality and reliable bond formation.
- Due to their high load capacity and ability to be installed at steep inclinations, MV piles are widely used for quay wall anchorage and as vertical or inclined tie elements.

12.6.5 Construction of additional new cut-off wall

Accommodating larger ships by additional dredging and deepening in front of a quay wall does indeed have significant implications for the port infrastructure, including the quay cranes and the jetty structure itself, as discussed in Chapter-14. Therefore, any such modifications should be carefully planned and executed to ensure the safety and longevity of the infrastructure. A thorough geotechnical and structural analysis should be performed to accurately assess the additional stresses and determine the most effective solutions.

The stability of a quay wall or jetty structure is influenced by several factors, including the type of soil, the depth of embedment, overall structural configuration and the design of the pile itself. When dredging is carried out around the piles, the surrounding soil is removed, reducing the lateral support that normally helps the pile stay stable. With less soil holding it from the sides, the pile has a longer unsupported (free-standing) length, making it more vulnerable to bending, sideways movement, or even structural failure.

A practical solution is to install a rigid cutoff wall to protect the dredged soil layer below the existing mud level. This wall restores lateral support to the piles and prevents further erosion around them. For the cutoff wall to function effectively, it must:

- have sufficient rigidity to withstand lateral pressures,
- extend deep enough below the design dredge level to ensure stability,
- be installable in front of the existing quay wall with minimal disturbance to the current structure during driving or installation, and
- be capable of integrating properly with the existing structural system.

These requirements can be satisfactorily addressed by adopting a suitable combi–sheet pile system, as illustrated below and explained in detail in Chapter 14.

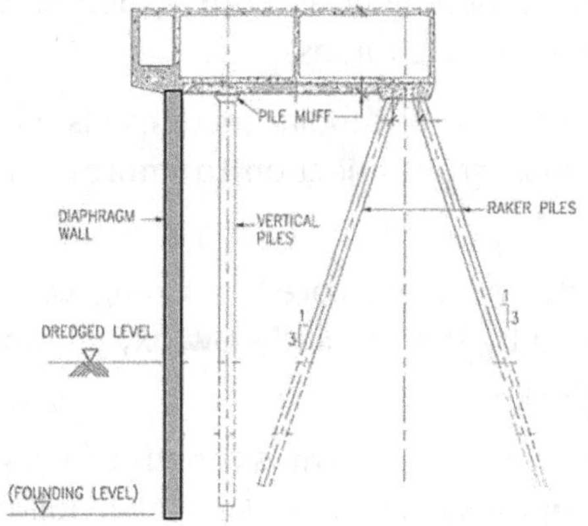

Figure 12.8 (a): Details of an Existing Quay Wall

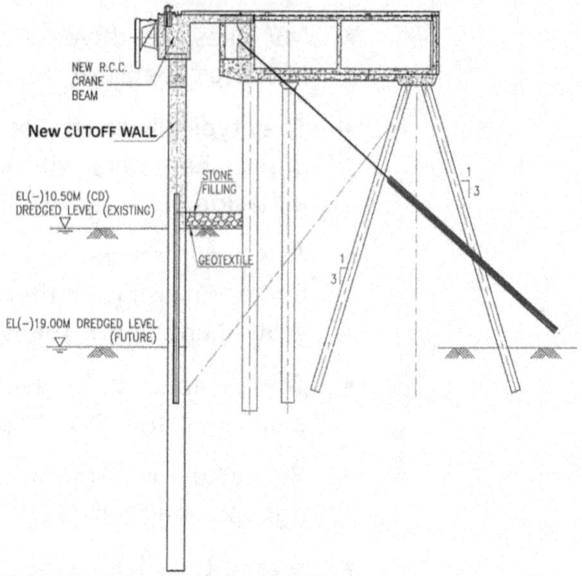

Figure 12.8 **(b)**: Integration of a Cut-off Wall with the Quay Wall for Harbour Basin Deepening

Thus, a structure stays safe only when the soil, piles, and loads are in balance. If dredging reduces soil support, this balance is disturbed, and the structure may become unstable. So, before dredging or making changes near a jetty, engineers must check how the reduced soil support will affect the structure. This ensures the structure stays safe, balanced, and able to do its job.

The modified quay structure must fulfil the following basic functional and technical requirements:

i) Retaining function

The modified structure has to safely retain soil and water (if required) after deepening of the port basin. Typically, for this aspect, the retaining height for the integrated structure must be agreed for the established dredge depth.

ii) Bearing function

The quay wall must safely sustain the new additional imposed loads by cranes, vehicles and the stored goods. It is essential that the handling equipment for the goods is safe. Due to the requirement for high handling speeds, extra horizontal forces may be generated due to the acceleration of the cranes. In some cases, additional storage areas and a transhipment facility need to be incorporated in the design to meet the functional requirements.

iii) Mooring function

The upgraded structure must be able to ensure that the larger proposed ships can be safely moored at their berths.

12.6.6 Enhancement of Deck Strength through RCC Overlay

In marine and infrastructure applications, integrated deck slabs often require strengthening due to increased operational loads, progressive material deterioration, or updated design standards. When an existing deck slab lacks adequate strength or stiffness for new service requirements, adding a suitable RCC overlay becomes a practical strengthening solution. A reinforced concrete (RCC) overlay is widely used because cutting and removing the old concrete for replacement can compromise structural integrity and may be environmentally or operationally unfeasible. By placing a new overlay that is properly bonded and integrated with the existing slab, the overall structural performance, stiffness, and durability of the deck can be significantly enhanced.

12.6.6.1 Mechanism of Stiffness Enhancement

The key benefit of an RCC overlay comes from the increased stiffness provided by the improved composite action between the new layer and the existing slab.

Once the new concrete layer is adequately bonded to the existing slab, both layers act as a single structural unit with a larger moment of inertia, thereby increasing flexural stiffness and reducing service-load deflections. The increased section depth also contributes to improved resistance against bending and shear forces, while the addition of new concrete assists in redistributing stresses more uniformly, limiting crack widths and enhancing long-term durability.

Alongside the stiffness benefits, the RCC overlay contributes to a substantial increase in load-carrying capacity. The introduced reinforcement supplements the existing steel and allows the section to accommodate greater ultimate loads. With increased rigidity, load distribution across the slab becomes more uniform, reducing localised stresses and improving overall structural integrity. This enhancement also improves fatigue resistance—a critical factor for jetty and deck slabs subjected to repetitive dynamic loads from berthing, mooring operations and operational vehicle traffic.

i) Design Considerations

To achieve the intended structural performance, key design considerations must be addressed. The bonding interface between the old and new concrete requires careful preparation through surface roughening, cleaning, and the application of suitable bonding agents to ensure monolithic behaviour. Potential issues related to differential shrinkage or thermal movements between the two layers must also be considered to prevent cracking or delamination. The thickness of the overlay should be determined through structural analysis, balancing serviceability requirements with ultimate strength needs.

ii) Practical Benefits in Jetty Structures

For jetty structures, an RCC overlay offers several practical benefits. The increase in slab stiffness enhances the overall rigidity of the jetty framing system, improving its capacity to resist lateral loads from wind, waves and berthing impacts. The overlay also provides a protective barrier to the original concrete, extending the service life of the structure by limiting exposure to harsh marine environments. Importantly, this method of strengthening is cost-effective and can typically be executed with minimal disruption to ongoing operations, making it a preferred solution compared to full slab replacement.

In summary, the application of an RCC overlay is a reliable and efficient method for improving both the stiffness and load-carrying capacity of existing deck slabs. By creating a composite section with increased depth and supplementary reinforcement, it enhances structural resilience and operational performance, making it particularly valuable for strengthening jetty and marine infrastructure.

12.6.7 Use of Voided Slab as an Alternative to Thick RCC Overlay

In situations where the level of the deck slab is to be raised or requires significant strengthening, a conventional RCC overlay may lead to considerable additional dead load, which can adversely affect the efficiency of the structural system and impose higher demands on existing supports and foundations. As an alternative, the adoption of a voided slab system provides a structurally effective solution that achieves the required stiffness and load-carrying performance while substantially reducing the self-weight of the slab.

12.6.7.1 Concept of Voided Slab

A voided slab incorporates hollow formers made of lightweight materials within the concrete volume, removing concrete from regions that contribute minimally to structural capacity. This reduction in non-critical concrete allows the slab to maintain its overall depth and flexural efficiency while decreasing dead load, often by 20–35% compared to an equivalent solid slab. Such weight reduction is particularly advantageous in retrofit or rehabilitation projects, where additional load from a thick overlay may overstress the existing framing system or foundations.

Structurally, the voided slab behaves similarly to a ribbed or waffle slab, enabling efficient load distribution while optimising material use. The preserved effective depth ensures that flexural stiffness remains high, and the placement of voids within the lower-stress middle zone ensures that moment capacity is not compromised. Local reinforcement around the voids can be strategically provided to maintain adequate shear and punching resistance, ensuring that global and local performance criteria are met. From a design standpoint, proper placement and alignment of void formers during casting are essential to achieving uniformity and structural integrity. In retrofit applications, the new voided slab must also be bonded effectively to the existing slab to ensure composite action, requiring careful surface preparation and detailing. Additionally, considerations related to construction feasibility, formwork requirements, and compatibility of deformation characteristics must be addressed.

12.6.7.2 Application in Jetty Structures

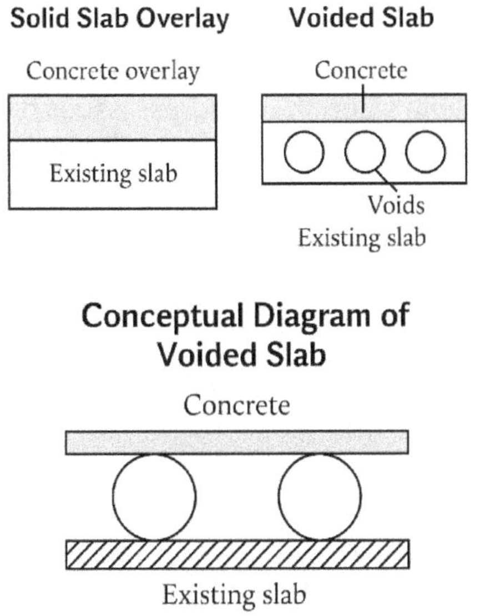

Figure 12.9: Concept of Voided Slab

For jetty structures, the voided slab system offers several noteworthy advantages. The increased structural depth contributes to improved lateral stiffness without introducing excessive weight, thereby enhancing the deck's capacity to resist lateral actions from berthing, mooring, wind, and wave forces.

The reduction in dead load also leads to more efficient and economical foundation design, particularly relevant for pile-supported marine structures where axial and lateral demands are sensitive to superimposed loads. Furthermore, the approach aligns with durability and sustainability objectives by reducing concrete consumption and lowering long-term stress levels within the slab.

In summary, the voided slab system represents a technically robust and weight-efficient alternative to a thick RCC overlay for enhancing deck slab performance. It provides comparable improvements in stiffness and load-carrying capacity while mitigating the disadvantages associated with increased dead load, making it especially well-suited for strengthening jetty decks and other marine infrastructure where both structural demands and foundation constraints are critical.

12.6.8 RCC Jacketing to Enhance Structural Performance

RCC jacketing is a well-established strengthening and rehabilitation technique used extensively in marine, industrial, and civil infrastructure to improve the performance of structural elements such as beams and piles. The method involves encasing the existing member within an additional layer of reinforced concrete, accompanied by supplementary steel reinforcement, to enhance its load-carrying capacity, stiffness, and durability. This approach is particularly useful in structures that have deteriorated due to ageing, environmental exposure, corrosion, or increased operational demands, and is also widely applied to upgrade seismic resistance and restore safety margins.

12.6.8.1 Objectives of RCC Jacketing

For beams, RCC jacketing is employed to strengthen members that are under-designed or have experienced material degradation over time. By increasing the effective cross-sectional area and adding new longitudinal and shear reinforcement, the technique significantly improves both flexural and shear capacities. The enhanced stiffness helps reduce excessive deflections and vibrations, contributing to improved serviceability. The process typically includes surface preparation to ensure effective bonding, placement of new reinforcement, installation of formwork, and careful concreting and curing to achieve monolithic behaviour between the old and new sections. The resulting section exhibits increased moment capacity, better load distribution characteristics, and improved durability and fire resistance.

In the case of piles—particularly those functioning in harsh environments such as marine structures—RCC jacketing provides a reliable solution to address issues related to corrosion, scour, or increased axial and lateral load requirements. Depending on site conditions, jacketing may be executed in dry conditions or underwater, using high-strength concrete and corrosion-resistant reinforcement to ensure long-term durability. The encasement restores or enhances axial and bending strength, improves lateral stiffness, and provides superior resistance to environmental deterioration. In seismic regions, the improved confinement offered by the jacket also enhances ductility and cyclic load performance.

12.6.8.2 Structural Implications

The structural implications of RCC jacketing are significant. Increasing the cross-sectional dimensions results in a higher moment of inertia, which not only raises the strength but also reduces stress concentrations and improves global stiffness. The enhancement in stiffness leads to lower deflections and improved dynamic response under service loads. Additionally, the increase in strength enables the structure to safely resist higher loads, thereby restoring or upgrading the structural performance to meet current design demands and safety standards. RCC jacketing has been successfully applied across a wide range of infrastructure and pile-supported marine structures requiring rehabilitation. Its adaptability, effectiveness, and cost-efficiency make it a preferred method for strengthening and extending the service life of existing structural members.

Jacketing of Marine Piles: For strengthening and protecting marine piles, RCC jackets are commonly installed by first wrapping hessian cloth supported on coiled wires around the pile, creating a uniform annular gap. The lower edge of the wrapping is securely tied to the pile to form proper containment around the deteriorated section. Once this enclosure is prepared, micro-concrete or free-flowing fibre-reinforced concrete is poured into the gap between the pile surface and the hessian wrapping. This method provides an effective way to form an RCC jacket around marine piles, restoring their strength and improving durability.

This technique is also highly effective for protecting steel piles in the intertidal and splash zones, where corrosion rates are typically severe. The RCC jacket acts as a durable barrier, significantly reducing exposure to seawater and aggressive marine conditions.

12.6.9 Installation of Tiebacks/Anchors for Quay Wall Stabilisation

Quay walls are important waterfront structures that hold back soil and withstand sideways forces from ships during mooring and berthing, along with pressure from water and the extra loads placed on the land behind the jetty. As ports upgrade to handle deeper-draft vessels, heavier quay cranes, and higher operational and storage loads, the height of soil retained behind the wall—and the pressure acting on it—often increases. This increase in lateral pressure can put additional stress on the quay wall and may affect its overall stability.

To deal with these additional enhanced pressures, tiebacks are used as a dependable way to help the quay wall handle the extra sideways forces. These are tensioned elements anchored into the ground behind the wall, and they work by pulling back against the increased lateral loads. This helps balance the forces acting on the wall and improves its overall stability and performance.

i) Purpose of Tiebacks

Tiebacks serve multiple structural and geotechnical functions, including:

- **Counteracting Lateral Earth and Hydrostatic Pressure**

  Tiebacks resist the horizontal forces exerted by soil and water, reducing the load directly acting on the quay wall.

- **Enhancing Structural Rigidity**

  By anchoring the wall to stable ground, tiebacks increase its stiffness and reduce its susceptibility to deformation.

- **Minimising Lateral Deflection**

  Tiebacks restrain the wall from excessive movement, maintaining its vertical alignment and serviceability.

- **Reinforcing the Backfill Zone**

  Tiebacks stabilise the soil behind the wall, preventing settlement and lateral displacement of the backfill material.

The fundamental principle behind tieback installation is the **transfer of lateral forces** from the quay wall into the **passive zone** of the soil. The passive zone is the region where soil resists movement and provides anchorage. By drilling inclined holes through the wall and anchoring steel rods or strands into this zone, the lateral pressures are redistributed away from the wall. This significantly reduces the bending moment and shear forces acting on the structure, thereby improving its stability and load-bearing capacity.

Anchors or Tiebacks offer several structural advantages:

- **Reduction of Overturning Moment and Sliding Forces**: By anchoring the wall, tiebacks counteract destabilising forces and improve overall equilibrium.
- **Alignment and Serviceability**: Tiebacks maintain the horizontal alignment of the wall, ensuring functional performance under operational loads.
- **Dynamic Response Control**: The increased rigidity reduces vibrations and dynamic movements caused by berthing impacts or crane operations.
- **Backfill Stabilisation**: Tiebacks restrain the soil behind the wall, preventing lateral movement and settlement, which enhances the long-term stability of the structure.

The installation of tiebacks is a strategic and effective engineering intervention for quay wall stabilisation. By transferring lateral pressures to stable ground zones, tiebacks reinforce both the wall and the surrounding soil, ensuring structural integrity under demanding environmental and operational conditions. Their application reflects advanced geotechnical and structural engineering practices aimed at safeguarding critical waterfront infrastructure.

ii) Selecting Appropriate Anchors

Choosing the right type of anchor depends on the interaction between the wall, the ground, and the operational environment. Anchor design is not a standardised process; it requires a detailed understanding of site-specific physical and geotechnical conditions. The topography and surface layout influence the anchor orientation, depth, and load path. Soil characteristics—including composition, strength parameters, stiffness, and stratification—determine the anchor's bond performance and long-term behaviour. Adequate clearance behind the wall is essential for drilling or installing tie-backs, and the anchor zone must be free from underground utilities, debris, or hard obstructions. Groundwater conditions also influence installation, especially for tie-rods or dead-man anchors, where water pressure affects excavation stability and anchor performance. Finally, serviceability requirements—including allowable wall movement and long-term deformation limits—guide the choice of anchor type and capacity.

iii) Anchor Location – Practical Understanding for Deadman Anchors

When using anchors in a retaining wall—like deadman anchors—it's really important to place them outside the active failure zone behind the wall. If the anchor ends up inside this zone, it won't work properly. It can interfere with how the soil resists movement and reduce the overall strength of the system. Basically, it stops the soil from developing its full passive resistance, which is what helps hold the wall in place.

The passive resistance of the soil—how much it pushes back against the wall—depends on things like:

i) The shear strength of the soil (how well it holds together),

ii) Its density, and

iii) How the wall itself deforms under pressure.

To improve the stability of an anchored wall, engineers can:

1) Keep the soil in front of the wall intact, especially during dredging or excavation.
2) Improve the soil's shear strength without disturbing the backfill—methods like injection grouting are useful here.
3) Install a new anchor wall and tie-rods, based on a full analysis of how the system behaves together.

Now, depending on where the anchor block is placed, its effectiveness changes:

i) Anchor A: Located inside the active failure zone—offers no resistance.

ii) Anchor B: Sits in a zone where the active and passive wedges overlap—resistance is reduced.

iii) Anchor C: Positioned to develop full resistance.

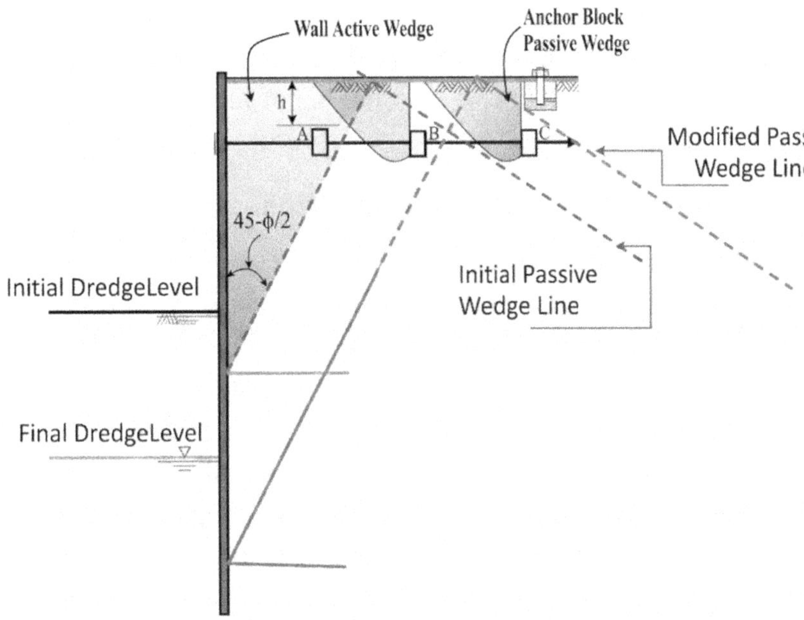

•

Figure 12.10 Shows the Effect of Anchorage Location on the Resistance Developed

For optimal performance, anchor blocks shall be founded against firm, undisturbed natural soil and installed in a manner that prevents settlement under service and ultimate load conditions. A minimum factor of safety of 2.0 is recommended to ensure structural reliability and satisfactory long-term performance. Dredging activities undertaken in front of the wall may significantly modify the in-situ stress conditions of the soil and lead to a reduction in passive earth resistance, as illustrated in the figure above. In particular, deepening of the ground level in front of an existing anchored wall causes an expansion of the active failure wedge. As a result, anchor blocks that were originally positioned within the passive resistance zone may become located within the active zone, thereby reducing or impairing the effective anchor resistance.

These effects shall be carefully evaluated during the design stage and rigorously verified during construction to ensure continued stability and performance of the anchored wall system.

12.6.10 Strengthening of Structure Adding Portals

12.6.10.1 Principle and Efficacy of the Portal System

Adding an extra portal frame to the existing jetty bents is basically a smart upgrade to make the jetty better at handling sideways forces. Marine structures mainly face lateral loads from mooring, berthing impacts, wind, waves, currents, and even earthquakes. To keep the jetty safe, stable, and working well over the long term, strengthening its framing system is important.

By adding this new bay, the jetty becomes stiffer as a whole, spreads lateral loads more efficiently, and gains extra redundancy. This means the structure is less likely to get overstressed in any one spot.

Besides improving strength, the additional portal frame also gives the jetty more usable width. This extra space can be used for new equipment, upgraded berth facilities, or other operational needs. So, this upgrade is not just about strengthening—it also improves the jetty's functionality and helps support future expansion.

12.6.10.2 Engineering Rationale and Application

Upgrading existing jetties is often necessary to accommodate evolving operational requirements, larger vessel dimensions, updated design standards, and the need to prolong structural service life. Introducing an additional portal frame is an effective enhancement strategy that strengthens both the structural capacity and functional performance of the jetty.

The added frame contributes to greater lateral stability by distributing lateral forces more evenly across the bent system. It also improves overall stiffness, reducing lateral displacements and controlling frame sway under environmental and operational loading. Furthermore, the modification increases structural redundancy, enabling the system to better withstand localised overstress without compromising the integrity of the entire bent. By creating additional load paths, the upgraded framing system becomes more resilient to higher berthing and mooring demands. Importantly, this form of strengthening is also cost-efficient, as it can be implemented locally without requiring major reconstruction of the existing jetty.

In practical terms, the addition of a new bay enhances the jetty's capability to safely handle larger vessels, higher operational loads, and more rigorous design conditions typical of modern marine infrastructure.

12.6.10.3 Technical Basis: Portal Method

To understand how a multi-bay jetty frame behaves under lateral loads, engineers often use the **Portal Method** during the early design stage. It is a quick and simple way to estimate how forces move through the structure. More detailed checks are done later using advanced tools like 3D finite-element analysis, but the Portal Method helps give an initial understanding—especially when assessing the effect of adding a new bay or portal frame.

Key Assumption

The Portal Method makes one main assumption:

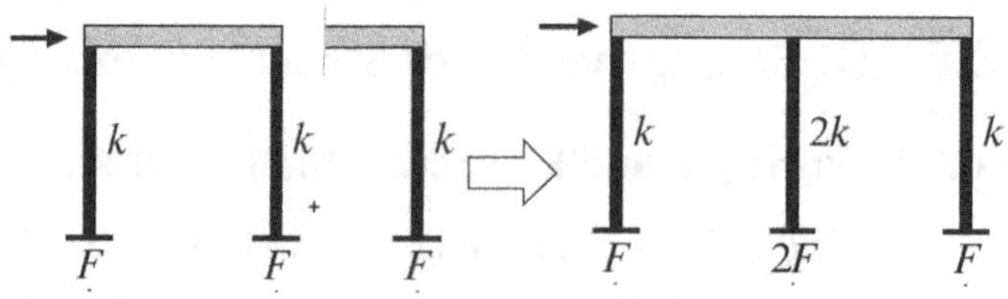

Figure-12.6: Lateral Stability Improvement by additional Portal action

Interior columns carry about twice as much lateral shear as exterior columns. This is because:

i) Interior columns are located between two bays,

ii) They receive load from both sides,

iii) Their central position usually makes them stiffer.

This assumption simplifies calculations while still giving a realistic picture of how the frame behaves under lateral forces. The simplified model results in a clear load-sharing pattern. The Load Distribution Logic is as follows;

a) Exterior Columns (typically of similar size) sharing gravity loads from above, each supports axial force from one side only

b) Interior Columns formed by combining two adjacent portal frames must carry axial forces from both sides (left and right); therefore, they are designed to be twice as strong as exterior columns.

This approach helps estimate lateral shear and bending forces quickly. It is especially useful when checking how the structure will behave after adding an extra bay to the existing jetty, as it shows how the additional frame will help improve stiffness and redistribute loads.

12.6.11 Stabilisation of Quay Wall During and After Dredging

Dredging near a quay wall—especially when deepening the water for ship access—can disturb the soil support and increase the risk of wall movement or failure. To manage these risks, engineers use several smart and practical techniques:

- Don't Over-Dredge Near the Wall: Digging too deep or too close can remove the soil that helps push back against the wall (called passive resistance). Keeping enough soil in front helps the wall stay balanced.
- Add More Anchors or Tie-Backs: Reinforcing the wall with extra support—like ground anchors or tie-rods—helps it resist the added pressure from soil and water, especially after dredging.
- Make the Backfill Stronger: Improving the soil behind the wall, either by compacting it or using better materials, helps reduce the sideways pressure pushing on the wall.
- Strengthen the Soil in Front of the Wall: Techniques like grouting (injecting cement-like material into the soil) or replacing weak soil with stronger material can boost stability without disturbing the backfill.
- Install a Relieving Platform: This is a structure that spreads out the weight from vehicles or equipment near the wall, so the wall doesn't take all the pressure directly. It helps reduce stress from live loads.
- Build a Rigid Cutoff Wall in Front: Adding a solid wall in front of the quay wall helps block soil movement and reduces pressure on the main structure, improving overall stability.
- Review the Structural Design: After dredging or any major change, it's important to reassess the design. New soil conditions and loading patterns might require updates to keep the wall safe and functional.

Wharf Functional Upgradation and Capacity Enhancement

Chapter Summary

This chapter addresses the functional upgradation of existing wharf and quay wall structures to meet evolving port operational requirements driven by larger vessels, deeper drafts, and high-capacity ship-to-shore (STS) cranes. It outlines the limitations of legacy marine structures when subjected to increased ship dimensions, elevated berthing and mooring forces, and higher vertical and horizontal operational loads, highlighting the need for targeted functional and structural enhancements. The chapter explains how functional upgradation extends beyond structural integrity to include operational performance, focusing on quay load-carrying capacity, crane rail systems, berthing energy absorption, and long-term serviceability. The impact of modern terminal equipment on quay geometry, crane beam systems, rail alignment, and foundation behaviour is systematically addressed.

Upgradation strategies range from minor repairs to full reconstruction, categorised as very soft, soft, and hard interventions. The chosen approach depends on structural condition, foundation capacity, remaining service life, and future operational needs, guided by a structured decision framework.

Special emphasis is placed on ensuring wharf compatibility with modern STS cranes through strengthening of crane beams, rails, anchorage systems, and upgrading marine fenders to handle higher berthing energies from larger vessels. Related upgrades to deck slabs, mooring systems, and utilities form an integral part of functional upgradation.

Learning Objectives

- Upgradation of existing wharves is essential to accommodate larger vessels and higher-capacity quay cranes.
- Renovation strategies follow a structured approach based on site conditions, operational demands, and future load requirements.
- New vessel size, Crane geometry and wheel load-path behaviour are critical design parameters for the quay wall upgradation.
- Fender systems must be enhanced to meet the higher berthing energy of larger vessels, with suitable spacing and alignment.

13.1 Introduction

The maritime industry has indeed undergone significant advancements, leading to the emergence of highly specialised vessels with unique designs tailored to maximise efficiency and meet specific navigational requirements. The development of ships in recent years indicates that ships are highly specialised. Modern ships, particularly those with larger longitudinal surface areas, have a significant impact on port infrastructure. Larger vessels, such as the Post-Panamax, Cape size, and Valemax type ships, exert greater force on mooring devices due to current, wave, wind and berthing force on the structure. Consequently, ports need upgradation in their facilities, including berthing structures, to meet the requirements of ships of these classes, ensuring efficient, safe, and sustainable operations.

Functional Upgradation of harbour facilities for new upcoming ships is a complex task. It involves not only strengthening structures for heavier operating loads, larger quay cranes, but also deepening the waterfront and updating quay structures to support higher capacity, modern cranes, and other operational demands. This requires careful consideration of structural stability, including adjusting existing harbour facilities to ensure enough water depth, proper mooring systems, and strong, stable structures for increased loads. These adjustments must be both theoretically sound and practically tested to ensure reliable port operations.

To adapt berthing structures to evolving ship requirements, four main points may be followed:

i) Deepening the harbour bottom at the berth to accommodate deeper-draft vessels and associated equipment.

ii) Strengthening and upgrading the existing structure, integrated with modern materials, to develop a deeper pier or jetty capable of supporting higher deck loads, including loads from large quay cranes and increased operational demands.

iii) Structural reconstruction and modification to cater for increased mooring and berthing loads

iv) Application of appropriate analysis and design methods, using data obtained from geotechnical, geophysical, hydrodynamic investigations, and condition surveys, to validate the selected upgradation approach.

Together, these methods enable the upgrading of berthing structures to maintain functionality, comply with current safety standards, and accommodate anticipated vessel and cargo requirements. As marine transport continues to evolve, designers must plan proactively to ensure structures can adapt to future needs. This approach supports safe, efficient, and sustainable port operations over the long term

13.2 Upgradation and Stabilisation of Existing Structures

Many older harbour structures require repairs and strengthening due to evolving operational demands and long-term development plans. Rather than reconstructing these facilities entirely, ports often choose to upgrade existing structures to meet modern shipping requirements. In addition to reconfiguring quay structures to accommodate larger vessels and high-capacity cranes, the key reasons for undertaking such upgrades include:

i) **Limited Space Availability:**

Where additional land or waterfront space is unavailable for new wharf or harbour basin construction, upgrading existing structures becomes the most practical and viable option.

ii) **Cost Efficiency:**

Completely dismantling old structures and constructing new ones can be significantly more expensive. Targeted repairs, customised modifications, and structural reinforcements offer a cost-effective way to extend the service life of existing facilities.

iii) **Environmental Consideration:**

Upgrading and reusing existing structures is a more environmentally sustainable approach compared to demolishing old facilities and building new ones.

Overall, these factors underline the importance of strategic planning and sound engineering to ensure harbour structures continue to meet the evolving demands of the maritime industry. In this context, renovation and upgradation form an integral part of repair, rehabilitation, and service-life extension, and the following points shall therefore be carefully considered.

13.3 Repair and Rehabilitation

i) **Repair (or Patching)**

Repair refers to localised actions taken to address small or isolated areas of deterioration, such as concrete spalling in quay decks or minor damage to piles. These repairs are generally short-term solutions, may need to be repeated over time, and are mainly intended to restore serviceability rather than structural strength.

Repairs are usually part of a broader maintenance and asset management program and typically include:

- **Preventive measures**, such as applying marine-grade protective coatings to reduce exposure to aggressive environmental conditions;
- **Corrective measures**, including localised structural repairs or minor retrofitting to accommodate small changes in usage or loading.

ii) Rehabilitation

Rehabilitation is a more extensive and invasive process than localised repair. It involves the systematic removal of deteriorated or damaged material and the restoration of structural integrity using new or improved materials. Rehabilitation often includes strengthening measures to restore, and sometimes enhance, the original design strength of the structure.

A successful rehabilitation program follows a structured and systematic approach, which includes:

- Detailed condition assessment using visual inspection and non-destructive testing;
- Identification of the root causes of deterioration or structural deficiency;
- Evaluation of the extent and severity of damage;
- Determination of the appropriate level of intervention;
- Selection of repair and strengthening materials compatible with the existing structure and environmental exposure;
- Proper surface preparation, execution, and curing of repair works to ensure long-term durability and performance.

13.4 Enhancement of Service Life in Concrete Structures

The service-life model explains how a reinforced concrete structure behaves and gradually deteriorates over time. Due to environmental exposure and natural ageing, the structure slowly loses strength and functional performance. Ideally, this deterioration occurs in distinct stages—starting from a sound condition, progressing to the serviceability limit, and eventually reaching a stage where structural failure may become possible, as shown in *Figure 13.1*.

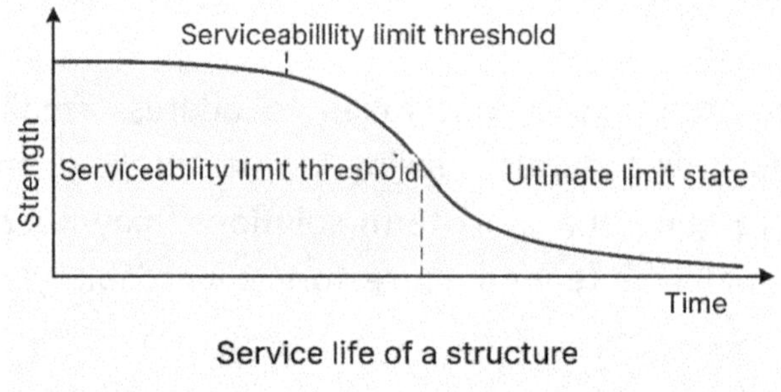

Figure 13.1: Service Life Model of RCC Structure

However, shortcomings during construction or later use—such as poor-quality materials, improper concrete placement, inadequate curing, poor detailing, lack of maintenance, or loading beyond design limits—can significantly reduce structural reliability and shorten the service life. Poor construction and site practices introduce defects that adversely affect durability.

To control the progression of material deterioration, a systematic and periodic maintenance program is essential, as illustrated in *Figure 13.2*.

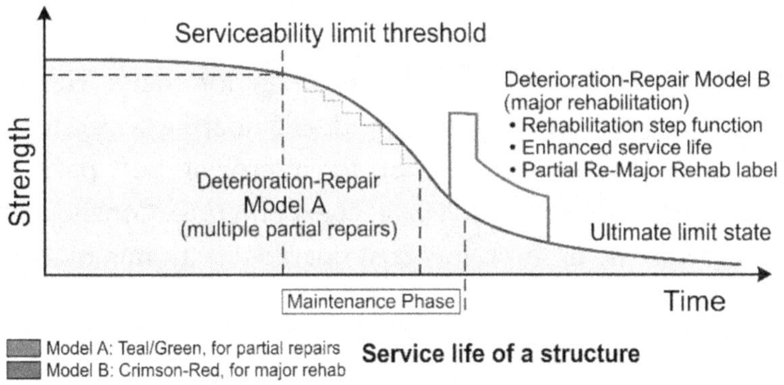

Figure 13.2: Service Life Model of Structure under Periodic Repair and Maintenance Program

When maintenance is delayed or neglected, deterioration accelerates and may require a highly extensive and invasive repair and strengthening program to restore the original design strength. Such an intervention is known as rehabilitation, as shown in *Figure 13.3*. Rehabilitation may be carried out in single or multiple stages over the remaining design life of the structure.

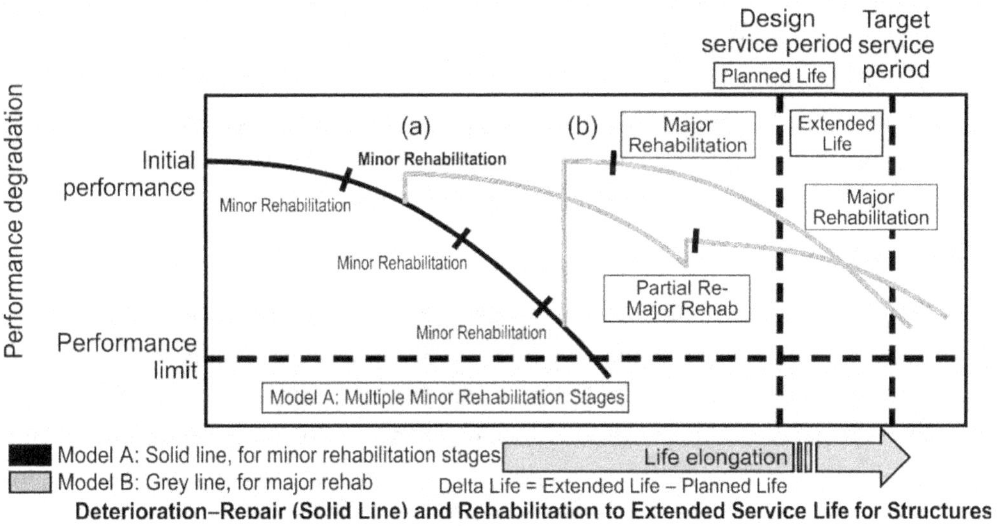

Figure 13.3: Service Life Extension Model of Structure after Rehabilitation

13.5 Practical Overview of Port Renovation Techniques

While employing a structured approach, renovation methods generally fall into three categories: **very soft**, **soft**, and **hard** techniques—each with its own approach, cost, and impact on operations.

i) Very Soft Renovation Techniques

Very soft renovation techniques represent the least disruptive approach for upgrading port infrastructure. These methods include repairs, injection, drilling, grouting, and the replacement of unsuitable fittings and accessories. Such interventions allow port operations to continue with minimal interruption, making them particularly suitable for rapid maintenance actions and urgent upgrade requirements.

These techniques are cost-effective, environmentally considerate, and sufficiently flexible to address localised issues without the need for major construction activities. When supported by appropriate planning and engineering assessment, very soft renovation methods provide an efficient solution for improving port performance while minimising operational downtime and overall project expenditure. Consequently, they are best applied in situations requiring prompt, low-cost upgrades with minimal disruption to ongoing port functions.

ii) Soft Renovation Techniques

Soft renovation techniques involve moderately invasive methods for strengthening and upgrading existing harbour structures. These include the installation of anchors to reinforce quay walls, localised strengthening of members through jacketing or carbon-fibre composite (CFC) wraps, and the use of injection and grouting to improve structural integrity. They also cover the replacement of outdated deck fittings—such as crane rails, fenders, and mooring arrangements—with higher-efficiency components capable of meeting updated load requirements without necessitating full reconstruction.

These methods enhance structural performance and increase load-carrying capacity while requiring significantly less time and cost compared to hard renovations. Although they may introduce some operational disruptions, these impacts are generally manageable with proper planning and sequencing. Soft renovation techniques are particularly advantageous in areas with spatial constraints and provide an effective middle ground between minimal intervention and complete rebuilding. As such, soft methods offer a balanced solution that delivers improved performance at a reasonable cost.

iii) Hard Renovation Techniques

Hard renovation techniques involve major construction activities and substantial structural intervention when existing harbour facilities can no longer meet performance, safety, or operational requirements through minor or moderate upgrades. These measures may include partial demolition and replacement of outdated structural components—such as crane-rail beams and heavily loaded deck elements—as well as the installation of additional tie-backs, anchor piles, barrettes, reinforced concrete (RCC) overlays, or relieving platforms to strengthen or replace deteriorated sections of the quay.

In situations where significant deepening or dredging is planned, hard renovations may also involve constructing cutoff walls or similar retaining systems to maintain the bearing capacity and stability of existing jetties and quay structures under the updated hydraulic and geotechnical conditions.

Hard renovation methods provide durable, long-lasting infrastructure capable of accommodating larger vessels, heavier cranes, deeper drafts, and higher cargo volumes. While these interventions require greater investment and may disrupt operations during construction, they ensure full compliance with modern safety standards and deliver comprehensive improvements in structural performance and service life.

iv) When Hard Renovation Is Required

Hard renovation techniques are selected when structural or operational demands exceed what very soft or soft renovation methods can safely address. These conditions typically include:

- **Significant Structural Deterioration or Damage:** When core elements such as beams, piles, decks, quay walls, or crane-support systems are severely deteriorated or have reached the end of their service life, minor strengthening measures are insufficient. Hard renovation becomes necessary to restore structural integrity.
- **Need to Support Much Higher Loads, Deeper draft and Dredging:** If the port must accommodate larger vessels (e.g., Post-Panamax, Capesize, Valemax), heavier cranes, automated equipment, or increased cargo throughput, the existing structure may lack adequate load-bearing capacity. Hard renovation provides the required reinforcement or reconstruction.

  Substantial berth deepening or continuous dredging may compromise the stability of existing foundations. Installing cutoff walls, new piles, or relieving platforms is often required to ensure safe operations under the new conditions.
- **Major Functional Upgrades or Layout Changes:** Reconfiguration of port layouts—such as upgrades for mechanisation or automation, changes in equipment positioning, or integration with new cargo facilities—may necessitate partial demolition and redesign of structural components.
- **Compliance with Updated Engineering Codes and Safety Standards:** Older structures designed to outdated standards may not meet current seismic, structural, or operational requirements. Hard renovation ensures full alignment with modern codes and extends overall design life.
- **Long-Term Reliability Requirements:** For ports targeting 30–50 years of reliable service, hard renovation provides durability and future-proofing that lesser interventions cannot achieve.
- **Insufficient Capacity After Soft or Very Soft Methods**: If non-destructive tests (NDT) and structural assessments indicate that soft renovation methods cannot achieve the required factor of safety, load capacity, or risk mitigation, hard renovation becomes the only viable option.

13.5.1 Logical Sequence for Selecting Renovation Technique

i) Structural Condition Assessment

As outlined in the assessment chapter, the strength and adequacy of the existing quay structure shall be established using verified material properties, prevailing loading conditions, geometric characteristics, and the overall structural response.

Structural defects resulting from static and dynamic actions—such as deformations, cracking, crushing, and localised damage—along with non-structural defects caused by environmental factors (humidity, temperature variations, chemical or biological influences) shall be systematically identified and quantified.

A detailed structural analysis shall subsequently confirm the global integrity of the quay, evaluating its load-bearing capacity and overall system behaviour. Based on the verified strength, durability, and the required target reliability level, the appropriate rehabilitation strategy shall be defined. Interventions shall be categorised as follows:

- **Minor defects** (surface corrosion, hairline cracks, localised spalling)

  → Proceed to **Step A**
- **Moderate degradation** (section loss, significant cracking, damaged fenders, minor settlement)

  → Proceed to **Step B**
- **Severe structural distress** (major settlement, instability, large deformations, extensive reinforcement loss, unsafe structural conditions)

  → Proceed to **Step C**

ii) Very Soft Technique (Step-A): Applicability Check

Very soft renovation shall be adopted when:

- The load-bearing capacity and global structural integrity remain largely intact;
- Only superficial, localised, or non-structural repairs are required to maintain serviceability;
- Quay operations are not significantly affected;
- No urgent strengthening or invasive upgrades are necessary.

Generally, very soft renovation comprises patch repair, surface coating, anti-corrosion treatment, localised concrete repair, and replacement of berth fittings and accessories.

If deterioration exceeds these thresholds → proceed to **Step B**.

iii) Soft Technique (Step-B): Applicability Check

Soft renovation is appropriate when:

- Structural capacity is partially reduced but remains recoverable and adequate for future operational loads;
- Damage is localised rather than systemic, allowing strengthening without major reconstruction;
- Future service loads are comparable to current design loads;
- The foundation system remains structurally sound;
- Environmental or operational constraints favour minimally intrusive methods where feasible.

Soft renovation technique may include localised RCC jacketing, FRP or steel plate strengthening, sheet-pile reconditioning, anchor or tie-rod enhancement, and moderate deck repairs.

If deterioration is more extensive → proceed to **Step C**.

iv) Hard Technique (Step – C): Applicability Check

Hard renovation shall be undertaken when:

- The structure exhibits severe deterioration or poses safety risks under intended loading;
- Global stability or foundation integrity is compromised;
- Significant settlement, deformation, or instability is present, requiring structural upgrading for higher future loads;
- Rehabilitation necessitates major structural inclusions or system-wide interventions;
- Geotechnical, dredging, or hydraulic conditions warrant a more robust solution.

Hard renovation technique may involve partial rebuilding of quay wall, strengthening and replacement of major load-carrying beams, installation of new cutoff sheet-pile wall or diaphragm walls, full deck replacement, strengthening of backfills, and major foundation reinforcement.

v) Evaluation of Operational and Future Requirements

- If future vessel size, berthing energy, or operational loads are expected to increase significantly → **Hard Renovation is preferred**.
- If operational downtime must be minimised and structural conditions permit, → **Soft or Very Soft Renovation is preferred**.

Table-13.1: Decision Rule Renovation Techniques

Structural Condition	Operational Requirement	Foundation Condition	Future Load Increase	Recommended Renovation
Minor defects	No major changes	Good	No	**Very Soft**
Moderate damage	Functional but weakened	Adequate	Minor	**Soft**
Severe or unsafe	Major operations affected	Poor/ compromised	Significant	**Hard**

13.6 Upgrading Existing Wharves for New Crane Systems

Over time, ports have seen big improvements in container crane technology, especially in how much they can lift and how large they are. This is mainly because container ships are getting bigger, which means ports need better equipment to handle them. While ship size is limited by water depth and the shape of the approach channel, newer ships are getting wider more quickly than they're getting deeper. Wider ships are more stable and can carry more containers stacked high, which helps reduce shipping costs by moving more containers at once.

13.6.1 Understanding Crane Geometry for Optimal Wharf Design

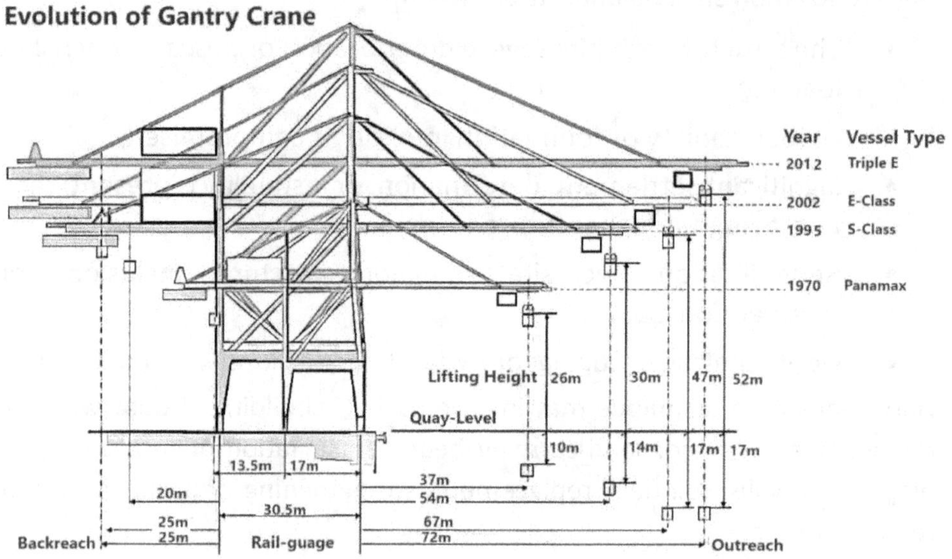

Figure-13.4: Growth of Crane Dimensions in Relation to Increasing Ship Size

To handle these larger ships, the cranes that serve them also need to grow—reaching farther, lifting higher, and handling heavier loads. With older, smaller ships being replaced by much larger ones, ports have to install taller cranes with longer outreach. On top of that, recent supply-chain pressures have pushed ports to increase capacity. Instead of building completely new terminals, many ports are upgrading their existing ones. But placing larger, heavier cranes on older docks is challenging because these cranes put much higher forces on the structure. That's why it's crucial to assess the quay properly and strengthen it as needed, ensuring it can safely support the added load.

As ship-to-shore (STS) cranes get bigger and reach farther, they bring major changes to port infrastructure. Longer outreach affects crane stability, so ports either need to widen the crane base or add counterweights to prevent tipping. But space on the berthing platform is limited, making this a tough design challenge. Bigger cranes also mean heavier wheel loads, which put more pressure on rails, support beams, and piles—raising concerns about structural safety. Over the past 20 years, dock loading capacity has jumped from around 70 to 130 tons per meter—a 185% increase—showing how fast cargo handling tech is evolving. While these cranes boost efficiency, they also demand stronger, well-planned infrastructure to keep port operations safe and reliable.

The wharf designer needs a basic understanding of the geometry and layout of the quay crane. One important aspect is how the crane wheels are spaced, because this spacing affects how the load is distributed onto the wharf structure. The distance between wheels is mainly based on the crane's overall width (bumper-to-bumper length) and on any additional parts, such as a stowage bracket located at the centre of the crane. Traditionally, using eight wheels has worked well to keep stresses on both the wheels and the rails within safe limits. As cranes have become heavier over time, the wheels and rails have also grown larger and stronger to handle the increased loads.

When evaluating loads on the wharf, both the weight of the wharf structure and the crane itself are treated as *dead loads*. These loads are predictable and can be estimated very accurately—usually within a few percent once the crane is assembled. In contrast, containers, trucks, and yard equipment are considered *live loads* because they move and change frequently. Although these live loads do vary, their effect on the main wharf girders is generally small. Understanding the difference between dead and live loads helps engineers design wharf structures and cranes that remain safe and stable under all operating conditions.

13.7 Crane Dimensional Characteristics

Rail-mounted quay cranes are key equipment for cargo handling at ports. Their dimensions are carefully designed to ensure safe operation, stability, and smooth movement along the quay. The main dimensional factors include the rail gauge, distance between crane corners, number and spacing of wheels, outreach, and lift height.

The rail gauge—the distance between the rails—and the centre-to-centre distance between crane corners determine how stable the crane is and whether it fits the port's rail layout. These dimensions directly influence how the crane carries loads and maintains balance during operations.

The number of wheels at each corner and their spacing help distribute the crane's weight and the weight of the containers it lifts. Proper wheel arrangement ensures smooth travel on the rails and prevents excessive stress on both the crane structure and the rail tracks.

The outreach defines how far the crane can reach over a ship, typically between 30 m and 70 m, depending on vessel size. The lift height indicates how high the crane can lift a container above the quay—usually 21 m to 51.5 m above quay level—and how deep it can lower a container into the ship's hold—typically 10 m to 20 m below quay level.

Together, these dimensions ensure that RMQCs can safely handle modern container ships while meeting the operational needs of busy marine terminals.

13.7.1 Basic Crane Loads

All crane components other than the rated lifting load are treated as dead load. Designers must also account for inertia loads arising from hoisting, trolley travel, and crane movement. Overload scenarios such as collisions, snagging, and motor stalls must be included in the design. Environmental actions—particularly wind and seismic forces—are also critical, with the stowed wind load often governing crane design.

To align crane and wharf design, port engineers should provide the supplier with:

- **Service load combinations**
- **Factored load combinations**

To ensure consistent coordination between the crane and wharf design, port engineers should provide the crane supplier with both service load combinations and factored load combinations, along with a basic load table for the supplier to complete.

Further, it shall be noted that quay or STS cranes impose large vertical and horizontal loads, requiring verification of crane-specific load data before quay design, considering the following Crane-Induced Loads on Quay Structures.

It should further be noted that quay or STS cranes impose significant vertical and horizontal loads on quay structures. Therefore, crane-specific load data must be verified before finalising structural design, considering the following crane-induced loads, which shall include;

13.7.1.1 Wheel Loads

Wheel loads are transmitted through bogie wheels during crane movement. It include Vertical loads (crane weight + container), horizontal loads (travel + wind), and impact loads during lifting.

The track load per meter must be calculated for appropriate rail selection and foundation design.

13.7.1.2 Corner Point Loads

These are concentrated vertical loads at the crane's four support corners, which can be up to 1,0000 kN for modern cranes. Corner loads are critical for checking deck strength and foundation capacity. It's important to remember that these loads aren't fixed—they change based on site conditions. Things like cargo type, crane usage frequency, and environmental factors (like wind) can affect how much load is applied and where. That's why ports need site-specific load studies and dynamic simulations to make sure the quay design is safe and efficient.

Table 13-2: Example reference of loads on quay by different cranes from two trending manufacturers

Crane type	Loads by Supplier-A	Loads by Supplier-B
Panamax	35–45 tons/wheel	30–45 tons/m
Post Panamax	45–70 tons/wheel	40–55 tons/m
Super Post Panamax	46–90 tons/wheel	60–80 tons/m

Based *on 8 Wheels per Corner at 1.00m Spacing*

Since crane loads vary depending on site conditions, cargo type, operational frequency, and wind exposure, site-specific load studies and dynamic analyses are essential for ensuring safe quay performance. A sample wheel-load distribution from past projects may be provided for reference.

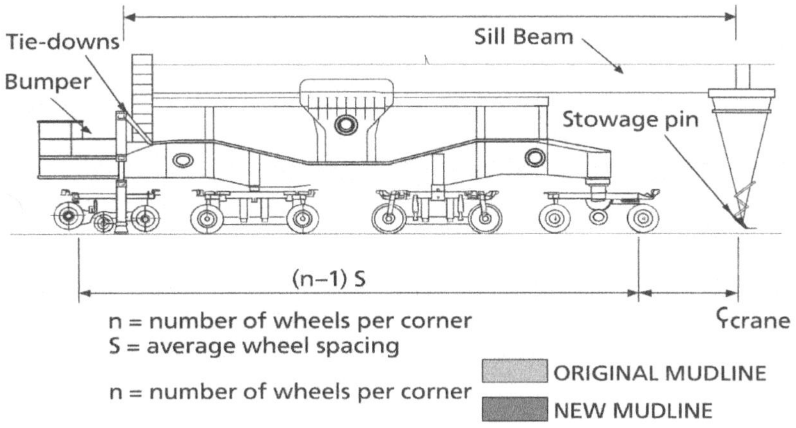

Figure-13.5: General Arrangement of Crane Wheels at Corners

13.7.2 Crane Base Configuration for Better Load Management

- **Wider Corner Spacing:** Increasing the centre-to-centre distance between crane corners creates a larger base, allowing more wheels and improving how loads are spread across the structure.
- **More Wheels per Corner:** Adding extra wheels at each corner reduces the load on individual wheels and decreases wear on the rails.
- **Proper Wheel Spacing:** Keeping suitable spacing between wheels ensures smooth, even load transfer and helps avoid local stress on the rails, improving overall stability.

13.7.2.1 Defining Safe Wheel Loads for Modern Cranes

Setting safe wheel load limits is essential to ensure that quay structures can support the operational demands of modern, high-capacity cranes. This begins with establishing realistic wheel load values based on the crane's lifting capacity, outreach, and operating conditions. Specialist crane consultants assess these loads under various scenarios—including wind, seismic effects, and dynamic forces generated during container handling—to determine suitable design thresholds.

When the quay's structural capacity is marginal, reinforcing the dock often proves more practical and cost-effective than reducing the crane's weight. Strengthening the quay not only ensures immediate operational safety but also allows for future upgrades and heavier equipment. As ports transition to larger cranes, designing for higher wheel load capacities becomes a strategic requirement. If load limits cannot be increased, crane manufacturers must verify compliance through accurate measurements and load assessments. Engaging experienced consultants helps ensure that crane systems meet current performance needs while remaining flexible for future increases in ship size and handling demands.

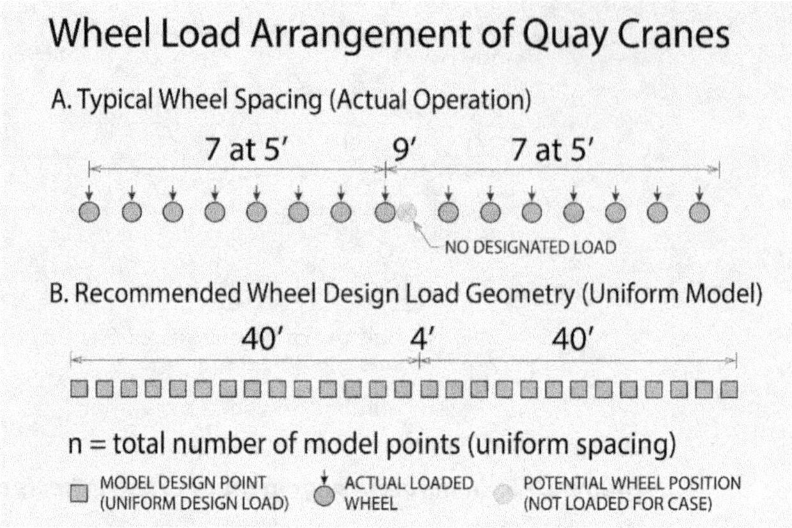

Figure-13.6: Typical Wheel Load Arrangement of Quay Cranes

13.7.2.2 Storm Anchors and Stowage Arrangements

Crane design data usually comes from the manufacturer, but it often doesn't match exactly what wharf designers need. One key detail is the factored uplift load at each crane corner, which helps assess how much force the tie-downs must resist. Load sharing between multiple tie-downs can vary a lot—sometimes one tie-down takes the full load due to crane movement, construction tolerances, or deformation. The direction of the uplift force also matters, as even small misalignments can change how bolts are loaded and how force is spread across the tie-down hardware. To reduce these issues, it's helpful to design the tie-down link plate so it can rotate around its weak axis, and to chamfer the pinhole in the ear plate to reduce eccentric loading.

Designers also need to understand the crane stop load provided by the manufacturer. This is usually given as the rated bumper reaction, which is the maximum force when the crane hits the stop at full gantry speed. This reaction depends on how fast the crane is moving. At low speeds, the oil in the bumper flows slowly through a small opening, causing little resistance. At high speeds, the oil flow creates a strong reaction as the gas spring compresses, which can significantly impact the structure.

13.8 Design Methodology

The design and evaluation of piers and jetties require estimating all major loads acting on the structure—dead loads, live loads, seismic forces, and marine forces such as waves and currents. Structural components such as piles, beams, slabs, and walls must be designed to safely resist these loads, and the overall stability of the jetty must be checked under the worst-case loading conditions.

Upgrading a general-cargo berth to a mechanised cargo-handling berth often reduces deck surcharge because mechanised systems distribute loads more evenly. This uniform load distribution lowers stress concentrations and improves the deck's structural performance.

Modern cranes also help reduce structural demand. Newer models use more wheels to spread loads over a wider area, decreasing pressure on the wharf deck and reducing forces transferred to the supporting piles. This improves operational safety, protects the structure, and allows the facility to handle heavier loads without compromising long-term integrity.

13.8.1 Evaluating Dock Strength for Crane Upgrades

When upgrading a wharf to handle higher-capacity cranes, the first step is to check whether the existing crane rails can support the new loads. This assessment uses detailed load data from crane manufacturers and includes reviewing the rail geometry, material properties, and supporting beam strength. In many cases, stronger rail profiles and reinforced beams are required. Additional piles may also be needed beneath the rail beams—on both the waterside and landside—to safely support the increased crane loads.

The crane design engineer is key to this process. By reducing the crane's self-weight during design, the engineer can significantly lower the wheel loads transferred to the quay. A lighter yet strong crane not only protects the dock structure but also decreases steel usage and overall construction cost, making it a practical and economical approach for port upgrades.

13.8.2 Design of Crane Rails

Crane rails are the interfacing elements between the crane and the supporting structure. The section properties of the rail should be properly verified for a trouble-free crane operation. A poor rail installation can result in significant ongoing maintenance costs and downtime resulting from the replacement of wheels or the wearing of rail. A rail installation consists of rail, clips, rail pads, bearing plates, holding down bolts, and grout, which needs proper detailing and workmanship. Other critical factors in the installation of rails are welding procedures, expansion joint design, and installation tolerances. When selecting crane rails, consideration should be given to the operating load, wheel diameter, the width of the rail head, and the hardness of the elements. Undersized rails result in premature wear and may require replacement within five to ten years, while a proper rail installation should last 20 years or more.

STS cranes/ Wharf cranes/ Yard cranes have higher wheel loads, but slow traveling speed, rails with 80mm wide flange (CR-80 rails) may be used. For large STS cranes and fast-running wide-span cantilever RMG cranes, a practical rail should have a wearing surface 100 mm wide or more.

To verify the crane rails' size, the information about design wheel loads, crane wheel diameter, and material properties of rails is very important. With service/ operating loads, appropriate design safety factors should also be applied in accordance with design specifications and the standard code of practice.

For designing crane rail interfaces, it is important to consider the following:

- Wheel loads, wheel diameter and spacing, and equaliser centre distance
- Rail type, geometry, gauge, and tolerances

Operating wheel loads are a combination of dead, wind, lifted, and other loads. The wind angle and all possible positions of the load, as well as the simultaneous operation of different crane drives, must be considered. The loads are added in different combinations reflecting actual operations. Crane wheel load calculations also consider overload and extreme load conditions such as snag events, storm wind, and earthquakes.

13.8.3 Design Verifications for Crane Rails

There are various rail selections that may be used for wharf cranes, and it is necessary to check whether a particular rail will be suitable for applicable operating and extreme loads. Some of the rail sections commonly used for cranes are shown below.

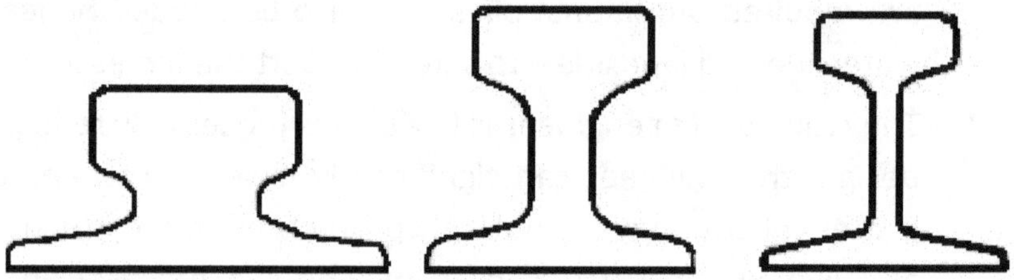

Figure-13.7: Typical Cross-sections of Crane Rails

i) Grade of Steel for Rail

Crane Rails are generally made in steel grades from 690 N/mm^2 to 1100 N/mm^2. The most commonly available crane rails in India have 880 to 1080 N/mm^2 tensile strength. It is important to have rails that are compatible with the crane wheels. An increased strength grade is highly desirable in heavy-duty applications or where the project is concerned with a long life.

13.8.3.1 Crane Rail Selection and Design

The national design standards typically call for the radii at the edge of the rail head to be subtracted from the total width. For practical purposes, it is advised that one should use 95% of the total width as the effective head width when designing and selecting the rails for a particular application.

ii) Rail Head Width Required

The simplest method for calculating rail head width is to allow 2.5 mm head width per tonne for vertical wheel load for heavy-duty cranes. Allow 1.7 mm for normal and light-duty cranes.

More conservative figures can be calculated by using the following formula: (Derived from Kempe's Engineers' Year Book).

$$H = \frac{(1580\ x\ W)}{D} \langle for\ heavy - duty\ cranes \rangle$$

Where;

H = full rail head in mm

D = wheel diameter in mm

W = wheel load in tonnes force

Example:

W = 40 Tonnes

D = 680 mm

H = Rail Head Width =?

$$H > \frac{(1580 \text{ x } 40)}{680} = 92.95 \text{ mm}$$

ISCR – 100 Rail is O.K

iii) Web Crushing Stress

Wheel loads due to wharf cranes are very high, and the thickness of the web for crane rails is relatively thin. Therefore, the web crushing strength shall also be checked. For web stress checking, it is assumed that the load spreads from the crane wheel into the rail through a strip of rail along the head. The length of this along the rail may be adopted as 10% of the head width. The load then passes down through the rail, spreading out at 45º as shown in the figure below.

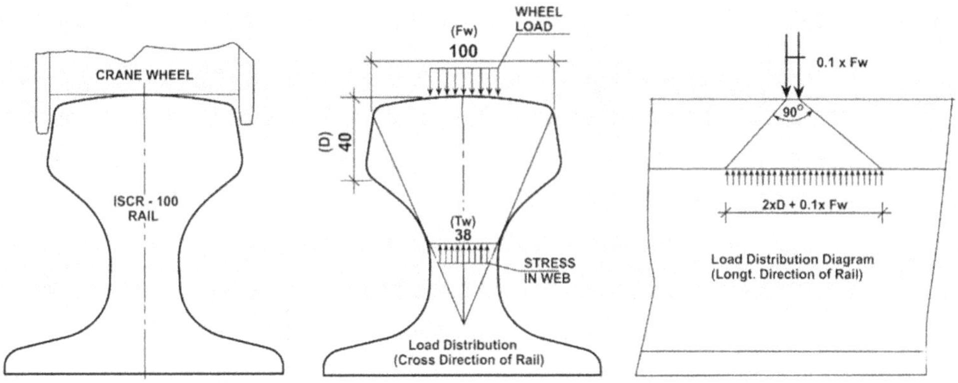

Figure-13.8: Web Stress in Crane Rails

Considering the combined stress in the web due to web crushing and minor bending effect, it is generally limited to $130 N/mm^2$.

Example

Crane Wheel Load = 400 KN

Rail Designation = ISCR – 100

***Rail* Dimensions**

Flange Width (Fw) = 100 mm

Flange Depth (D) = 40 mm

Web Thickness (Tw) = 38 mm

Total web area under stress = Tw x (0.1 x Fw + 2.D)

= 38 mm x (0.1 x 100 mm + 2 x 40 mm)

= 38 mm x 180 mm

$$= 6840 \text{ mm}^2$$

$$\text{Web stress} = \frac{400 \times 1000 \text{ N}}{6840 \text{ mm}^2}$$

$$= 58.48 \text{ N/mm}^2 < 130 \text{ N/mm}^2$$

The other precautions that need to be considered for the longevity of crane rails are as follows;

- Rails that are out of tolerance in the horizontal plane, out of gauge, will result in premature wear of rails and wheels—only time will show if the rail will wear the wheel flanges or the wheels will wear the rail.
- With a good rail installation, there should be little or no discernible wear of rail or wheel, assuming the crane is also properly aligned and does not rack excessively while running on rail.
- Rails improperly aligned vertically can induce racking loads in cranes, which may impair the crane structure prematurely.
- ISO 12488-1 may be used as a practical guide for rail installation tolerances.

13.9 Upgradation of Marine Fender Systems on Quay Walls

Fenders serve as the critical interface between a ship and the berth, acting primarily as a safety buffer to protect vessels, infrastructure, and personnel. When a ship makes contact, the fender compresses, dissipating energy and distributing the resulting forces through the system's components—such as panels, anchors, and chains—ultimately transferring them to the supporting structure via a defined load path.

When upgrading existing quay walls or jetty structures to accommodate larger vessels, the selection and installation of appropriate fender systems becomes essential for ensuring operational safety and structural integrity. As vessel dimensions increase, the lateral forces generated during berthing and mooring also rise, placing greater stress on the quay infrastructure. To safely manage these forces, existing fenders may need to be replaced or supplemented with higher-capacity units designed to absorb increased berthing energy. Suitable fender types for such upgrades include **cell fenders**, known for their high energy absorption and low reaction force; **cone fenders**, which offer excellent performance under large compression angles and are ideal for high tidal variations; and **super arch fenders**, which are robust and suitable for heavy-duty applications. In cases where space constraints exist, **foam-filled fenders** or **pneumatic fenders** may be used due to their flexibility and ease of installation.

Installation of upgraded fenders must follow precise engineering guidelines to ensure performance and safety. Key considerations include:

- **Correct positioning** based on tidal range and vessel draft to ensure consistent contact during berthing.
- **Anchoring and mounting** systems must be designed to transfer loads effectively into the quay structure without overstressing it.

- **Alignment and spacing** should follow standards such as those from PIANC or British Standards, typically limiting spacing to less than 15% of vessel length.
- **Corrosion protection** and **marine-grade materials** should be used to ensure long-term durability in harsh marine environments.
- **Load path verification** through structural analysis is essential to confirm that the quay wall can safely absorb and distribute the reaction forces from the fender system.

By integrating suitable fender types with proper installation practices, upgraded quay walls can safely accommodate the largest designed vessels while maintaining structural integrity and operational efficiency.

The following guidelines are provided to support the fender design and installation process, ensuring alignment with international standards and promoting long-term structural durability.

13.9.1 Fender Design and Selection Criteria

In line with Newton's Second Law of Motion, the force exerted on a wharf when a moving vessel is brought to a stop is proportional to the rate of change of momentum. A sudden halt results in a high-impact force, akin to a hard collision. However, using a soft, compressible medium like a rubber fender reduces the rate of momentum change, thereby minimising the force and protecting both the vessel and the berth.

Effective fender design requires a multidisciplinary approach, combining theoretical knowledge with practical insights from real-world berthing operations. Design codes and standards emphasise the importance of a sound understanding of these principles. This guide aims to support designers and specifiers by outlining key input parameters, guiding the calculation of berthing energy, and helping in the selection of suitable fender types.

To accurately design and select marine fenders, the calculation of berthing energy must consider several factors, including:

- Vessel type, class, and dimensions
- Approach velocity of the vessel
- The vessel's approach direction during berthing
- Type and configuration of the berthing structure
- Harbour basin exposure conditions
- Availability of docking assistance
- Tidal range, which influences fender positioning

Selecting suitable marine fender systems for upgrading a quay structure shall follow a systematic engineering procedure to ensure compliance with international best practices, operational requirements, and long-term durability.

The method adopted for future projects shall typically include the following stages:

i) Determination of Future Berthing Requirements

Fender upgrading shall be aligned with projected marine operations. This includes Design vessel size, displacement, Berthing velocity, angle, Environmental loads (wind, current, tide) and operational requirements (berthing frequency, contact locations, berthing approach)

These parameters form the basis for selecting upgraded fender systems suitable for future terminal use.

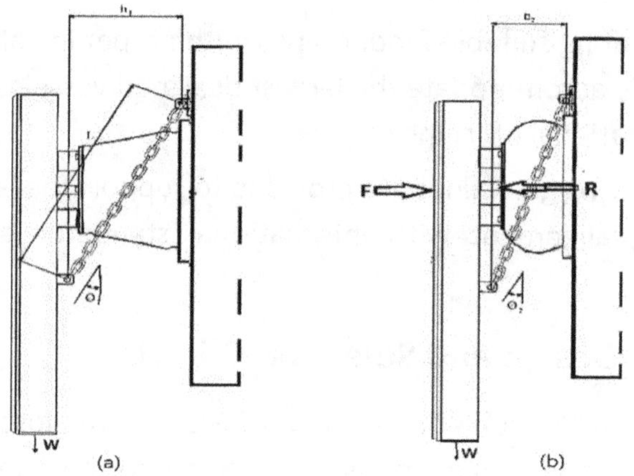

Figure-13.9: Application of Load in Fender

ii) Fender Type Selection and Performance Design

Based on updated berthing energy calculations, the appropriate fender type shall be selected (e.g., cone fenders, cell fenders, arch fenders, pneumatic fenders, composite systems).

Selection shall ensure Sufficient energy absorption, Acceptable reaction forces on the quay wall, Corrosion-resistant and low-maintenance components and Compatibility with quay structural capacity

iii) Structural Verification of the Quay Wall

Before installation of upgraded fenders, the structural capacity of the quay wall shall be checked for Increased reaction forces from the new fender system. If required, structural strengthening (anchor upgrades, backing plates, embedment strengthening) shall be incorporated into the upgradation program.

iv) Installation of Upgraded Fender System

Installation shall follow approved manufacturer guidelines and project specifications, which may include drilling and placement of new anchors or chemical anchor systems, Installation of backing plates, chains, brackets, and fixing hardware and Alignment checks to ensure correct fender projection and berthing line.

Thus, a judicious fender selection shall be based on updated berthing energy requirements and operational conditions to ensure optimum performance and long-term reliability. The design process compares vessel berthing energy with the capabilities of available fender types—such as cone, cell, arch, pneumatic, and composite fenders. Cone and cell fenders are suited for heavy-duty, high-energy applications, while arch and pneumatic fenders are appropriate for medium-duty berthing or where greater deformation flexibility is required. The selected fender must provide adequate energy absorption without imposing excessive reaction forces on the quay wall and must be compatible with the structural capacity of the quay. Durable, low-maintenance components and optimized fender panels with UHMW-PE pads shall be used to ensure uniform load distribution and prevent localized wall damage, thereby ensuring safe and efficient berthing performance.

13.9.2 Performance-Based Comparison of Marine Fender Types

13.9.2.1 Cone Fenders

Followings are the performance characteristics of cone fenders;

- Very high energy absorption with low reaction force, making them suitable for large vessels.
- High angular performance—efficient even under large approach angles.
- Stable deformation pattern ensures predictable load transfer to the quay wall.
- Long service life due to robust rubber geometry and minimal stress concentration.

Cone fenders are generally best suited for deep-water berths, Container terminals, heavy-duty applications.

13.9.2.2 Cell Fenders

The following are the performance characteristics of Cell Fenders

- High energy absorption and moderate reaction forces, though slightly lower efficiency than cone fenders.
- Excellent shear and torsional resistance, suitable for berthing with chain systems.
- Stable footprint allows use with large frontal panels.
- Performs well under low-to-moderate berthing angles.

These are best suited for: oil & gas berths, bulk terminals, general cargo quays.

13.9.2.3 Arch Fenders

Arch fenders possess the following Performance Characteristics

- Moderate energy absorption with higher reaction forces compared to cone and cell fenders.
- Simple, durable, and cost-effective design.
- Performs well under frequent, low-impact berthing.
- Suitable for locations with limited space, as they do not require large frontal panels.

These fenders are suitable for small to medium vessels, tug berths, and fishing harbours.

13.9.2.4 Pneumatic (Yokohama-type) Fenders

Performance Characteristics of Pneumatic Fenders are;

- Very high energy absorption with extremely low reaction forces due to air-based compression.
- Excellent performance under varying drafts, tidal ranges, and berthing angles.
- Portable and not structurally anchored to the quay wall, avoiding wall overload.
- Requires periodic pressure maintenance but offers superior cushioning.

These are best suited for ship-to-ship (STS) operations, temporary berthing, offshore terminals, and LNG berths.

13.9.2.5 Extruded / Cylindrical / D-Type Fenders

Performance Characteristics

- Lower to moderate energy absorption depending on size.
- High adaptability to curved or complex structures.
- Easy to install and replace.
- Best for low-speed, low-energy impacts.
- Best suited for pontoon edges, small craft terminals, marinas, piers, and workboats.

Table-13.3: Performance Comparison of Fenders

Fender Type	Energy Absorption	Reaction Force	Durability	Angular Performance	Structural Demand on Quay	Best Application
Cone	Very High	Low	Very High	Excellent	High panel load spread needed	Large vessels, heavy terminals
Cell	High	Moderate	High	Good	Requires good anchorage	Bulk, tanker, general cargo
Arch	Moderate	High	High	Moderate	Minimal	Small–medium vessels
Pneumatic	Very High	Very Low	Moderate–High	Excellent	Very low	STS, offshore, temporary berths
Cylindrical/ D-Type	Low–Medium	High	Moderate	Good	Minimal	Marinas, small craft

13.9.3 Fenders Alignment and Spacing for Safe Berthing Operations

During berthing, the vessel shall be carefully aligned to ensure gradual contact with the fenders, thereby preventing any direct impact on the quay wall. The following conditions must be met to ensure safe and effective berthing:

- Fender Condition: All fenders shall be in good working order, free from wear, damage, or deformation, and capable of delivering their specified energy absorption capacity.
- Fender Alignment: Fenders must be properly positioned and aligned to uniformly absorb the berthing energy as per design requirements.
- Multi-Point Contact: The vessel shall make contact with multiple fenders simultaneously to distribute impact forces evenly and minimise localised stress on the berth structure.
- Fender Design Parameters: The size and spacing of the fenders are critical and shall be selected based on vessel type, berthing energy calculations, and structural limitations of the quay wall.

13.9.4 Fender Arrangement and Spacing In Between

Once the appropriate type and size of fenders are selected, the next essential step is determining the quantity required, which is largely influenced by the spacing between units. Proper fender spacing is critical to the overall performance and safety of the system. Excessively wide spacing—often aimed at reducing costs—can compromise protection, increasing the risk of vessels making direct contact with the quay structure during berthing. To prevent such incidents, British Standards recommend that for continuous quay walls, the installation pitch should not exceed 15% of the vessel's length. However, this spacing must be suitable for the full range of vessels expected to berth at the facility. Specifically, the arrangement should accommodate the smallest vessels in terms of spacing, while the fender sizes must correspond to the largest designed vessel to ensure adequate energy absorption and structural protection.

The maximum spacing between fenders (S) can be calculated with this equation:

$$\boldsymbol{S} \leq 2.\sqrt{\boldsymbol{R_B^2} - (\boldsymbol{R_B} - \boldsymbol{P_U} + \boldsymbol{\delta_F} + \boldsymbol{C})^2}$$

Where,

S = Maximum spacing between fenders

$\boldsymbol{R_B}$= Bow radius of the vessel's broadside (in meters or feet)

$\boldsymbol{P_U}$ = Uncompressed height of the fender, including the panel

$\boldsymbol{\delta_F}$ = Fender deflection allowance

$\boldsymbol{C}$ = Clearance allowance between the vessel and the fender system

Note:

If radius info is not available, one can use this estimation to find out the info:

$$R_B \approx \frac{1}{2}.\left[\left(\frac{B}{2}\right) + \left(\frac{L_{OA}^2}{8.B}\right)\right]$$

Where,

B = Beam (width) of the vessel

L_{OA} = Overall length of the vessel

This approximation provides a practical method to estimate the bow radius when

13.9.5 Fender (or Panel) Surface Pressure

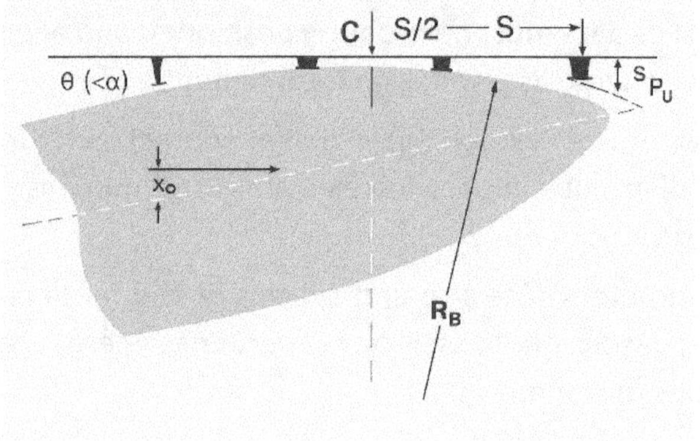

Surface pressure value of the fender has to be less than the vessel's allowable hull surface pressure. For certain fenders, like super cell type and cone type, they most commonly come with frontal frames/panels that distribute the pressure. So to decrease the surface pressure value, one can increase the surface area of the panel.

Figure-13.10: Spacing between Fenders for Angular Berthing

The hull form or curvature of the ship is important. The bow radius influences where a ship contacts the fender relative to its centre of mass. It also influences the number of fenders compressed depending on their spacing. Bow flares may push the upper edges of the fender closer to the structure, so the upper edges of the panel, chain brackets, etc., need to be checked for clearance.

13.9.6 Choosing a Suitable Frontal Panel

To choose a suitable panel, one has to consider hull pressures allowed for the berthing vessels. The following table shows a rough guide of allowable hull pressures of certain popular type of vessels. (just for reference):

Table-13.4: Limiting Hull Pressures on Frontal Panels for various types of Ships

Vessel Type	**Hull Pressure (KN/m^2)**
Tankers	150-250
ULCC & VLCC(Coastal Tankers)	250-350
Product & Chemical Tankers	300-400
Bulk Carriers	150-250
Post-Panamax Container Ships	200-300
Panamax Container Ships	300-400
Sub-Panamax Container Ships	400-500
General Cargo	300-600
Gas Carriers	100-200

13.10 Crane End Buffer and Tie-Down Arrangements

A crane end buffer is required on the quay wall to safely absorb and dissipate the kinetic energy of the crane in the event of over-travel along its rail track. It serves as a critical safety device that prevents structural damage to the quay, protects crane components from impact loads, and ensures controlled deceleration of crane movement.

13.10.1 Technical Requirements of Crane End Buffers

A crane end buffer is required on the quay wall to safely absorb and dissipate the kinetic energy of the crane in the event of over-travel along its rail track. It functions as a critical safety device that prevents structural damage to the quay, protects crane components from impact loads, and ensures controlled deceleration of crane movement.

Technical Requirements of Crane End Buffers

- **Energy Absorption Capacity**: The buffer shall be capable of absorbing the maximum kinetic energy generated by the crane at its rated travel speed and load condition, in accordance with manufacturer design and relevant standards (e.g., EN 13155, FEM/ISO guidelines).
- **Rated Maximum Reaction Force**: The buffer's reaction force must remain within the permissible limits of the quay wall structure and crane end carriage. Structural verification of supporting elements is required prior to installation.
- **Mounting and Alignment Requirements**: The buffer shall be accurately aligned with the crane rail and buffer stop frame to ensure uniform impact and proper load transfer.

Figure-13.11: Image of End Buffer for RMQC

The crane end buffer shall be installed on the quay wall at the designated location by setting the anchor bolts and securing the buffer unit to ensure effective absorption of crane travel impact loads. The installation is completed with functional checks to verify smooth crane movement and proper energy absorption performance.

13.10.2 Stowage Pin

A crane stowage pin on the quay wall is provided to securely restrain the rail-mounted crane in its designated parking or stowage position during idle periods or extreme weather conditions. The stowage pin engages with a corresponding slot or locking plate on the crane end-carriage, preventing unintended longitudinal or transverse movement caused by wind loads, surge forces, or operational vibrations.

Figure-13.12: Image of Stowage Arrangement for RMQC

It is typically fabricated from high-strength steel and anchored to a reinforced concrete block on the quay wall using heavy-duty mechanical or chemical anchors designed to withstand uplift, shear, and lateral loads. Proper alignment, embedment depth, corrosion protection, and load-capacity verification are essential to ensure reliable locking performance and to safeguard both the crane and quay infrastructure during non-operational periods.

13.10.3 Crane Tie-Down System

A crane tie-down system is installed on the quay wall to hold rail-mounted quay cranes securely during strong winds, cyclones, or when they are not in use. The system includes steel tie-down lugs or anchor points fixed into a reinforced concrete block, which connect to matching locking sockets on the crane. When the tie-down is engaged, it stops the crane from lifting, sliding, or tipping over due to extreme wind forces.

This setup ensures the crane stays safely locked in its parking position and protects both the crane and the quay during severe weather or emergency conditions.

Figure-13.13: Image of Tie-Down Arrangement for RMQC

13.10.3.1 Design Requirements for Crane Tie-Down System

The tie-down system shall be designed to withstand the maximum storm wind loads acting on the crane, as defined by the crane manufacturer and relevant design standards (e.g., EN 1991-1-4 for wind actions, FEM/ISO crane design guidelines). The tie-down anchors must safely resist uplift, shear, and longitudinal forces generated under extreme wind conditions. Tie-down lugs or pins shall be manufactured from high-strength structural steel with appropriate corrosion protection suitable for marine environments, and the embedded anchorage block shall be designed with adequate embedment depth and shear keys to ensure effective load transfer into the quay structure.

13.10.3.2 Compatibility and Alignment

The tie-down interface shall be fully compatible with the crane's tie-down socket geometry and locking mechanism. The tie-down positions must allow complete engagement without introducing eccentric forces on either the crane or the quay structure. Accurate vertical and horizontal alignment of the tie-down components is essential to prevent misalignment and ensure reliable performance under load.

Modernising Maritime Infrastructure

Steel Sheet Piles: Application in Wharf and Waterfront Structures

Chapter Summary

This chapter presents a concise yet comprehensive overview of how steel sheet piles are applied in wharf and waterfront structures, with particular focus on their role in quay wall strengthening, deepening, and bulkhead construction under varied geotechnical and marine conditions. Whether used as standalone retaining elements or integrated into high-capacity combi-wall systems, steel sheet piles offer notable advantages in structural efficiency, drivability, durability, and rapid construction within port environments.

Further, the chapter describes the geotechnical and structural considerations for various anchorage types and outlines the stability criteria and safety factors relevant to each. It also highlights key construction and durability requirements, such as drivability, hammer and equipment selection, and overall construction control. Together, these topics form a complete engineering basis for the effective use of steel sheet piles in high-performance wharf and waterfront projects.

Learning Objectives

- Steel sheet piles are available in different profiles with varying stiffness and bending capacity, while combi-walls use large king piles for heavy marine applications.
- Anchorage systems for quay walls must be selected according to ground conditions, space constraints, and required stability.
- The durability of steel sheet piles depend on effective corrosion control and suitable protection measures for long service life in marine environments.
- Drivability of sheet pile is controlled by pile strength, soil resistance, and proper hammer selection.

14.1 Introduction

Steel sheet piles are widely regarded as highly efficient and versatile structural elements used extensively in marine and waterfront infrastructure. Their inherent material properties—such as high strength, durability, and ease of installation—make them particularly well-suited for the development and upgrading of wharf structures, quay walls, and various waterfront retaining systems. This chapter presents an overview of the key characteristics, benefits, and specific applications of steel sheet piles in wharf development and port infrastructure, with particular emphasis on strengthening and upgrading existing quay walls and constructing bulkhead systems under diverse site-specific conditions and marine environments. Prior to addressing application-related and practical considerations, the fundamental characteristics of steel sheet piles are presented below.

14.2 Forms of Steel Sheet Piles and Their Configurations

Steel sheet pile systems are available in a wide range of profiles and configurations, enabling engineers to address diverse retaining wall requirements and sub-soil conditions. Their versatility allows effective solutions for both shallow and deep waterfront structures.

14.2.1 Z-Type Sheet Piles

Z-type sheet piles consist of a profile shaped like an elongated "Z," with a continuous web and interlocks positioned symmetrically on both edges, far from the neutral axis. This arrangement provides a high section modulus, making Z-piles one of the most efficient sheet pile profiles in terms of flexural resistance.

Due to their unsymmetrical geometry, individual Z-piles may twist during handling or installation; therefore, they are typically driven as crimped or welded doubles. Z-piles are widely used for intermediate to deep retaining walls, including cantilever and tied-back systems.

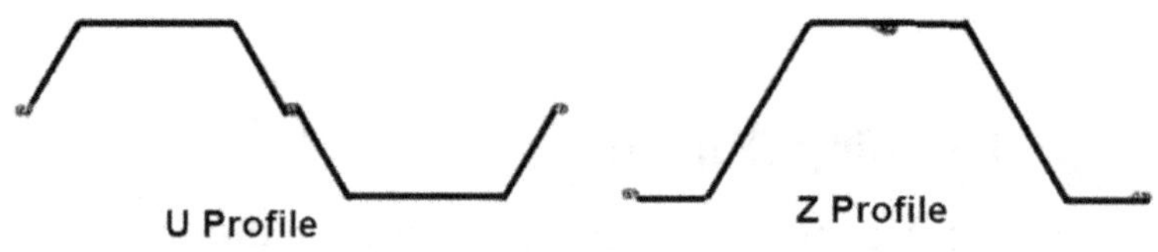

Figure 14.1 U-type and Z-type steel sheet piles

14.2.2 U-Type Sheet Piles

U-type sheet piles have a U-shaped cross-section with interlocks located on both sides, forming a continuous wall with a central axis through the double-pile configuration. While their bending resistance is lower than comparable Z-profiles due to the interlock position at the neutral axis, U-piles generally offer higher section modulus.

Narrow U-piles are suitable for harder ground conditions, while wider U-piles are preferred where drivability is easier. Compared to U-profiles, Z-piles offer greater system width, fewer interlocks per metre of wall, and improved watertightness.

14.2.3 Contiguous O-Piles

O-piles are pipe-shaped sections fitted with welded connectors, allowing each pipe to interlock with the next to form a continuous steel wall. The primary load-bearing element is the circular pipe itself, giving O-pile walls high stiffness and axial capacity.

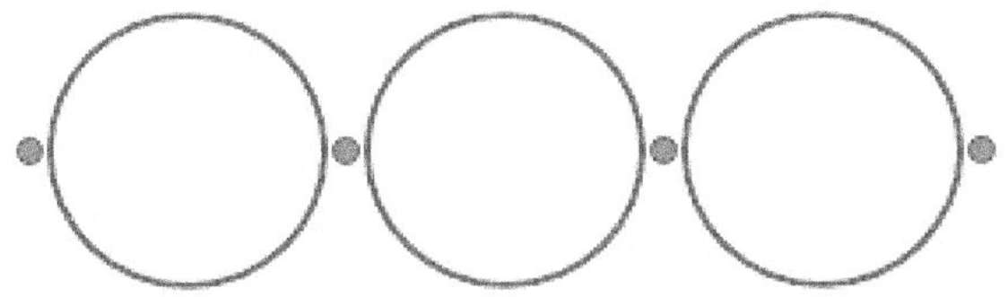

These systems are suitable for applications requiring significant bending resistance and wall rigidity.

Figure 14.2 O- type steel sheet pile wall

14.2.4 High Modulus Piles

High-modulus piles are formed by combining Z-profile sheet piles with deep universal beams (H-beams) welded in a soldier-pile arrangement. This configuration substantially increases the bending moment capacity and wall stiffness. The welded assembly is driven as a unified panel, providing enhanced structural performance for deep or heavily loaded waterfront structures.

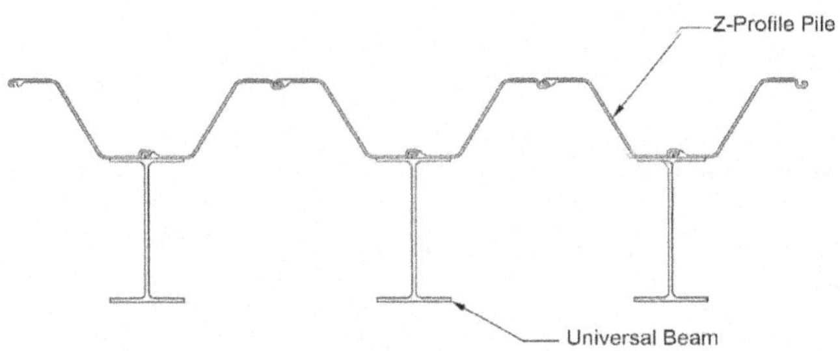

Figure 14.3 Z-type sheet pile wall + H-Beam System

14.2.5 Combined (Combi) Walls

Combined walls are used where conventional sheet piles do not provide sufficient stiffness or bearing capacity. They are a combination of primary king piles acting as the main load-bearing members and secondary sheet piles serving as infill elements that transfer lateral loads.

Combi-walls may utilise tubular piles, H-piles, or box piles as the primary elements. Tubular piles, in particular, provide high vertical capacity and wall stiffness and are commonly used in heavy marine works and deep-draft quay structures where greater strength and deeper penetration into competent strata are required.

King piles carry the majority of the vertical and horizontal forces, while secondary sheet piles act as diaphragms, spanning between king piles and transferring lateral loads. This system provides increased wall stiffness, reduced horizontal movement, and cost-effective alternatives to secant walls. The various forms of Combi-wall structures are as follows;

14.2.6 H + Z Combi-Wall

In this system, H-piles (king piles) are connected to Z-sheet piles using specially designed connectors. The interlocks provide adequate free play to facilitate installation while ensuring structural engagement under design loads. This configuration forms a strong, continuous retaining wall suitable for deep and heavily loaded quay structures.

Figure 14.4: H+Z combi-wall

14.2.7 O + Z Combi-Wall

O + Z systems incorporate O-shaped tubular piles as primary load-bearing elements, connected to Z-piles or U-Piles through welded sheet-pile connectors. These structures offer high axial and bending resistance and are used in marine environments requiring significant wall stiffness.

Other variations include O + U and HZ-combi walls, depending on structural and geotechnical requirements.

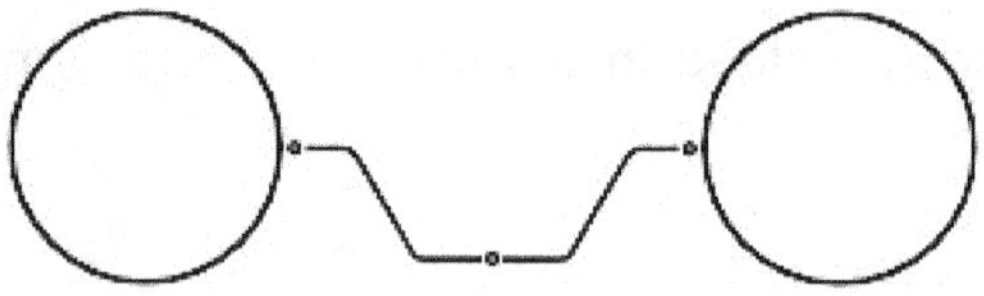

Figure 14.5 O+Z combi-wall

The other combined sheet pile systems are shown in Figure 14.6.

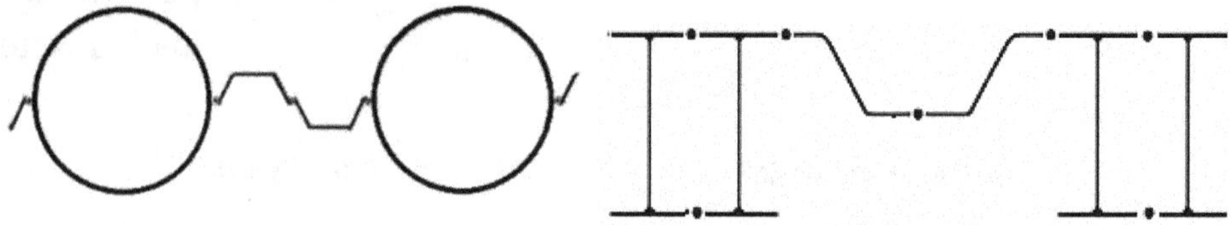

O+U Combi-wall   H+Z Combi-wall

Figure 14.6 Other combi-walls

14.3 History of Sheet Piles in Port Construction

Advancements in steel manufacturing enabled the production of hot-rolled I, Z, and U sheet pile profiles, and in 1914 the first steel sheet pile with a hot-rolled interlock was developed. This innovation provided sheet pile elements with improved drivability, enhanced soil-tightness, and reduced weight. The first steel sheet pile wall in the Port of Rotterdam was constructed in 1927, thirteen years after the introduction of the Larssen-type interlocked sheet pile. Following the Second World War, combined systems consisting of steel bearing piles and sheet piles were introduced, and by 1985, these systems became standard practice in the Port of Rotterdam, forming essential components in deep-draft wharf construction. The primary wall shapes and elements used in combined wall systems are described in the following section.

14.3.1 Key Historical Dates in Sheet Pile Development

- **Late 1800s** – Early timber and iron sheet piles with primitive reverted interlocks used for quay and river structures.
- **1902** – Karl Larssen patents continuous interlocking steel sheet piles, marking the beginning of modern sheet-pile technology.
- **1906** – Trygve Larssen introduces the first hot-rolled sheet pile section.

- **1914** – Hot-rolled I, U, and Z profiles with factory-rolled interlocks introduced, improving drivability, water tightness, and structural capacity.
- **1923–1931** – Industrial expansion of Larssen sheet piles in Europe; Japan begins domestic production by 1931.
- **1927** – First major steel sheet-pile quay wall constructed at the Port of Rotterdam.
- **1945** – Post–World War II expansion drives demand for deeper and heavier port structures; combined wall (combi-wall) systems using king piles and sheet piles are introduced.
- **1950s** – Tubular king piles adopted for combi-walls, enabling deeper embedment and higher bending resistance.
- **1960s** – Straight-web sheet piles were introduced for cellular cofferdams in large gravity-type marine structures.
- **1970s** – Mechanisation of pile driving through diesel hammers, vibratory hammers, and hydraulic systems enables installation of longer and heavier sections.
- **1980s** – Modern H + Z and O + Z combi-wall systems were adopted widely in deep-draft port construction, including Rotterdam.
- **1996** – EN 10248-1 formalises European sheet-pile steel grades S270GP, S355GP, and S430GP.
- **2000s** – Cold-formed sheet piles gain prominence as economical options for lighter retaining structures; ball-and-socket interlocks developed.
- **1990s–2005** – API undertakes extensive research upgrading tubular joint and pile design formulations for offshore and nearshore applications.
- **2005** – Nippon Steel introduces Hat-Type Sheet Pile 900, improving drivability and structural reliability.
- **2006–2007** – **Eurocode 3 Part 1-6** introduces unified limit-state design rules for cylindrical/tubular shells, applicable to large-diameter tubular piles.
- **2014** – **API RP 2A-WSD, 22nd Edition** released; becomes a key global reference for tubular pile design, widely applied in wharf and quay-wall foundation engineering.
- **2020s** – Digitalisation advances: FEM, BIM, and automated pile-driving systems widely adopted.
- **2023** – High-strength sheet-pile grade **S460GP** introduced in updated EN 10248-1:2023.

This integrated chronology captures a 120-year progression from timber piles and rudimentary interlocks to advanced hot-rolled, cold-formed, HAT-type, and high-strength steel sheet piles, supported by modern Eurocode (EC3) and API RP 2A tubular-pile design frameworks. The evolution reflects continuous enhancements in structural efficiency, drivability, and load-carrying performance. Future developments in high-strength metallurgy, precision installation systems, advanced corrosion protection, and fully integrated digital design and monitoring platforms will further increase structural reliability, extend service life, and optimise whole-life performance in marine, geotechnical, and urban

14.4 Selection of Steel Sheet Piles for Cut-off Walls

Cut-off walls constructed for quay walls, berthing structures, bulkheads, and waterfront retaining systems are primarily intended to retain soil and water, support operational surcharge loads, and safely transmit vertical and horizontal forces to the underlying foundation strata throughout the service life of the facility.

Steel sheet pile walls are widely adopted for quay wall construction due to their high structural efficiency, suitability for marine environments, rapid installation, and adaptability to variable ground conditions. These systems are designed to perform satisfactorily during construction, normal operation, and under extreme loading conditions. In general, the design and selection of steel sheet pile systems involve the following key considerations:

- Selection of appropriate steel sheet pile geometry;
- Selection of sheet pile steel grade and material properties;
- Assessment of durability requirements for steel sheet piles;
- Evaluation of suitable anchorages for wall
- Verification of sheet pile drivability;
- Assessment of wave tranquillity within the basin.

i) Selection of Steel Sheet Pile Geometry

For a given structural purpose and site condition, the selection of sheet pile profile and geometry is a critical aspect of the design and is governed by structural, geotechnical, drivability, economic, and construction considerations. Various types of primary sheet piles are available, including U, Z, H and O sections, as well as combined systems such as H-Z, O-Z and O-U profiles, as described in the relevant sections.

Steel sheet piles forming quay walls are primarily subjected to bending moments, shear forces, and axial forces. The combined action of bending, shear, and axial load shall be assessed in accordance with the governing design codes.

ii) Sheet Pile Steel Grades and Material Properties

The structural efficiency of sheet pile walls may be enhanced through the adoption of steels with higher yield strength. Higher-grade steels, notably S390GP and S430GP, are widely available, providing minimum yield strengths of 390 N/mm^2 and 430 N/mm^2, respectively. A further increase in capacity is anticipated with the introduction of S460GP, having a minimum yield strength of 460 N/mm^2, in the forthcoming revision of EN 10248.

As the bending resistance of sheet piles is directly proportional to the yield strength of the steel, the use of higher-strength grades allows lighter pile sections to be selected for a given design action. This results in reduced overall steel tonnage and improved economy, while maintaining satisfactory constructability and installation performance.

Steel materials for sheet piles specified in accordance with EN 10248 shall satisfy the following mechanical performance requirements:

- **Strength**, sufficient to meet ultimate and serviceability limit state requirements;
- **Ductility**, adequate to maintain load-bearing resistance under large deformations associated with soil–structure interaction;
- **Toughness**, sufficient to prevent brittle fracture under impact loading and stresses induced during pile driving;
- **Weldability**, appropriate to ensure reliable fabrication and site construction.

Steel pipe piles are supplied in accordance with EN, ASTM, or API standards and are manufactured as seamless, LSAW, or SSAW carbon or high-strength steel pipes. SSAW piles are available in various diameters with typical wall thicknesses up to approximately 25 mm and are generally suitable for economical foundation applications. LSAW piles, fabricated from rolled steel plates, can be supplied with wall thicknesses of 100 mm or greater and are intended for heavy-load or critical foundation conditions.

a. Steel Grade Designation

Steel grade designations are defined in EN 10027-1 for hot-rolled products and structural steels. Grades are primarily classified by their specified minimum yield strength at ambient temperature, supplemented by additional identifiers related to mechanical properties and delivery conditions.

S 355 J R M H	
Principal Symbol	
S	denotes Structural Steel
Mechanical Properties	
355	denotes a minimum yield strength in N/mm² for thickness < 16 mm
J	denotes minimum impact energy value (toughness) at 27 J for Charpy test
K	denotes minimum toughness at 40 J
R	denotes a test temperature at +20°C
0	denotes a test temperature at 0°C
2	denotes a test temperature at −20°C
Delivery Conditions	
N	denotes Normalised rolled steel
NL	denotes Normalised rolled weldable steel
M	denotes Thermo-mechanically rolled steel
H	denotes Hot rolled hollow section

iii) Durability of Steel Sheet Piles

Steel sheet piles located in marine or saline environments shall be designed for durability, considering corrosion as a time-dependent deterioration mechanism influenced by moisture, oxygen availability, and chloride exposure.

Marine structures are generally subjected to five distinct exposure zones: atmospheric, splash, tidal, submerged, and mud zones. The most severe corrosion commonly occurs in the splash zone and slightly above the Mean High Water Level (MHWL), where cyclic wetting and drying combined with high oxygen availability accelerate material degradation. The relative severity of corrosion across the different exposure zones is illustrated in Figure 8.3.

Corrosion mitigation measures shall be selected on a project-specific basis. Where corrosion rates are low, additional protection may not be required; however, in aggressive environments, appropriate durability measures are essential to achieve the specified design life. Common approaches include:

- **Sacrificial steel allowance**, enabling anticipated section loss due to corrosion to be accommodated within structural design limits;
- **Higher-strength steels**, subject to verification of deflection and local buckling behaviour;
- **AMLoCor micro-alloyed steel**, offering significantly reduced corrosion rates in marine immersion and low-water zones;
- **Protective organic coatings**, requiring careful specification due to potential damage during handling and installation;
- **Cathodic protection systems**, effective below the half-tide level and requiring specialist design;
- **Concrete encasement**, providing robust protection in splash and tidal zones, subject to interface and concrete durability considerations.

iv) Suitability of Anchorage Arrangements

Excavation in front of embedded walls or placement of backfill results in stress redistribution within the soil mass, which may lead to lateral wall deformation and increased earth pressures requiring structural restraint. Accordingly, anchorage systems are introduced at the head of sheet pile walls to control deflections, enhance stability, and improve design economy by reducing bending moments and embedment depths. Anchorage systems transform cantilever walls into propped or anchored systems, enabling lighter structural sections and more economical solutions. The selection of anchorage type directly influences wall behaviour, construction methodology, and long-term performance.

System selection shall be based on site-specific considerations, including ground conditions, depth and extent of competent strata, groundwater regime, availability of space behind the wall, construction sequence, and permissible wall deformations. Anchorage systems shall be positioned outside potential active failure zones to ensure full mobilisation of resistance and long-term stability.

v) Drivability of Sheet Piles

Steel piles and sheet piles shall possess adequate drivability to overcome soil resistance during installation and achieve the required penetration depth. The limit of drivability is defined as the maximum soil resistance that can be overcome without structural damage or premature refusal when driven using a suitably sized hammer. Inadequate assessment of drivability is one of the most common deficiencies in driven pile design.

The pile driving system comprises three interacting components:

- the pile;
- the pile driving hammer; and
- the surrounding soil.

Pile Requirements

To ensure adequate drivability:

- The pile shall have sufficient stiffness to transmit driving forces effectively;
- The pile shall have adequate structural strength to withstand driving stresses, with total induced stress limited to 0.9 × fy, verified in accordance with EN 1993-5.

  Installation of a deep cut-off wall using a combi-wall system in front of existing quay structures typically requires high driving energy, generating significant ground vibrations. In the presence of clean, loose to medium-dense granular soils at the toe level, vibration-induced compaction may occur, leading to seabed subsidence and potential outward wall movement or tilting. Accordingly, ground susceptibility to vibration-induced instability shall be assessed in advance, and suitable mitigation measures implemented.

vi) Wave Tranquillity after Installation

Reconfiguration of waterfront structures shall be undertaken without compromising harbour basin tranquillity. Wave-absorbing wall systems are preferred over conventional vertical walls as they significantly reduce wave reflection and associated wave-induced loads.

A porous combi wall configuration, formed by installing intermediate sheet piles of limited length between tubular king piles, permits partial transmission and dissipation of wave energy. When combined with a rock revetment behind the combi wall, the system functions as an effective wave-absorbing arrangement, enhancing basin tranquillity and supporting safe and efficient port operations.

The above factors are discussed in detail in the subsequent sections.

14.4.1 Advantages of Steel Sheet Piles as Cut-off Wall

Steel sheet piles provide several benefits that directly support wharf and waterfront construction installing a robust cut-off wall:

i) Steel sheet piles are manufactured with predetermined shapes and dimensions, allowing full inspection before and during installation to ensure consistent quality.

ii) The design, installation, and quality assurance associated with driven piles eliminate uncertainties and deliver a reliable, cost-effective foundation system suited for varied subsurface conditions.

iii) A wide range of steel grades and pile profiles is available, enabling high structural strength and increased working load capacity, resulting in economical solutions for marine foundations.

iv) Driven steel piles can be engineered to resist compression, tension, and lateral loads according to project requirements, site constraints, and budget considerations.

v) Steel sheet piles are well-suited for marine and near-shore environments and typically do not require special casing systems for offshore or riverfront installations.

vi) When driven through water, steel piles can be used immediately, allowing construction to progress without curing delays.

vii) Steel piles eliminate curing time, enabling significantly faster construction schedules and minimising overall project timelines.

14.5 Functional Role of Combi-Sheet Piles as Cut-off Wall

Combi- sheet piles are flexible systems that can be used in many soil conditions and for different types of quay wall constructions. They help to build and upgrade a strong and cost-effective quay wall and bulkhead that can carry heavy loads and resist large bending forces. They create straight vertical cutoff walls that allow ports to build upgraded waterfront areas for enhanced cargo handling, storage, and general wharf operations. Sheet pile walls also provide safe and stable berthing edges that can withstand the forces from ships during berthing and mooring. They support heavy equipment such as cranes, trucks, and utility lines. In deeper waters, high-modulus sheet pile systems are used, where large main piles provide most of the strength, and smaller secondary piles act as stiffeners to improve overall performance. The very specific functional advantages relevant to the upgradation and reconstruction of waterfront engineering are as follows;

i) **High Strength-to-Weight Ratio:** Their favourable strength-to-weight ratio reduces overall structural self-weight, which is especially advantageous in marine and offshore environments where load efficiency is essential.

ii) **Rapid Installation and Off-Site Fabrication:** The ability to fabricate sheet piles off-site reduces construction time and minimises risks associated with weather, tidal variations, and other site-dependent constraints.

iii) **Superior Driveability:** Steel sheet piles possess excellent resistance to impact, bending, lateral loading, and heavy compression, making them suitable for driving through varied soil layers, including cohesive, granular, and underwater strata.

iv) **Predictable Material Properties:** Manufactured under controlled conditions, steel sheet piles provide consistency in strength, stiffness, and durability, essential for long-term waterfront performance.

14.5.1 Comparative Study of Combi-Wall Systems

A typical combi-wall system is a very strong and effective cutoff wall that shall ensure structural continuity and prevent soil loss through the joints. The interlock connections between primary and secondary elements must be robust. As combi-steel sheet piles have various forms, the behaviour and performance of these combi-wall connections were thoroughly assessed by de Gijt and Heijndijk (1996), who evaluated three key components:

- the primary king pile,
- the intermediate sheet pile, and
- the interlock or connector system.

Their analysis considered bending stiffness in both principal directions, deformation characteristics, ultimate capacity, dimensional stability, installation practicality, and cost. Each parameter was scored on a scale of 1 (poor) to 5 (excellent), and the comparative results are illustrated in Figure 14.7.

Type of Pile	Rigidity		Torsion	Buckling	Dimensional Stability	Installat-ion	Costs
	Ix	Iy					
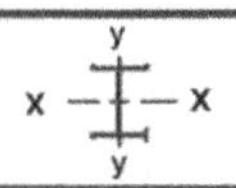	3	1	1	3	4	3	2
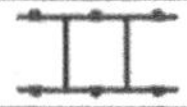	5	3	3	4	4	4	1
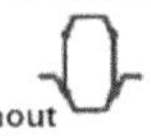 Without Reinforcement Rib	4	3	3	2	3	3	3
 With Reinforcement Rib	5	3	3	3	4	3	3
 Welded Along Seam	5	5	5	5	5	5	4
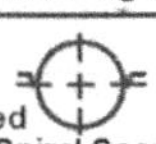 Welded With Spiral Seam	5	5	5	5	4	4	5

***Figure 14.7** Evaluation of the types of combi-wall systems*

The study concluded that steel tubular piles generally provide superior overall performance. However, H-beams combined with Z-piles may be preferred where specific site constraints or geometric considerations dictate alternative configurations. Large-diameter open-ended tubular piles used as primary elements in combi-walls offer several engineering advantages

14.5.2 Use of Combi–Sheet Pile Wall for Upgradation

For deepening and structural upgradation of quay walls, combi-sheet pile walls—consisting of large-diameter tubular king piles combined with intermediate sheet piles—are increasingly adopted worldwide. When configured as a cutoff wall in marine environments, combi-walls offer distinct engineering, geotechnical, and hydraulic advantages. The key benefits are summarised below.

v) **Efficient Driving Characteristics**: Tubular piles, with their closed cross-section and uniform stiffness, exhibit excellent drivability. Their geometry minimises risks of buckling or deviation during installation, thereby ensuring improved verticality and alignment control.

vi) **High Vertical and Horizontal Load Resistance**: The large, uniform moment of inertia of tubular sections provides high flexural stiffness essential for resisting berthing impacts, crane loads, soil pressures, and hydrodynamic forces. Their uniform buckling resistance in all directions enhances axial capacity, making them particularly suitable for carrying large vertical loads and resisting horizontal loads from soil and differential water pressure.

vii) **Efficient Foundation**: During driving, open-ended tubular piles naturally develop a soil plug that increases both shaft resistance and end-bearing capacity. After installation, tubular piles can be further strengthened internally by placing compacted granular fill inside the pile. Such infill enhances end bearing resistance in the founding strata and reduces buckling risks for the free-standing portion of the pile.

viii) A reinforced concrete plug cast at the top of the pile stabilises the compacted fill and provides a robust ‘t pile head for integration into the existing structure.

14.5.3 Porous Combi-Wall for Reduced Wave Reflection

Using infill sheet piles of limited length between tubular king piles creates a porous wall system that offers significant hydraulic and operational benefits:

i) Wave Energy Dissipation

The gaps between the intermediate sheet piles allow partial transmission of wave energy instead of reflecting it entirely. This reduces wave load on piles.

In combination with a rock revetment placed behind the combi-wall, the porous configuration serves as an effective wave-absorbing system, lowering wave reflection and improving overall tranquillity within the berthing basin. The improved calmness enhances safety, reduces vessel motions, and facilitates more efficient cargo-handling operations.

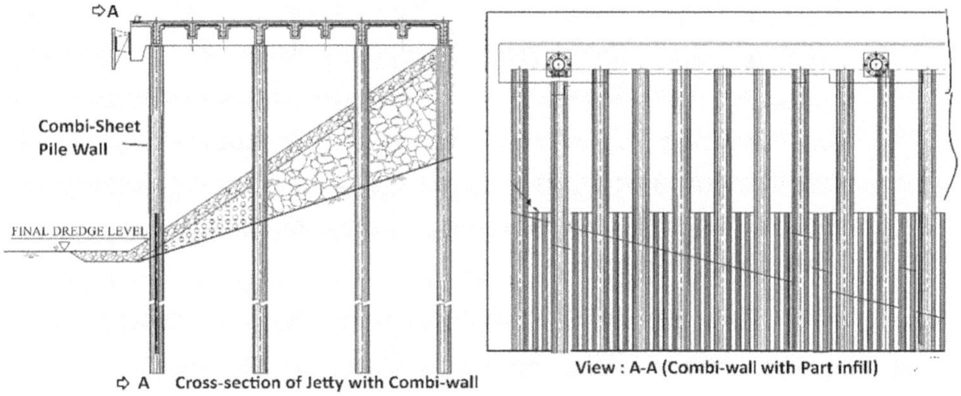

Figure 14.8: Comb—Wall Arrangement for Wave Absorption

Reference – Vlorë Harbour:

A technically comparable example to the Vlorë Harbour wave-absorbing combi-wall system is described in the paper *"Innovative Combiwall Quay-Wall with Internal Rubble Mound Chamber: Numerical Tools Supporting Design Activities"* presented at the *26th International Ocean and Polar Engineering Conference*.

Although this study focuses on Vlorë Harbour itself, it provides generalised evidence of similar applications in other port environments, noting that:

i) Quays built over piles combined with absorbing rubble-mound slopes are used to *enlarge or rebuild structures in existing docks*, especially where reduction of wave reflection inside harbour basins is required.

This rubble-mound configuration acts similarly to a rock revetment, providing wave energy dissipation behind a structural retaining system such as a combi-wall.

ii) Such rubble-mound absorbing systems are highlighted as common practice in situations where low-reflection performance is essential for mooring and manoeuvring conditions. They reduce wave reflection into the basin in much the same way as the rock revetment behind the porous combi-wall proposed for Vlorë.

iii) The paper notes that several vertical dissipative solutions—including Jarlan-type wave-absorbing chambers—have been adopted globally to reduce wave reflection by combining porous or open vertical structures with internal rubble-mound energy-dissipating layers. This configuration is conceptually equivalent to a porous combi-wall plus rock revetment system.

14.6 Case Study: Wharf Upgradation Using Combi Sheet Piles

To gain a clear understanding of the application of sheet piles as cut-off walls in quay wall and wharf upgradation projects, particularly for deepening works and additional dredging, typical scenarios encountered in existing marine structures were examined. In the executed projects, sheet piles were introduced as cut-off walls to facilitate deepening and dredging activities, control soil movement, and enable the installation of anchors or tie-back systems for enhancing the lateral stability of the modified quay structures. Simultaneously, modifications were carried out to the deck structures to integrate the new cut-off walls with the existing structural systems, ensuring compliance with the revised operational and loading requirements.

This section presents three representative case studies illustrating the use of sheet pile cut-off walls for the upgradation of existing wharves and quay structures. Each case study was characterised by a distinct structural configuration and specific engineering challenges. The typical upgradation scenarios considered were as follows:

i) **Upgradation of a wharf supported by a gravity retaining wall**, where a tubular-type combi sheet pile wall was constructed as a cut-off wall and supplemented with tie-back anchors. This approach enhanced the structural capacity of the wharf, improved overall stability against lateral forces, and extended the functional service life of the existing retaining system.

ii) **Upgradation of a quay wall constructed with an RCC diaphragm wall**, where the works focused on achieving greater dredging depths, improving structural strength, increasing load-carrying capacity, and accommodating higher berthing and operational demands. For the modification of the existing quay wall, a tubular-type combi sheet pile cut-off wall with tie-back anchors was installed to provide the required structural support and stability.

iii) **Upgradation of an open berthing structure with a revetment**, where a cut-off sheet pile wall composed of HZ-type piles was installed to provide effective earth retention, protection against scour, and additional structural support for expanded port operations. To ensure adequate lateral stability, anchoring systems were installed behind the sheet pile wall. Additionally, the deck structure was strengthened, and crane rail beams were modified to safely accommodate increased operational loads and modern handling equipment.

14.6.1 The Fifth Dock at Port of Antwerp

This case study described the upgradation of an existing wharf supported by a masonry-type gravity retaining structure. At the Port of Antwerp, the Fifth Dock quay wall had originally been constructed in the early nineteenth century to accommodate vessels with drafts of less than 7 m. With the subsequent increase in vessel sizes and the need to accommodate Cape-size vessels, it became necessary to deepen the dock and upgrade the existing quay wall structure. Accordingly, the dock wall was upgraded in 1994 to meet the revised operational and navigational requirements.

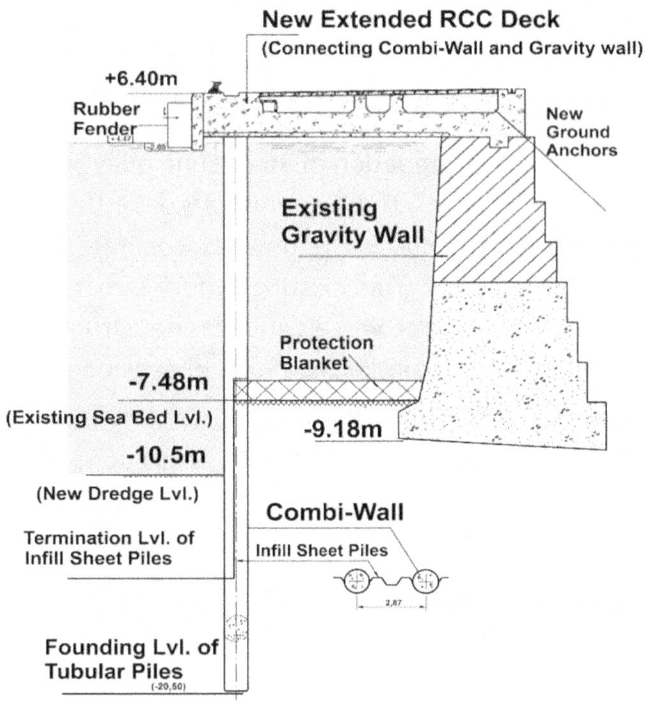

Figure 14-9 Typical cross-section of Quay Wall Renovation at Antwerp

The typical schematic arrangement of the upgraded quay wall was illustrated in Figure 14-9. As part of the adopted solution, the existing masonry gravity quay wall was retained. To facilitate deepening and structural augmentation, a new cut-off wall was introduced in front of the existing structure. This cut-off wall comprised a tubular-type combi-wall, designed to safely retain soil at the founding strata and ensure stability under increased dredging depths. The tubular piles, acting as king piles, were extended up to the deck level, while the infill sheet piles were terminated at the level of the protection blanket provided at the mudline of the existing quay wall. The infill sheet piles were deliberately terminated at a lower level to create a partially open and porous wall system, thereby mitigating the effects of hydrostatic pressure induced by tidal variations on the newly constructed cut-off wall. Based on a detailed assessment of the geotechnical conditions and their spatial variability, an open combi-wall arrangement with infill sheet piles of limited penetration was adopted, positioned at the level of the scour protection.

To meet the revised operational requirements at the jetty deck level and to integrate the new cut-off wall with the existing masonry quay wall, a new deck structure was constructed. This deck structure effectively connected the new cut-off wall system with the retained existing quay wall, ensuring structural compatibility and continuity.

As a consequence of increased dredging depths, modified load conditions, and enhanced operational demands on the upgraded quay wall, new grout anchors were installed. These anchors were designed to reduce the horizontal forces acting on the wall due to earth pressures and increased bollard loads, as well as to minimise eccentricity and inclination of the resultant forces at the foundation level. Although this solution for revamping the wharf was technically demanding, it proved to be less costly and less time-consuming than alternative solutions that would have required demolition or complete replacement of the existing quay wall.

14.6.2 Augmentation of General Cargo Berths at Paradip Port

This case study documented the upgradation of an existing quay structure supported by an RCC diaphragm wall at Paradip Port. The General Cargo Berths were more than three decades old and comprised an anchored reinforced concrete diaphragm wall with a thickness of approximately 1.10 m. The existing wharf structure had originally been designed to accommodate Panamax-size vessels and was not structurally adequate to berth the larger Cape-size vessels proposed under the port's expansion plan.

To address the revised operational requirements, the existing quay wall was planned for reconfiguration and deepening to enable berthing of vessels up to 180,000 DWT. As part of the long-term development strategy, the berthing basin was envisaged to be deepened from the existing level of –11.5 m to –19.8 m.

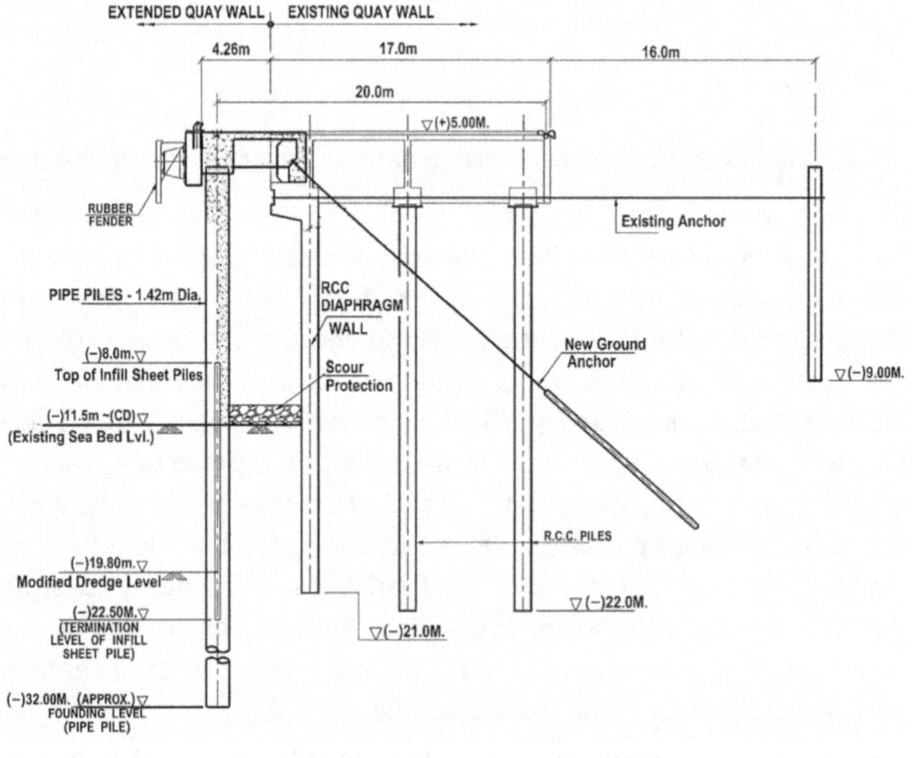

Figure 14.10 Typical cross-section of Augmented Quay Wall

A comprehensive assessment of the existing berthing structures, along with a detailed evaluation of the sub-soil conditions across the port area, was carried out by the author. This study supported the planning, design verification, and structural reconfiguration of the existing quay wall to ensure that the upgraded system satisfactorily met the new performance criteria. Regular technical assistance was also provided to the site execution team to ensure compliance with the approved design and construction requirements.

The approved augmentation scheme involved the installation of a cut-off wall in front of the existing quay wall, comprising tubular piles and infill sheet piles forming a combi-wall system, as schematically illustrated in Figure 14-10.

The augmentation process involved the following key construction stages:

i) A new cut-off wall was constructed using deep tubular piles combined with infill sheet piles. The tubular piles functioned as primary load-bearing members as well as the main retaining elements, while the infill sheet piles provided retention of the soil mass at the interface, thereby enabling safe dredging up to the final level of –19.80 m.

ii) Tie-back anchors were installed at each mooring location and at intermediate points along the berth to resist additional lateral loads arising from the berthing and mooring of larger vessels.

iii) The alignment of the combi-wall was set approximately 3.26 m seaward of the existing quay face to facilitate berth widening, accommodate mechanisation requirements, and provide additional deck space.

iv) The newly constructed cut-off wall was structurally integrated with the existing berth deck by extending the deck slab up to the cut-off wall alignment. Fender systems and bollards were relocated along the modified quay face to suit the new operational configuration.

The adopted construction methodology proved to be economical and significantly reduced the project execution time. The project represented one of the first successful applications of a tubular combi-wall system in India, and, following completion, the augmented quay wall was placed into full operational service.

14.6.3 Upgradation of Port Elizabeth Marine Terminal

This case study described the upgradation of an existing wharf characterised by an open berthing structure with a revetment at the Port Elizabeth Marine Terminal, located near New York City. The wharf structures along Berths 82 through 98 had originally been constructed in the early 1970s to serve as general cargo berths and were designed to accommodate Panamax-size vessels. With the growing demand for containerised cargo, the terminal was subsequently equipped to handle container ships. In response to continued growth in vessel size and the introduction of next-generation container vessels, the Port Authority initiated a program to upgrade the approximately thirty-year-old wharf structures by increasing the harbour draft.

As part of the development plan, the berthing basin was deepened from the original level of –10.0 m to –15.20 m to accommodate vessels of up to 150,000 DWT. The original berthing platform comprised a reinforced concrete deck supported on timber piles. Lateral stability of the berthing and mooring system was originally achieved through tie rods and anchor piles connecting the pile heads to the anchoring system behind the wharf.

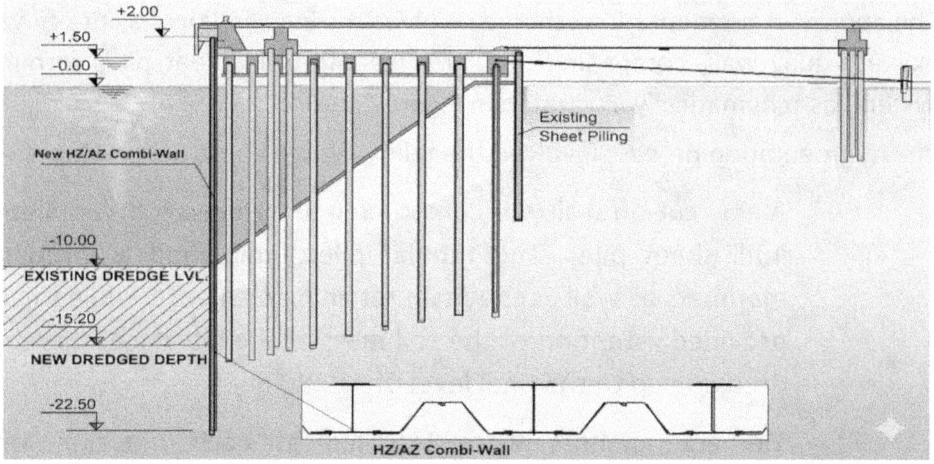

Figure 14.11 Typical cross-section of Modified Wharf at Elizabeth Terminal

The deepening and improvement works for the wharf included the following key elements:

i) A combi-sheet pile cut-off wall, consisting of a combination of H-beams and Z-type sheet piles, was installed along the face of the wharf. This cut-off wall enabled the dredging depth in front of the wharf to be safely increased to –15.20 m while maintaining stability of the existing foundation system.

ii) The structural deck elements were rehabilitated and strengthened to support larger container-handling cranes. In addition, the existing fender systems were replaced with higher-capacity fenders capable of resisting the increased berthing energy associated with the proposed 150,000 DWT vessels.

iii) Additional anchor systems incorporating deadman anchors were installed to reduce deflections and stresses in the berthing structure arising from increased mooring forces and deeper dredging levels.

The selected cut-off sheet pile wall system offered the advantage of retaining the existing founding strata beneath the wharf while allowing dredging in front of the toe of the structure to the required depth. The primary load-carrying members, consisting of H-beams, were installed from the wharf deck along the berthing line, while intermediate short Z-type sheet piles were driven between the H-beams using follower-pile techniques to complete the cut-off wall.

This solution proved to be highly cost-effective for the renovation and upgradation of the wharf, as it minimised disruption to the existing structure while achieving the required deepening. In addition, the deck structure was strengthened, and crane rail beams were modified to safely accommodate increased operational loads and modern container-handling equipment, ensuring long-term operational reliability of the upgraded terminal.

14.7 Drivability of Steel Sheet Piles

Ensuring the drivability of steel sheet piles is a critical design and construction consideration, as the ability of a pile to reach the required embedment depth directly influences the performance of the completed structure. Drivability refers to the capability of the pile to penetrate soil under driving forces without damage and is governed by the interaction between soil resistance and the structural stiffness and strength of the pile section. The pile must efficiently transmit driving energy from the hammer to overcome soil resistance while simultaneously withstanding high dynamic stresses, which may approach up to 90% of the steel's minimum yield strength. These driving stresses are significantly higher than service-level stresses and therefore demand adequate stiffness for energy transfer and sufficient strength to resist impact-induced forces during installation, as discussed below.

i) Pile Driving System

A pile driving system comprises the pile, the driving hammer, the surrounding soil, and supporting equipment such as cranes, leads, and templates. The effectiveness of pile installation depends on the proper interaction and compatibility of these components, which control the efficiency of energy transfer from the hammer to the pile. Inappropriate matching of the pile, hammer, and soil conditions can result in inefficient penetration, excessive driving stresses, or structural damage to the pile.

The maximum stresses in a driven pile generally occur during the installation phase rather than during its service life. Consequently, driving-induced stresses must be kept within prescribed allowable limits to prevent damage. For steel piles, the permissible maximum compressive stress during driving is commonly restricted to approximately ninety percent of the minimum yield strength of the material, ensuring safe installation under severe dynamic loading conditions.

ii) Influence of Soil Stratification and Hardness

Soil stratification and stiffness have a direct influence on pile drivability. Soft and very soft soils such as peat, silt, and soft clays generally allow easy driving, while loosely deposited sands and gravels without cementation also present favourable conditions. Moderately dense sands, fine gravels, and firm clays result in moderate driving resistance. Difficult to very difficult driving conditions occur in densely packed sands and gravels, stiff to very stiff clays, cemented layers, cobbles, and weathered or weak rock. These conditions are commonly evaluated using borehole logs and standard penetration test data.

Accordingly, the stiffness of pile section and hammer or pile driver shall be selected.

iii) Selection of Driving Method

A pile driving system comprises the pile, the driving hammer, the surrounding soil, and supporting equipment such as cranes, leads, and templates. The effectiveness of pile installation depends on the proper interaction and compatibility of these components, which control the efficiency of energy transfer from the hammer to the pile.

Inappropriate matching of the pile, hammer, and soil conditions can result in inefficient penetration, excessive driving stresses, or structural damage to the pile.

The maximum stresses in a driven pile generally occur during the installation phase rather than during its service life. Consequently, driving-induced stresses must be kept within prescribed allowable limits to prevent damage. For steel piles, the permissible maximum compressive stress during driving is commonly restricted to approximately ninety percent of the minimum yield strength of the material, ensuring safe installation under severe dynamic loading conditions.

iv) Management of Difficult Driving Conditions

In challenging ground conditions where high driving resistance is expected, several techniques can be employed to improve drivability and prevent pile damage. Driving shoes may be fitted to strengthen the toe and facilitate penetration into hard or cohesive strata. Predrilling or augering helps reduce point resistance and shaft friction, particularly in dense or cemented soils. In non-cohesive granular soils, water jetting at the pile toe can significantly lower penetration resistance by loosening and lubricating the soil. Toe modifications for H-piles can be used to navigate hard strata or boulder layers more effectively. Additionally, when an obstruction is encountered, it is often beneficial to proceed with driving adjacent piles first and return later, once soil displacement has reduced localised resistance. These methods collectively assist in overcoming difficult layers while protecting the structural integrity of the pile.

v) Dynamic Effects, Setup, and Restrike

Pile driving induces substantial disturbance in the surrounding soil, generating excess pore pressures that subsequently dissipate with time, leading to either setup (an increase in capacity) or relaxation (a decrease in capacity). In cohesive soils, where pore pressure dissipation is slow, immediate post-driving resistance may not reflect true long-term capacity. Therefore, restrike tests—either set-checks performed shortly after initial driving or re-drives conducted after a longer rest period—are essential to verify final capacity and ensure an accurate assessment of drivability. Such tests allow the engineer to account for time-dependent soil behaviour and confirm whether the driven pile has achieved the required bearing capacity.

14.7.1 Driving System Components and Alignment Control

Reliable drivability and accurate sheet pile installation depend on the proper configuration of the driving system and alignment controls. Hammer cushions, pile helmets, and follower systems must be correctly matched to the hammer and pile section to ensure efficient energy transfer and prevent head damage under high driving stresses.

Cranes must provide adequate lifting capacity and boom height to maintain safe handling and alignment of long piles. Leads—whether swinging, semi-fixed, or fixed—play a critical role in keeping the hammer and pile in proper verticality throughout driving. Similarly, templates or guide frames are often indispensable, especially in panel driving operations, as they maintain alignment, ensure correct positioning, and reduce risks of declutching or pile misalignment during installation.

i) Driving Methods and Installation Sequence

Appropriate driving methodology is crucial for achieving the required penetration while maintaining alignment and protecting the sheet pile interlocks.

- The pitch-and-drive method, although simple, is limited to short piles in loose soils and carries a higher risk of misalignment.
- Panel driving provides better control over verticality and minimizes drifting or declutching, making it suitable for denser or obstructed soils. In very dense sands, gravels, or rocky layers, staggered driving is preferred to reduce cumulative soil displacement and maintain vertical alignment by alternating the order of pile penetration.
- A properly planned driving sequence helps reduce soil heave, avoid premature refusal, and prevent interlock damage, thereby ensuring efficient and continuous progress of piling works.

14.8 Installation of Combi-Sheet Pile Wall

A combined (combi) sheet pile wall consists of high-stiffness primary structural members (king piles), such as tubular piles or H-sections, installed at regular intervals and connected by lighter, more flexible intermediate sheet piles. The installation of king piles is a critical operation and must be executed with a high degree of accuracy to ensure that each pile is driven perfectly vertical or to the specified inclination (batter). This precision is essential to maintain parallel alignment between king piles and to achieve the designed centre-to-centre spacing. Once the king piles are successfully installed and verified for position and alignment, the intermediate sheet piling sections are subsequently positioned and driven between them.

The successful installation of a combined wall requires meticulous attention to the selection of equipment, the use of appropriate driving templates, and strict control of the installation methodology, as outlined below.

14.8.1 Installation Process

When constructing a combined wall system incorporating tubular or H-section king piles, installation shall always begin with the king piles, as these form the primary load-bearing elements of the wall and exhibit the highest stiffness.

To ensure positional accuracy and alignment, king piles must be installed using a rigid driving template. Only after completion and acceptance of the king pile installation should the more flexible intermediate sheet piles be installed, as illustrated in the accompanying figure showing the recommended sequence of installation.

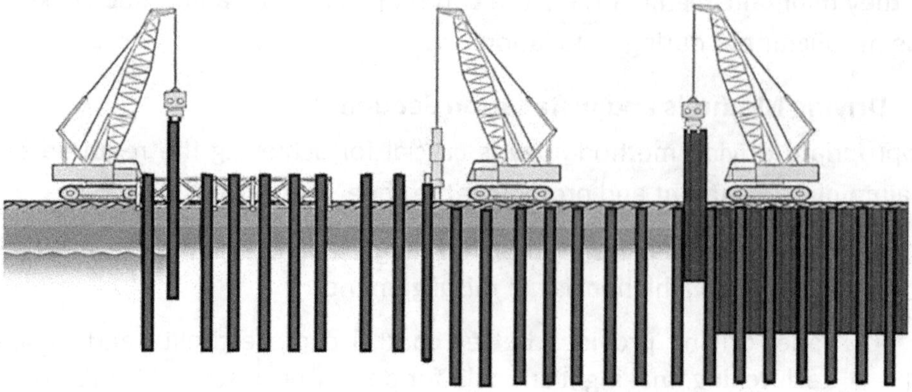

Figure 14.12: Order of Installation of Combined Wall

It is recommended to employ a crawler crane equipped with a leader, or alternatively, a two-level driving template to precisely control pile alignment during driving. The type of supplementary equipment required is dependent on whether installation is carried out from land or from water. For marine installation, the crane must be mounted on a stable platform such as a pontoon or jacket structure. In addition, a secondary pontoon or barge may be connected to the main working platform to facilitate the storage and supply of king piles and intermediate sheet piles, ensuring efficient and continuous installation operations.

14.8.2 Order of Installation

The driving of piles into the ground results in soil densification and lateral displacement in the surrounding strata. Consequently, the sequence of pile installation is of critical importance in order to minimise cumulative soil movement and associated positional inaccuracies. Adopting an appropriate installation order ensures optimal control over horizontal alignment and verticality of the piles, as detailed below.

i) Installation of King Piles

Prior to pitching and driving heavy tubular king piles, it is essential to install a rigid and well-aligned driving template. The primary function of the template is to accurately control the horizontal positioning of the piles. Driving templates are typically fabricated from structural beams and/or hollow sections. Short transverse beams are arranged to maintain the specified spacing between adjacent piles. Small welded "notches" are provided on these cross beams to engage and retain the pile interlocks, thereby preventing rotation of the piles during driving. The clearance between the interlocks and the notches shall be limited to only a few millimetres to maintain alignment within acceptable tolerances. A typical driving template configuration should incorporate an odd number of pile positions, as illustrated in the reference figure.

Once all king piles within a panel or template have been pitched, priority shall be given to driving the first and last piles, followed by the remaining piles in an alternating sequence. This installation order is critical to achieving the best possible results in terms of horizontal positioning and verticality. The recommended driving sequence, indicated numerically in the reference figure, generally progresses from the outer piles towards the centre. This approach is governed by geotechnical considerations: driving tubular piles in dense or firm strata causes soil compaction due to vibration and induces a plugging effect. If piles are driven consecutively in a linear sequence, uneven soil compaction on either side of adjacent piles may occur, resulting in lateral displacement and loss of verticality. Site experience has demonstrated that the installation sequence shown in Figure 11.8 provides superior control of pile alignment.

Where the driving template accommodates five (5) piles, the first stage involves driving the two outer piles. These piles provide lateral confinement for the remaining piles within the panel, thereby minimising differential soil compaction and reducing the risk of lateral pile movement during subsequent driving operations. Following this, the central pile is driven, and finally, the second and fourth piles are installed. This staged driving sequence is particularly critical in dense sand and stiff or hard clay soils, where soil displacement effects are more pronounced.

Figure 14.13: Sequence of King Pile Driving

Upon completion of driving all five piles within the panel, the driving template shall be advanced to the right-hand outer pile. This pile then serves as the initial pile for the next installation section, ensuring continuity of alignment and spacing for the combined wall system.

ii) Installation of Intermediate Sheet Piles

Pitching and driving of the intermediate sheet piles shall be carried out only after the successful installation of the king piles. The installed king piles provide a fixed reference and act as guides for the alignment and positioning of the infill sheet piles.

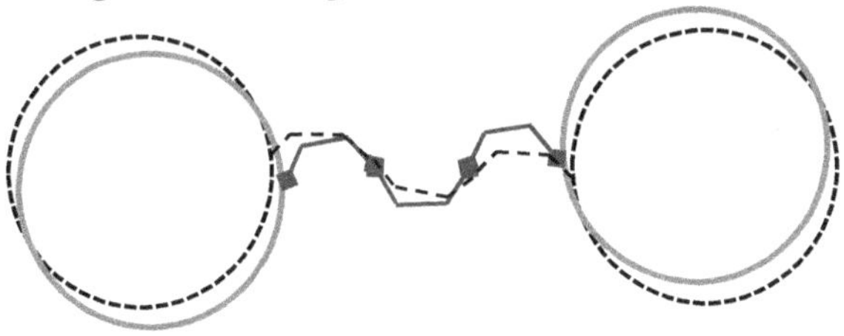

Figure 14.14 : Driving Tolerances in Combi-wall

As intermediate sheet piles are comparatively more flexible than the king piles, they are capable of accommodating minor deviations in verticality or orientation of the king piles. This flexibility allows the intermediate piles to follow these small imperfections while remaining fully engaged within the interlock system, thereby ensuring proper clutching and structural continuity of the combined wall, as illustrated in Figure 14.14.

iii) Driving Tolerances

Driving tolerances for intermediate sheet piles are generally stringent and depend on soil conditions as well as the length, shape, and size of the piles. As standard practice, pile verticality should be maintained within 10 mm per metre of pile length. Horizontal positioning tolerances typically require the tubes to be within 10–15 mm of the combined wall centreline. During the pitching and driving of the king piles, continuous monitoring of alignment relative to the wall shall be carried out using surveying instruments such as theodolites.

14.8.3 Driving Equipment and Installation Method

Modern piling practices permit the use of either vibratory hammers or impact hammers for driving both king piles and intermediate sheet piles. Vibratory hammers are generally preferred, as they reduce the risk of structural damage. However, in dense or hard soil conditions, vibratory methods alone may be insufficient to achieve the required penetration depth, necessitating the use of impact hammers. In practice, a combination of both techniques is often adopted.

The typical installation sequence for a combined sheet pile wall is as follows:

1. King piles are initially driven to the maximum feasible depth using a vibratory hammer.
2. If required, an impact hammer is then used to advance the king piles to the design toe level.
3. Intermediate sheet piles, with connectors attached, are installed between the king piles and are preferably driven using a vibratory hammer, with an impact hammer used only at the final stage if necessary.

14.9 Durability Requirements in Marine Environments

The durability of steel sheet piles depends primarily on their ability to resist corrosion over the intended service life. Corrosion reduces steel thickness, section modulus, stiffness, and overall load-carrying capacity. Ensuring durability means making considerations for environmental exposure conditions, expected corrosion rates for each exposure zone, Design life of the structure and protective measures during design and installation. Steel does not deteriorate with age on its own — its condition depends on it.

i) Corrosion Behaviour of Steel Sheet Piles

Corrosion behaviour in steel is defined by a natural electro-chemical process that gradually destroys steel when exposed to moisture, oxygen, and aggressive chemicals.

Key Factors Driving Corrosion are;

- Moisture or humidity, which acts as an electrolyte, enabling electron flow between the anode and cathode areas on the steel surface.
- Oxygen accelerates the cathodic reaction. Corrosion increases in environments where oxygen is continuously replenished, such as tidal or splash zones. And,
- Chlorides (mainly from seawater) increase the electrical conductivity of water, allowing corrosion to occur more aggressively.

Table 14-1: Corrosion Rates And Behaviour Depending on Exposure Condition

Zone	Behaviour	Typical Rate (mm/year/side)
Splash Zone	Constant wetting + oxygen, rust layer removed	~0.075–0.08
Tidal Zone	Alternating wet/dry cycles	~0.035–0.04
Permanent Immersion	More stable, uniform corrosion	~0.035
Below Seabed	Very low oxygen → minimal corrosion	~0.012

The splash zone is the most critical area often governing corrosion design.

Steel sheet piles experience different corrosion rates depending on exposure, and their typical behaviour is as follows;

ii) Durability Requirements During Design

To ensure long-term performance, the design must include a corrosion allowance (sacrificial steel), providing sufficient additional thickness to account for anticipated corrosion loss so that the residual section remains structurally adequate over the design life. This allowance should recognise that corrosion rates vary along the pile length, with significantly higher deterioration occurring in the splash zone.

For all other exposure zones, sacrificial steel is generally more reliable than protective coatings, as coatings are prone to damage during pile driving. However, in the splash zone—where corrosion rates are exceptionally high—additional marine-grade protective coatings of suitable thickness are recommended to ensure adequate durability.

14.9.1 Use of Protective Systems to Improve Durability

Durability of steel sheet piles can be significantly enhanced through the use of suitable protective systems. These systems are selected based on the severity of corrosion exposure along different zones of the pile and the likelihood of coating damage during installation.

i) Protective Organic Coatings

Protective organic coatings such as epoxy, polyethene (PE), and polypropylene (PP) provide an additional barrier against moisture, oxygen, and chlorides. These coatings are especially beneficial in upper exposed sections, where atmospheric corrosion or splash effects can be more pronounced.

To be effective, the coatings must:

- Be applied to an adequate **thickness** to ensure long-term protection.
- Be capable of **resisting abrasion and mechanical damage** that occurs during pile driving, transport, and handling.
- Include appropriate **surface preparation** to achieve good adhesion and long coating life.

Coatings may require touch**-up repairs** at the site if damaged during installation.

ii) Cathodic Protection (CP)

Cathodic protection is widely used in marine environments to control corrosion by converting the steel surface into a cathode of an electrochemical cell. Two CP systems are typically adopted: i) Sacrificial anode systems and ii) Impressed current systems

Cathodic protection is most effective below the low-tide level, where the structure remains submerged. It is often used in combination with protective coatings, which reduces total current demand and improves long-term performance. It shall be designed by specialists to ensure adequate current distribution without affecting adjacent structures.

iii) Concrete Encasement

Concrete encasement provides a robust physical barrier around the steel, particularly valuable in splash and tidal zones where wave action, high oxygen availability, and chlorides accelerate corrosion. To ensure durability, the concrete must be high-quality, low-permeability, and chloride-free. Adequate cover thickness and proper placement and curing are essential, and encasement should be continuous across the high-risk zones and designed to avoid electrochemical corrosion at the steel–concrete interface.

Concrete protection is particularly useful for preventing Accelerated Low Water

iv) Corrosion-Resistant Steels (AMLoCor)

AMLoCor is a micro-alloyed, low-corrosion steel developed for steel sheet piles used in aggressive marine environments. Long-term in-situ testing has shown that AMLoCor provides significantly better corrosion resistance than conventional carbon steel, with up to a 3-fold reduction in corrosion rate in the permanent immersion zone and up to a 5-fold reduction in the low water zone, which is typically the most critical region. It also offers resistance to Accelerated Low Water Corrosion (ALWC). AMLoCor generally achieves the required service life without the need for additional protective coatings, and its mechanical properties meet standard design requirements, including compliance with EN 1993-5. Even when the design incorporates corrosion protection, durability can be compromised during installation if care is not taken.

v) **Preservation of Protective Coatings**

The preservation of protective coatings during installation is essential for ensuring long-term durability of steel sheet piles. When piles are driven through gravel, dense soils, or other abrasive materials, coatings are susceptible to scratching, chipping, or complete removal, which significantly reduces their corrosion-resistance. Therefore, the selection of protective coatings must account for expected driving stresses, ensuring adequate toughness and adhesion. Any damage observed during handling or driving must be repaired immediately to prevent early corrosion initiation. In highly abrasive ground conditions, sacrificial steel allowances may offer a more dependable protection strategy, as they remain effective even when coatings are compromised.

vi) **Ensuring High-Quality Backfill**

The type and condition of backfill surrounding the steel sheet piles have a major influence on long-term corrosion behaviour. Aerated or chemically contaminated fill materials promote higher corrosion rates due to increased oxygen availability and the presence of corrosive agents. Conversely, proper compaction reduces voids, limits oxygen ingress, and thereby slows the corrosion process. Where aggressive fill materials such as slag, industrial waste, or peat soils are present, additional protective measures—such as enhanced coatings, cathodic protection, or increased corrosion allowance—must be incorporated into the design to ensure the structural longevity of the pile system.

vii) **Avoid Steel–Concrete Interface Corrosion**

Special care is required when steel sheet piles are partially encased in concrete, as the interface between exposed steel and concrete can become a critical point for accelerated corrosion. This is primarily due to electrochemical potential differences between steel embedded in alkaline concrete and steel exposed to seawater or atmospheric conditions. Differential aeration at the transition zone can further create localized anodic regions, leading to rapid pitting and section loss. To prevent such deterioration, protection measures must be continuous across the steel–concrete interface, ensuring that there are no unprotected gaps or abrupt changes in exposure conditions. Proper detailing, high-quality concrete, and consistent coating systems across the interface are essential to maintaining long-term durability.

14.9.2 Driving System Components and Alignment Control

Reliable drivability and accurate sheet pile installation depend on the proper configuration of the driving system and alignment controls. Hammer cushions, pile helmets, and follower systems must be correctly matched to the hammer and pile section to ensure efficient energy transfer and prevent head damage under high driving stresses.

Cranes must provide adequate lifting capacity and boom height to maintain safe handling and alignment of long piles. Leads—whether swinging, semi-fixed, or fixed—play a critical role in keeping the hammer and pile in proper verticality throughout driving. Similarly, templates or guide frames are often indispensable, especially in panel driving operations, as they maintain alignment, ensure correct positioning, and reduce risks of declutching or pile misalignment during installation.

i) Driving Methods and Installation Sequence

Appropriate driving methodology is crucial for achieving the required penetration while maintaining alignment and protecting the sheet pile interlocks.

- The pitch-and-drive method, although simple, is limited to short piles in loose soils and carries a higher risk of misalignment.
- Panel driving provides better control over verticality and minimises drifting or declutching, making it suitable for denser or obstructed soils. In very dense sands, gravels, or rocky layers, staggered driving is preferred to reduce cumulative soil displacement and maintain vertical alignment by alternating the order of pile penetration.
- A properly planned driving sequence helps reduce soil heave, avoid premature refusal, and prevent interlock damage, thereby ensuring efficient and continuous progress of piling works.

14.10 Driven Piles in Weak Rocks

Design guidance for large-diameter driven piles in weak rock or hard clay is currently scarce. Onshore projects typically favour drilled and grouted piles to avoid uncertainties in capacity and installation associated with driven piles in such conditions.

While extensive experience and established standards like DNV (1992) and API (2000) exist for pile design in sands and clays, similar guidance for weak rock is lacking. The behaviour of driven piles in weak rock is not well understood, especially regarding capacity modelling and drivability.

Design methods for estimating skin friction and end bearing in weak rock vary significantly, leaving designers to select suitable approaches, define parameters, and plan site-specific investigations. This often results in excessive reliance on certain tests, overly conservative pile lengths, certification challenges, and pile refusal during construction.

14.10.1 Design Process

Optimal pile design should be based on a clear understanding of rock mass behaviour and the potential risks involved in the design, installation, and operation of driven piles in weak rock. The recommended process, drawn from experience with such conditions, includes:

1. Developing a ground model and geotechnical risk register.
2. Identifying rock mass properties.
3. Selecting the most suitable design method and related parameters.
4. Defining and specifying the required survey.
5. Establishing appropriate design parameters.
6. Performing pile design and drivability analysis.

7. Verifying the design through pile testing and/or pile driving monitoring (PDM).

To assess the design capacity and drivability of piles in soft rock, the following risks should be taken into consideration from the early stages of the project. The main challenges in designing and installing **driven piles in weak rock** include:

i) **Limited Design Guidance**: There is a lack of standardised guidelines for large-diameter driven piles in weak rock or hard clay, unlike sands and clays, which are covered by DNV and API standards.

ii) **Uncertainty in Pile Behaviour:** The mechanics of how driven piles interact with weak rock are not well understood, especially regarding **Pile capacity** (how much load the pile can bear) and **Drivability** (how easily the pile can be installed without refusal)

iii) **Inconsistent Design Approaches:** Significant variation exists in how skin friction and end bearing are calculated in weak rock, leading to inconsistent designs.

iv) **Site-Specific Complexity:** Designers must independently choose design methods, define geotechnical parameters, and plan investigations tailored to each site, increasing complexity and risk.

v) **Overreliance on Testing:** Due to uncertainty, there is often excessive dependence on specific testing techniques, which may not be fully reliable or representative.

14.10.2 Weak Rock Characterisation

Weak rocks lie within the transitional spectrum between soil and rock. Compared to soils, they are harder, more brittle, more dilatant, and more heterogeneous; whereas relative to stronger rocks, they are softer, less brittle, more compressible, and more prone to weathering. The primary characteristics influencing weak rock behaviour include:

- Uniaxial compressive strength
- **Rock mass strength and condition** (e.g., Geological Strength Index (GSI), Rock Quality Designation (RQD), Fracture Index)
- Grain size and mineral composition
- Moisture Content

  The strength of weak rock materials can vary considerably, making it challenging to define suitable lower and upper bound strength parameters for design and drivability evaluations. This variability may result in the need for longer piles to achieve capacity, while simultaneously creating difficult installation conditions when designs rely solely on rock strength test data. Weak rocks are generally classified into three main types: arenaceous (e.g., sandstone), argillaceous (e.g., hard clays, mudstone, siltstone), and calcareous (e.g., chalk, weak limestones, calcarenites), as outlined by Blyth and de Freitas (1991).

14.10.2.1 Type of Weak Rocks

Weak rocks can be broadly categorised based on their composition and behaviour under load and during pile installation:

- **Arenaceous (Sandstone)**

  Sandstones display a wide range of material and rock mass strength, as well as stiffness, with their mechanical properties being highly sensitive to moisture content. Under pile driving conditions, they typically behave in a brittle manner, often disintegrating into sand- or gravel-sized particles, which can affect drivability and end bearing performance.

- **Argillaceous (Mudstone, Siltstone)**

  Mudstones and siltstones generally exhibit low strength and stiffness, both at the material and rock mass level. They tend to behave like stiff, over-consolidated cohesive soils, and during pile driving, they often remould in a manner similar to clayey soils, influencing shaft friction and installation resistance.

14.10.2.2 Calcareous Rocks (Chalk, Limestone, Calcarenite)

Calcareous rocks, composed primarily of fine mineral grains cemented with calcite, exhibit varied behaviour during pile driving depending on their density and composition. Low to medium-density chalk tends to form a remoulded annulus around the pile, while higher-density chalk and limestone undergo complex processes including crushing, porosity reduction, and remoulding. Siliceous calcarenites with low carbonate content have been modelled using cohesive soil approaches for pile capacity; however, this method may not be universally applicable due to the potential for very low friction values. Design methods for these rocks should be validated through pile load testing. Specific recommendations exist for piles in chalk, accounting for crushing and compaction effects. Accurate assessment of rock properties, supported by testing, is essential to predict installation behaviour. Most references suggest limiting end bearing resistance based on allowable displacements rather than yielding, typically ranging from 2.5 qc to 5 qc, with normalised displacements (z/D) between 0.05 and 0.1, where z = settlement, D = Pile diameter.

Thus, Density significantly influences the behaviour of calcareous rocks during pile installation, particularly in terms of remoulding, crushing, and load-bearing characteristics:

Low-Density Calcareous Rocks (e.g., Chalk) tend to remould easily under driving stress, forming a soft annulus around the pile. As a result, this remoulded zone can reduce shaft friction and affect load transfer, requiring careful consideration in design.

The most widely accepted references for pile design in chalk are CIRIA (2002) and Carrington et al. (2011), particularly for onshore applications. CIRIA (2002) recommends limiting skin friction values ranging from 30 kPa to 50 kPa for large displacement pre-formed piles in low to medium-density chalk, and up to 120 kPa in high-density Grade A chalk with minimal discontinuities. Carrington et al. (2011) propose deriving limiting skin friction based on the peak post-cyclic shear strength, adjusted by an adhesion factor.

Design Implication: Lower end bearing resistance; pile capacity may be governed more by allowable displacements than material strength.

Medium to High-Density Calcareous Rocks (e.g., Dense Chalk, Limestone)

experience a combination of crushing, porosity reduction, and partial remoulding during driving. These processes can increase drivability resistance and complicate installation, but may also enhance end bearing if properly compacted.

Design Implication: Requires detailed assessment of crushing potential and stiffness; pile tests are essential to validate assumptions.

Siliceous Calcarenites may show very low friction due to low carbonate content and fine-grained structure. For such strata, cohesive soil models may not be suitable; frictional resistance can be underestimated or overestimated.

Design Implication: Load testing is critical to confirm design assumptions and ensure reliability.

14.10.2.3 Shaft Resistance Factor

In summary, density affects both the mechanical response of the rock and the interaction with the pile, influencing drivability, capacity, and the reliability of design models. Especially, the pile-rock interface along the shaft represents a major source of uncertainty in pile design within weak rock formations. Accurately estimating ultimate shaft resistance requires modelling the steel-rock interaction, which typically falls into one of three categories:

- **Complete disaggregation** of the rock at the pile wall, resulting in a frictional, cohesionless interface behaviour.
- **Complete remoulding** of the rock at the interface, leading to cohesive behaviour similar to that of clayey soils.
- **Limited disturbance**, where the pile remains in contact with the excavated rock mass, resembling a drilled socket condition. In such cases, oversized sockets may form, causing reduced contact stresses and lower frictional resistance.

Table 12.1: Behavior of Soft Rock

Rock Type	Interface Behaviour	Shaft Resistance Type	Key Design Consideration
Sandstone	Disaggregation	Frictional	Moisture-sensitive, brittle response
Mudstone/ Siltstone	Remoulding	Cohesive	Behaves like stiff clay, adhesion dominant
Chalk/ Limestone	Crushing/ Remoulding	Mixed (Cohesive/ Frictional)	Density-dependent, requires testing

Rock Type	Interface Behaviour	Shaft Resistance Type	Key Design Consideration
Calcarenite	Low friction potential	Frictional (variable)	Composition-sensitive, cohesive models may fail

14.10.2.4 Design Parameters Selection

Selection of design parameters should be grounded in a thorough understanding of rock mass behaviour, as emphasised by Hoek et al. (2005). Field analyses of pile installations in mudstone and sandstone have shown that in-situ measurements of rock mass properties correlate more reliably with actual pile capacity and drivability than laboratory-derived intact rock properties. To obtain meaningful rock mass strength data, geotechnical investigations must be carefully scoped and specified. At a minimum, these should include:

- Systematic rock logging using methods such as the Geological Strength Index (GSI)
- High-quality sampling of both weathered and intact rock for laboratory testing
- In-situ testing techniques like pressure-meter tests and/or downhole geophysical methods such as P-S logging

Given the current uncertainties in pile design methodologies, it is recommended to incorporate trial piles and static load testing into the design process. Additionally, Dynamic Pile Driving Monitoring (PDM) should be conducted during installation to verify as-built pile performance.

14.10.2.5 Weak Rock Characterisation for Pile Driving

For the purposes of pile design, *weak rocks* refer to geological materials with undrained shear strengths ranging from approximately 300 kPa to 10 MPa, as defined by ISO 1468-2:2003. These rocks may be inherently weak due to limited lithification or may result from the weathering of previously stronger formations.

The behaviour of weak rocks during pile driving is influenced by several key characteristics:

- **Material Strength**: Typically assessed through Uniaxial Compressive Strength (UCS), Point Load (PL), or Unconfined Undrained (UU) tests.
- **Rock Mass Condition**: Evaluated using classification systems such as the Geological Strength Index (GSI), Rock Quality Designation (RQD), Fracture Index (FI), and supported by in-situ methods like high-pressure dilatometer (HPD) testing, fracture orientation analysis, and P-S logging.
- **Grain Size and Mineral Composition**: These affect both strength and deformation behaviour.
- **Weathering Grade and Moisture Content**: These factors significantly alter the mechanical response of the rock during installation.

Field investigations often reveal substantial variability in UCS values, with strengths ranging from ~0.5 MPa to over 5 MPa within the same core run or rock layer. This variability highlights the challenge of designing piles based on conservative laboratory-derived values, which may not reflect the actual in-situ rock mass conditions encountered during driving.

14.10.2.6 Guidelines on Driving Resistance and Equipment Selection

Low UCS (<1 MPa)

- Rock behaves similarly to stiff cohesive soils.
- Standard impact hammers or vibratory drivers may be sufficient.
- Risk of remoulding or disaggregation is high, reducing shaft friction.

Moderate UCS (1–5 MPa)

- Requires more energy to penetrate.
- Hydraulic or diesel hammers with controlled energy settings are preferred.
- Monitoring is essential to avoid pile damage or refusal.

High UCS (>5 MPa)

- Driving becomes challenging; refusal is likely without pre-drilling.
- Pre-boring or socketing may be necessary to reduce resistance.
- Specialised equipment and techniques (e.g., down-the-hole hammers) may be required.

UCS affects the **degree of remoulding or crushing** at the pile-rock interface:

- **Lower UCS**: More remoulding → cohesive behaviour → adhesion-based design.
- **Higher UCS**: Less remoulding → frictional behaviour → higher shaft resistance.

Design Considerations

UCS variability within the same rock layer can lead to inconsistent drivability. Conservative UCS values may result in overdesign (longer piles) or underestimation of driving effort. Therefore, field validation through dynamic monitoring (PDM) and trial piles is essential.

Careful planning is required as High UCS zones may cause pile refusal, requiring contingency plans, and Low UCS zones may lead to unexpected settlement due to remoulding.

14.10.3 Backfilling Management for Retaining Walls

A. Placement of Backfill

Backfilling may begin once a sufficient length of the sheet-pile wall has been constructed and stabilised. Unless otherwise specified in the design, the backfill material should be clean, free-draining, and placed in uniform horizontal layers. For sands and gravels, uncompacted lift thicknesses may be placed up to 30 cm, while silts, clays, and mixed silty clays require reduced lift thicknesses not exceeding 20 cm to ensure consistent placement quality.

B. Compaction of Filling Material

Compaction is typically required to reduce compressibility, increase shear strength, control volume change, and minimise long-term settlement. When required by design, mechanical compaction should achieve approximately **92–93% of Standard Proctor maximum dry density**, with moisture-sensitive soils compacted within **±2%** of the optimum moisture content.

It is important to recognise that compaction decreases permeability and, if excessive, can significantly increase lateral earth pressures near the wall crest, which may adversely affect anchors or support systems. Densification of the soil in the passive zone and around anchorage elements is generally beneficial; however, compaction on the active side should be controlled in both method and degree. Heavy compaction machinery must be kept clear of newly installed bulkheads and within 2 m of the wall. Only light, hand-operated compactors should be used.

Hydraulic fill cannot be compacted during placement and should be treated in design as a temporary additional load acting on a potentially unstabilised wall until final consolidation and drainage have occurred.

C. Effect of Backfill Extent on Earth Pressure

When designing for active earth pressure, the anticipated lateral loads depend on the extent and characteristics of the backfill. Active pressure may develop either within the engineered backfill zone or along failure surfaces passing through adjacent native soils. Therefore, accurate prediction of earth pressures requires an understanding of the construction method, placement sequence, wall movement direction, and expected groundwater conditions.

The designer must have a complete understanding of the following aspects, along with the wall alignment:

i) **Existing site geometry**, including height and inclination of the retained soil, ground profile, soil distribution, adjacent loads, and nearby structures.

ii) **Soil bulk density**, typically ranging from 16–22 kN/m^3 and often increasing with depth.

iii) **Groundwater regime**, including final levels, pore-pressure distribution, seepage effects, and potential need for hydro-geotechnical modelling.

iv) **Shear strength parameters** (c', ϕ') and sub-soil stratification for evaluating stability and lateral loads.

v) **Soil stiffness**, required for deformation-based analyses using finite element or finite difference methods, obtained from in-situ tests, laboratory measurements, or back-analysis of comparable structures.

D. Special Considerations for Enhancing Wall Stability

The stability of sheet-piles for a quay wall can be significantly improved through the use of anchorage, particularly where large fluctuations between high and low water levels occur or where increased wall height requires greater lateral resistance. In such situations, additional tiebacks help optimise the wall section and enhance stability. Careful management of ground settlement around the anchorage zone is essential, as anchor blocks or deadmen must remain stable under service loads. While settlement is typically minimal in undisturbed soils, anchorage placed in uncompacted or recently filled ground may require pile-supported foundations or improved compaction of the passive wedge to at least 90% of Standard Proctor density. Tie rods installed above loose granular or soft cohesive soils may experience unanticipated tensile forces as underlying soils consolidate and induce sagging of the rods; mitigation measures include supporting tie rods on light vertical piles or encasing them within oversized conduits to reduce load transfer. During construction, soft or unstable material in the passive zone should be removed and replaced with compacted granular fill to ensure effective development of passive resistance and enhance overall wall performance.

Epilogue

1. Understanding the Changing Landscape of Maritime Infrastructure

I. Park, N. K., & Suh, S. C. (2019). *Tendency toward mega containerships and the constraints of container terminals*. **Journal of Marine Science and Engineering**, 7(131), MDPI.

II. Bassan, S. (2007). Evaluating seaport operation and capacity analysis—preliminary methodology. **Maritime Policy & Management**, 34(1), 3–19.

III. UNCTAD. (2014). *Review of Maritime Transport 2014*. United Nations Publications. ISBN: 978-92-1-112878.

2. Evolution of Maritime Trade in India

i. Mookerji, R. K. *Maritime History of India*. Asian Educational Services.

ii. Chandra, M. Trade and Trade Routes in Ancient India. Abhinav Publications.

iii. Alpers, E. A. *The Indian Ocean in World History*. Oxford University Press.

iv. Jain, S. Merchants, Trade and Commerce in Ancient India. New Royal Book Company.

3. Shipping, Logistics, and Essentials of Port Planning

I. Lloyd's Register. Ship Types and Sizes Explained.

II. Puertos del Estado. (1999). ROM 3.1-99: Design of Maritime Configuration of Ports, Approach Channels and Harbour Basins. Ministry of Public Works, Government of Spain.

III. PIANC. Guidelines for Harbour and Port Infrastructure.

IV. ISO 6346. Container Identification.

V. UNCTAD. Port Development: A Handbook for Planners.

VI. Tsinker, G. P. Port Engineering: Planning, Construction, Maintenance and Security. Wiley.

VII. Song, D.-W., & Panayides, P. *Maritime Logistics*. Emerald Publishing.

VIII. Das, J. K., Sinha, D., & Mukherjee, S. Analysis of Indian Port Performance Post Reforms: A Segmented Regression Approach.

IX. Shetty, K. D., Gurudev, V., & Dwarakish, G. S. (2021). *Factors affecting vessel turnaround time in a seaport*. **HYDRO 2020 Conference**.

4. Key Trends in Maritime and Port Industries

i. Liu, Q., & Ke, L. One Belt One Road: Policy implications on logistics route competition – China–Germany trade. University of Hamburg.

ii. USAID / Nathan Associates. Principles of Monitoring Port Performance: Cap-Haïtien Port Regulatory Strengthening Project.

iii. UNCTAD. (2014). Review of Maritime Transport 2014.

5. Coastal, Structural and Geotechnical Studies for Port Development

i. Davidson-Arnott, R. *Coastal Processes and Geomorphology*. Cambridge University Press.

ii. Kamphuis, J. W. Introduction to Coastal Engineering and Management. World Scientific.

iii. Pethick, J. Coastal Geomorphology: An Introduction. Edward Arnold.

iv. Holthuijsen, L. H. *Waves in Oceanic and Coastal Waters*. Cambridge University Press.

v. USACE. Coastal Engineering Manual (CEM).

vi. PIANC. Design of Marine Facilities for the Berthing, Mooring and Repair of Vessels.

vii. MoPSW (India). Guidelines for Port and Harbour Projects.

viii. CSIR-NIO & NIOT. Coastal and Oceanographic Technical Reports.

ix. MoEF&CC. CRZ Notification – Technical Guidelines.

6. Coastal Geoscience of India

i. Nayak, A. N., & Raju, S. V. K. Geomorphology of the Indian Coast. Geological Society of India.

ii. Wadia, D. N. Geology of India – Volume II. Tata McGraw-Hill.

iii. SAC–ISRO / GSI. Coastal Geomorphological Atlas of India.

7. Deterioration of Concrete Structures in Maritime Environment

i. Neville, A. M. Concrete Durability. CRC Press.

ii. Mangat, P. S., & Molloy, B. T. Durability of Concrete in Marine Environment. Chapman & Hall.

iii. Stewart, M. G., & Rosowsky, D. V. Fundamentals of Durable Reinforced Concrete. CRC Press.

iv. Broomfield, J. P. Corrosion of Steel in Concrete. CRC Press.

v. ACI 357R. Guide for the Design and Construction of Marine Structures.

vi. EN 206. Concrete – Specification and Performance.

vii. IS 456, IS 10262, IS 383. Bureau of Indian Standards.

8. Structural Condition Assessment Techniques

i. Neville, A. M. Properties of Concrete.

ii. Parrott, L. J. Carbonation in Reinforced Concrete. Concrete Association, UK.

iii. Geiker, M. R. et al. Durability-based design – European Perspective. Taylor & Francis.

iv. ASTM D6432-19. Guide for Ground Penetrating Radar (GPR).

9. Geotechnical Studies for Brownfield Port Upgradation

i. USACE EM 1110-1-1804. Geotechnical Investigations.

ii. Das, B. M. *Foundation Engineering*. Elsevier.

iii. Bowles, J. E. Foundation Analysis and Design. McGraw-Hill.

iv. Eurocode 7 – EN 1997-1.

v. BS 8002. Earth Retaining Structures.

vi. Tomlinson & Woodward. Pile Design and Construction Practice.

vii. Chanda, S., Kumar, M., et al. (July 2023). Assessment of liquefaction potential at Jaigarh Port using Standard Penetration Test data and associated consequences: A case study. In Proceedings of the 17th Symposium on Earthquake Engineering (Vol. 3).

viii. Chanda, S., Kumar, M., *et al.* (November 2023). *Evaluation of liquefaction-induced settlement in reclaimed ground: A case study.* Presented at the Sixth World Congress on Disaster Management.

10. Advanced Materials and Strengthening Techniques

i. Pilakoutas et al. FRP-Reinforced Concrete Design Guidelines. ICE Proceedings.

ii. ACI 440.2R. Externally Bonded FRP Systems.

iii. Phan et al. Strengthening Methodologies for Existing Structures. NIST.

iv. Bandara & Rajeev. *UHPFRC for Rehabilitation*. MDPI, 2023.

11. Repair, Rehabilitation and Retrofit of Wharf Structures

i. Kameda et al. (2017). *Pier Strengthening using Carbon Fibre Grid*. Nippon Steel Technical Report.

ii. Engindeniz et al. *Repair of RC Beam-Column Joints*. ACI Structural Journal.

iii. PIANC. Design Values for Berthing Velocity. World Congress 2018.

12. Wharf Functional Upgradation and Capacity Enhancement

i. Raza, S. et al. Strengthening of RC Columns by Jacketing. MDPI.

ii. World Bank / IFC. Terminal Handling Capacity Planning.

iii. Tsinker, G. P. *Port Engineering*. Wiley.

iv. Antonini, A. et al. (2016). *Wave simulation for quay wall design*. **NHESS**.

v. OCDI (Japan). Technical Standards for Port and Harbour Facilities.

vi. EAU (2004, 1996). Recommendations for Waterfront Structures.

13. Wharf Functional Upgradation and Capacity Enhancement

i. Raza, S., Khan, M. K. I., Menegon, S. J., Tsang, H.-H., & Wilson, J. L. Strengthening and Repair of Reinforced Concrete Columns by Jacketing: State-of-the-Art Review. MDPI.

i. American Concrete Institute (ACI). ACI 357R – Guide for the Design and Construction of Marine Structures. ACI, USA.

ii. World Bank / International Finance Corporation (IFC). *Terminal Handling Capacity Planning*. World Bank Group.

iii. United Nations Conference on Trade and Development (UNCTAD). *Port Development: A Handbook for Planners*. UNCTAD, Geneva.

iv. Tsinker, G. P.Port Engineering: Planning, Construction, Maintenance, and Security. Wiley.

v. Hammad, M., Bahrami, A., Khokhar, S. A., & Khushnood, R. A. A State-of-the-Art Review on Structural Strengthening Techniques with FRPs: Effectiveness, Shortcomings, and Future Research Directions. MDPI.

14. Steel Sheet Piles: Application in Wharf and Waterfront Structures

i. Manoj Kumar, A Practical Approach to Design of Steel Sheet Pile Structures. Notion Press.

ii. Antonini, A., Archetti, R., & Lamberti, A. (2016). *Wave simulation for the design of an innovative quay wall: The case of Vlorë Harbour*. **Natural Hazards and Earth System Sciences Discussions**, published June 2016.

iii. Douairi, M., & De Gijt, J. *Upgrading Techniques for Quay Walls*. ResearchGate publication.

iv. Committee for Waterfront Structures, Harbours and Waterways. *Recommendations (EAU 2004)*, 8th Edition. Ernst & Sohn, 2006.

v. OCDI, Japan. Technical Standards and Commentaries for Port and Harbour Facilities in Japan.

vi. Committee for Waterfront Structures (German Society for Harbour Engineering and Soil Mechanics). *Recommendations for Waterfront Structures, Harbours and Waterways (EAU 1996)*, 7th English Edition.

vii. Tomlinson, M., & Woodward, J. *Pile Design and Construction Practice*. Taylor & Francis.

viii. American Petroleum Institute (API). API Specification 2B: Specification for the Fabrication of Structural Steel Pipe.

ix. American Petroleum Institute (API). API RP 2A-WSD (2007): Recommended Practice for Planning, Designing and Constructing Fixed Offshore Platforms – Working Stress Design, 21st Edition, Washington, DC.

x. Lehane, B. M., & Jardine, R. J. (1994). *Displacement-pile behaviour in soft marine clay*. **Canadian Geotechnical Journal**, 31(2), 181–191.

xi. British Standards Institution (BSI), BS EN 12699:2015 – Execution of Special Geotechnical Works: Displacement Piles.

INDEX

U

V

W

Y

Z

www.ingramcontent.com/pod-product-compliance
Lightning Source LLC
LaVergne TN
LVHW082026200926
848481LV00010B/2928